Robert A. Guyer and Paul A. Johnson
Nonlinear Mesoscopic Elasticity

Robert A. Guyer and Paul A. Johnson

Nonlinear Mesoscopic Elasticity

The Complex Behaviour of Granular Media
including Rocks and Soil

WILEY-
VCH

WILEY-VCH Verlag GmbH & Co. KGaA

The Authors

Prof. Robert A. Guyer
UMASS – Physics Dept.
Hasbrouck Laboratories
Amherst, USA
guyer@physics.umass.edu

Prof. Paul A. Johnson
Los Alamos National Laboratory
Geophysics Group
Los Alamos, USA
paj@lanl.gov

Cover Picture
T. J. Ulrich, P. Johnson and R. A. Guyer,
Investigating Interaction Dynamics of Elastic
Waves with a Complex Nonliniear Scatterer
Applying the Time Reversal Mirror, Phys. Rev.
Lett., 98.10430 (2007)

Library of Congress Card No.: applied for
**British Library Cataloguing-in-Publication
Data:** A catalogue record for this book is
available from the British Library.
**Bibliographic information published
by the Deutsche Nationalbibliothek**
The Deutsche Nationalbibliothek lists this
publication in the Deutsche Nationalbib-
liografie; detailed bibliographic data are
available on the Internet at
<http://dnb.d-nb.de>.

© 2009 WILEY-VCH Verlag GmbH & Co.
KGaA, Weinheim

Printed in the Federal Republic of Germany
Printed on acid-free paper

Typesetting   le-tex publishing services
GmbH, Leipzig
Printing   betz-druck GmbH, Darmstadt
Binding   Litges & Dopf GmbH, Heppen-
heim

ISBN 978-3-527-40703-3

Contents

*Nonlinear Mesoscopic Elasticity: The Complex Behaviour of Granular Media
including Rocks and Soil.* Robert A. Guyer and Paul A. Johnson
Copyright © 2009 WILEY-VCH Verlag GmbH & Co. KGaA, Weinheim
ISBN: 978-3-527-40703-3

Preface

Nonlinear mesoscopic elasticity (NME) is the identifier of a collection of extreme/unusual elastic behaviors. The purpose of this book is to describe these behaviors as seen in particular physical systems, to suggest generalization beyond the particular based on a simple picture of the underlying physics, and to provide an analysis/theoretical framework for assessment of behavior and for the description of experiments. Thus we begin here with a brief (so that those who realize they are in the wrong place find that out sooner rather than later) description of the physical systems that are candidates for NME; six examples are shown. The behaviors that are associated with NME are many; eight examples are shown. The physical state of NME systems is specified in a multidimensional space of parameters, for example, length scale, time scale, the size of stress/strain fields, the strength of internal forces, etc. The boundaries of this space are set. At the end of the following overview we will provide an outline of the book.

Robert A. Guyer, Amherst
Paul A. Johnson, Los Alamos
2009

Nonlinear Mesoscopic Elasticity: The Complex Behaviour of Granular Media including Rocks and Soil. Robert A. Guyer and Paul A. Johnson
Copyright © 2009 WILEY-VCH Verlag GmbH & Co. KGaA, Weinheim
ISBN: 978-3-527-40703-3

Acknowledgements

We wish to acknowledge the contributions to this work by our many colleagues. These include our friends at Los Alamos National Laboratory, the Catholic University of Belgium, Turin Polytechnic Institute (Italy), University of Nevada at Reno (USA), Stevens Institute of Technology (USA), University of Paris East Marne-la-Vallée (France), University of Paris VI and University of Paris VII (France), the French Petroleum Institute, the Swiss Federal Institute of Technology (ETH), the Swiss Federal Laboratories for Materials Testing and Research (EMPA) Zurich (Switzerland), University of Le Mans (France), the National Oceanic Atmospheric Administration (USA), University of Mediterreanean (France), the Institute of Applied Physics (Russia), University of Maine (France), University of Stuttgart (Germany), and University of Massachusetts/Amherst (USA). We are grateful to the Institute of Geophysics and Planetary Physics at Los Alamos, the US Department of Energy, Office of Basic Energy Science and Los Alamos National Laboratory for their generous and ongoing support of this work.

Nonlinear Mesoscopic Elasticity: The Complex Behaviour of Granular Media including Rocks and Soil. Robert A. Guyer and Paul A. Johnson
Copyright © 2009 WILEY-VCH Verlag GmbH & Co. KGaA, Weinheim
ISBN: 978-3-527-40703-3

1
Introduction

1.1
Systems

Figures 1.1 to 1.6 show six examples of systems that have NME: powdered aluminum, thermal barrier coating, sandstone, cement, ceramic, and soil. For each figure there is a scale bar or caption that makes it clear that the systems of interest have noticeable inhomogeneities on a length scale smaller than the sample size, say 100 μm, but much larger than the microscopic scale, 0.1 nm. We imagine the physical systems that possess NME to have *very approximately* a *bricks-and-mortar* character. The bricks [quartz grains in the case of rocks, packets of crystallites (quartz, feldspar, ...) with clay particles in the case of soils, single crystals of aluminum in the case of powdered aluminum, ...] interface with one another across a distinctive, elastically different system, the mortar (a system of asperities in the case of rocks, a system of fluid layers and fillets in the case of (wet) soil, a layer of defective material in the case of aluminum powder, etc.). We are interested in these systems on a length scale that is large compared to that of their bricks. Systems built up to this length scale have important elastic features conferred by the geometry of the system that are strikingly different from those of their bricklike constituents.

For example, in the case of a Berea sandstone, the typical elastic modulus is an order of magnitude smaller than the corresponding modulus of quartz, that is, the

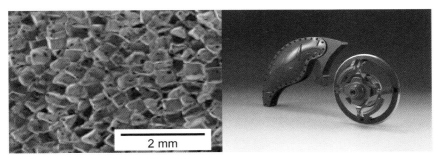

Fig. 1.1 Porous aluminum powder [9]. (Please find a color version of this figure on the color plates)

Nonlinear Mesoscopic Elasticity: The Complex Behaviour of Granular Media including Rocks and Soil. Robert A. Guyer and Paul A. Johnson
Copyright © 2009 WILEY-VCH Verlag GmbH & Co. KGaA, Weinheim
ISBN: 978-3-527-40703-3

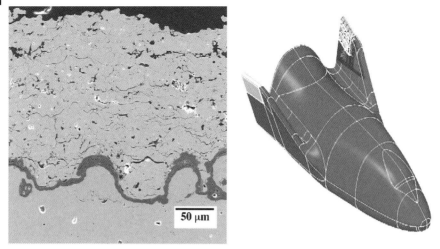

Fig. 1.2 Thermal barrier coating [10, 11]. (Please find a color version of this figure on the color plates)

Fig. 1.3 Sandstone (typical grain size 100 μm) [12]. (Please find a color version of this figure on the color plates)

bricks. This means that a given force, say across a sample, produces ten times as much displacement as it would if applied across the quartz alone. This displacement must reside in the mortar as the assembly process could not have altered the stiffness of the bricks. The mortar is a minor constituent of the whole comprising, perhaps, 10% of the volume. Ten times as much displacement due to 10% of the volume means that the mortar is very soft and that it carries strains approximately two orders of magnitude greater than those in the bricks. Accompanying the inhomogeneity in the structure is an inhomogeneity in the strain. There is a further important point. Ten percent by volume of soft material randomly distributed in otherwise hard material could not markedly modify the response of the assembly.

Fig. 1.4 Cement [13]. (Please find a color version of this figure on the color plates)

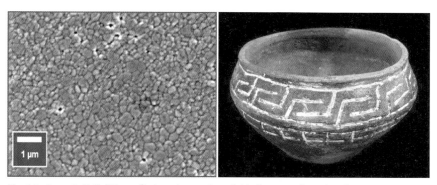

Fig. 1.5 Ceramic [14]. (Please find a color version of this figure on the color plates)

Fig. 1.6 Soil (sieved, typical grain size 1 mm) [15, 16]. (Please find a color version of this figure on the color plates)

The bricks-and-mortar picture captures an essential aspect of the way in which NME materials are constructed, that is, in such a way that the minority component (by volume) can effectively shunt the behavior of the majority component.

In identifying systems of interest with these simple ideas we cast a net that includes ceramics, soils, rocks, etc. But we do not pretend in any way to do justice to the disciplines of ceramic science, soil science, concrete science, ..., or even to elasticity in ceramics, soils, concretes, ... These are highly developed fields comprised of many subdisciplines. The discussion we present will be relevant more or less as dictated by the specific types of soil/ceramic/concrete/...

1.2
Examples of Phenomena

In Figure 1.7 we illustrate schematically eight examples of elastic behavior that we associate with NME. These include behavior that is quantitatively different from the usual behavior, behavior that is qualitatively different from the usual behavior, behavior that brings to the fore the importance of time scale and behavior in auxiliary fields. Not all NME materials possess these behaviors to the same degree. We sketch what is being illustrated schematically in each panel below. In the figure caption, information is given that locates an example of these experiments and characterizes them quantitatively.

1. The velocities of sound, c, of a sandstone are a factor of 2 to 4 less than those of the major constituent, for example, a quartz crystal. Thus the elastic constants of NME materials, K, $K \propto c^2$, might be less than the elastic constants of the parent material by an order of magnitude (even more for a soil).

2. When the pressure, P, is changed from 1 bar to 200 bar, the velocity of sound of a sandstone changes by a factor of 2. The same pressure change produces a 1% change in the velocity of sound in quartz (water, other homogeneous materials). Thus elastic nonlinearity, measured by $\gamma_c = \mathrm{d}\ln(c)/\mathrm{d}\ln(P)$, is very large for NME materials, often several orders of magnitude larger than that of the parent material.

3. When a sandstone (soil) is taken through a pressure loop, the strain that results is a hysteretic function of the pressure. In addition, when there are minor pressure loops within the major loop, the strain at the endpoints of the minor loop is "remembered". NME materials can have hysteretic quasistatic equations of state with endpoint memory.

4. A sample is subjected to a step in stress. Accompanying that step is a prompt step in strain followed by a slow further strain increase that evolves approximately as $\log(t)$. Recovery from the release of the step stress has a similar prompt step in strain and $\log(t)$ further reduction in strain. NME materials exhibit *slow dynamics* in response to transient loading.

5. The resonance of a bar of NME material is swept over at a sequence of fixed drive amplitudes. As the drive amplitude is increased, the resonant frequency shifts (to a lower frequency) and the effective Q of the system, measured by the amplitude at resonance, decreases. In a plot of the detected amplitude per unit drive, this is seen as a shift in the resonance peak accompanied by

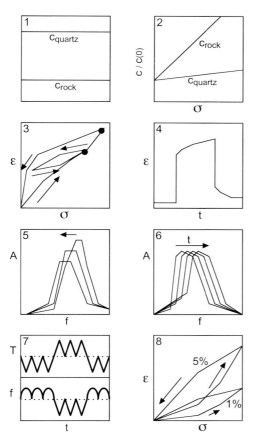

Fig. 1.7 Eight experiments. The eight experiments of interest are: (1) The velocity of sound, hence elastic constants, of a sandstone is a factor of 2 to 4 less than that of the major constituent, for example, a quartz crystal [1]. (2) When the pressure is changed, the velocity of sound of a sandstone changes by a factor of 2 for the application of 200 bar, whereas the same pressure change produces a 1% change in the velocity of sound in quartz (water, other homogeneous materials) [2]. (3) When a sandstone (soil) is taken through a pressure loop, the strain that results is a hysteretic function of the pressure and exhibits elastic endpoint memory [3]. (4) Accompanying the step in stress is a step in strain followed by a slow further strain response, that is, more strain, that evolves as $\log(t)$. Recovery from the release of the step stress has a similar strain step and $\log(t)$ further strain [4]. (5) The resonance of a bar of material is swept over at a sequence of fixed drive amplitudes. As the drive amplitude increases, the resonant frequency shifts (to lower frequency) and the effective Q of the system decreases [5]. (6) The slow evolution of the elastic state, brought about by an AC drive (compare to panel 4), can be seen in experiments in which the elastic state, once established, is probed by a low drive sweep over a resonance [6]. (7) When the temperature is changed slightly, the elastic response to that change involves a broad spectrum of time scales (compare to panels 4 and 6), suggesting $\log(t)$ behavior. In addition, the elastic response to temperature is asymmetric in the sign of the temperature change [7]. (8) A stress/strain loop similar to that in panel 3 is changed markedly by the configuration of fluid in the pore space [8].

a reduction of the amplitude at resonance. This behavior, which follows the fast motion of the drive, is an example of *fast dynamics.*

6. A bar of NME material is brought to steady state in response to a large-amplitude AC drive. The AC drive is turned off and the subsequent elastic state of the bar is probed with a low-amplitude drive that is swept over a resonance. The resonance, initially with resonance frequency shifted to a lower frequency as in panel 5, evolves back to a higher frequency approximately as log(*t*). The elastic state of the bar, established by a *fast dynamics drive*, relaxes once that drive is turned off by *slow dynamics.*

7. When the temperature of an NME material is changed slightly, the elastic response to that change, brought about by the temperature-induced internal forces, involves a broad spectrum of time scales (compare to panels 4 and 6), suggesting log(*t*) behavior at the longest times. In addition, the elastic response to temperature is asymmetric in the sign of the temperature change.

8. When an NME material is subjected to the internal forces of fluid configurations, a stress/strain loop similar to that in panel 3 is changed markedly. Much like a sponge, a rock is softer when wet.

The sequence of experiments sketched here call attention to the physical variables that are involved in the description of NME systems. The nature of a probe, the pressure, the temperature, the fluid configurations, the probe size, the duration of a probe, and the aftereffect of a probe having been present must all be considered and examined.

1.3
The Domain of Exploration

NME materials are probed in the complex phase space illustrated in Figure 1.8, that is:

1. Length. There are three length scales associated with NME materials, the microscopic scale (interatomic spacing) $a = 0.1\,\text{nm}$, the scale of inhomogeneity $b \approx 1\text{--}100\,\mu\text{m}$, and the sample size $L \gg b$. A quasistatic measurement is at $k \to 0$ ($k = 2\pi/\lambda$), whereas a resonant bar experiment is at wavelengths related to the sample size, $b \ll \lambda < L$.

2. Strain. There are judged to be two strain values of importance. At strains $\varepsilon < 10^{-7}\text{--}10^{-6}$, the nonlinear effects are small and have a more or less traditional behavior. At strains $\varepsilon > 10^{-3}\text{--}10^{-2}$ irreparable damage is done to a sample. The middle ground $10^{-7} < \varepsilon < 10^{-3}$ is the strain domain of NME.

3. Force. The standard for the strength of forces is the pressure given by a typical elastic constant, $K \approx \varrho c^2$, where ϱ is the density and c is the speed of sound, $K \approx 10^{11}\,\text{dyne/cm}^2 = 10^4\,\text{MPa}$ for a sandstone (1 atmosphere is $10^6\,\text{dyne/cm}^2 = 10\,\text{MPa}$). NME materials may be subject to a wide range of forces – applied forces, forces delivered to the interior of the systems from

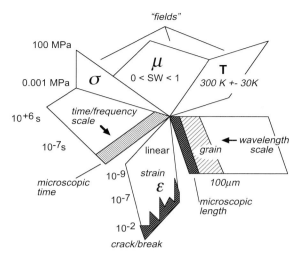

Fig. 1.8 Phase space. The materials of interest are probed on different time scales, length scales, and strain scales and with a variety of applied "fields".

the complex thermal response of constituents, or forces delivered to the interior of the system from arrangements of fluid in the pore space. The approximate strain consequence of a force (pressure) is found using $\varepsilon \approx P/K$, where P is the pressure. The strain range given above, $10^{-7} < \varepsilon < 10^{-2}$, implies 10^{-3} MPa $< P < 10^2$ MPa.

4. Time. The fastest time scale relevant to NME materials is approximately the time for sound to cross an inhomogeneity, $\tau \geq 100\,\mu\text{m}/c \approx 10^{-7}$ s. A resonant bar measurement is typically at 10^3–10^4 Hz (this scale is set by sample size L), a quasistatic measurement of stress/strain may last 10 min, and the strain response to a change in temperature may develop over a week. The range of time scales is enormous, 10^{-7} to 10^6 s.

All of these scales – length, time, and force – are far removed from the corresponding microscopic scales, for example, 0.1 nm is the microscopic length scale, 10^{12} Hz (a typical Debye frequency) is the microscopic time scale, and a microscopic energy per microscopic volume (say 0.1 eV$/(0.1\,\text{nm})^3 \approx 10$ GPa) is the microscopic force scale (stated here in terms of pressure since force alone means little).

1.4
Outline

Our interest is in the nonlinear elasticity of mesoscopically inhomogeneous materials. We will discuss the theoretical apparatus that is used to describe these materials, the phenomenology of the experiments conducted, and the large body of data that illustrates the behavior that characterizes these materials.

In Part I, Chapters 1–5, we give a theoretical introduction to traditional linear and nonlinear elasticity. We begin the discussion at the microscopic level. It is here that the basic structure of linear and nonlinear elasticity is established and the numbers that determine the magnitude of almost all quantities of interest are set. It is a short step from a microscopic description to the continuum description that corresponds to the traditional theory of linear/nonlinear elasticity. These topics are covered in Chapter 2, which is followed, in Chapter 3, by a series of illustrations of the consequences of the theory. To get to the domain of elasticity of mesoscopically inhomogeneous materials we must jump a gap. Across this gap, where we will work, we start with a theoretical apparatus, having the same form as the traditional theory of linear/nonlinear elasticity, to which we will add a collection of *ad hoc* ingredients that have no immediate source in the domain we have left behind. A variety of mesoscopic elastic elements, contacts, interfaces, etc. are described in Chapter 4. So also is an effective medium scheme for turning mesoscopic elastic elements into elastic constants suitable for a theory of elasticity. The coupling of the elastic field to auxiliary fields, particularly temperature and saturation, is taken up in Chapter 5.

In Part II, Chapters 6–9, we introduce hysteretic elastic elements, or strain elements with an elaborate stress response, Chapter 6. The dynamics of elastic systems carrying these elastic elements can be complex because of an internal field that responds to stress slowly in time. A discussion of the resulting *fast* and *slow* dynamics is given in Chapter 7. A set of practical matters related to data analysis and modeling of data sets is taken up in Chapter 8. This is followed by a description in Chapter 9 of a wide variety of considerations that relate to using data on elastic systems for characterization (spectroscopy) and for location (tomography).

In Part III, Chapters 10–13, we discuss experiments. Quasistatic measurements, including coupling to auxiliary fields, are described in Chapter 10. Dynamic measurements, dynamic/quasistatic to dynamic/dynamic, are described in Chapter 11. The current picture of fast/slow dynamics is given a full airing. In Chapter 12, field experiments that touch on NME are described. The final chapter, Chapter 13, contains a description of a wide variety of nondestructive evaluation applications of NME.

References

1 Bourbie, T., Coussy, O., and B. Zinszner (1987) *Acoustics of Porous Media*, Butterworth-Heinemann, New York.

2 Gist, G.A. (1994) Fluid effects on velocity and attenuation in sandstones, *JASA*, **96**, 1158–1173.

3 Boitnott, G.N. (1997) Experimental characterization of the nonlinear rheology of rock. *Int. J. Rock Mech. Min. Sci.*, **34**, 379–388.

4 Pandit, B.I. and Savage, J.C. (1973) An experimental test of Lomnitz's theory of internal friction in rocks. *J. Geophys. Res.*, **78**, 6097–6099.

5 Guyer, R.A., TenCate, J.A., and Johnson, P.A. (1999) Hysteresis and the dynamic elasticity of consolidated granular materials. *Phys. Rev. Lett.*, **82**, 3280–3283.

6 TenCate, J.A., Smith, D.E., and Guyer, R.A. (2000) Universal slow dynamics in granular solids. *Phys. Rev. Lett.*, **85**, 1020–1023.

7 Ulrich, T.J. (2005) (thesis), University of Nevada, Reno.

8 Carmeliet, J. and van den Abeele, K. (2002) Application of the Preisach–Mayergoyz space model to analyse moisture effects of the nonlinear elastic response of rock. *Geophys. Res. Lett.*, **29**, 48.1–48.4.

9 Baumeister, J., Banhart, U.J., and Weber, M. (1996) Damping properties of aluminium foams. *Mater. Sci. Eng.*, **A205**, 221–228.

10 Rejda, E.F., Socie, D.F., and Itoh, T. (1999) Deformation behavior of plasma-sprayed thick thermal barrier coatings. *Surf. Coat. Technol.*, **113**, 218–226.

11 Eldridge, J.I., Zhu, D., and Miller, R.A. (2001) Mesoscopic nonlinear elastic modulus of thermal barrier coatings determined by cylindrical punch indentation. *J. Am. Ceram. Soc.*, **84**, 2737–2739.

12 Guyer, R.A. and Johnson, P.A. (1999) Nonlinear mesoscopic elasticity: evidence for a new class of materials. *Phys. Today*, **52** (4), 30–36.

13 Brandt, A.M. (2009) *Cement Based Composits*, 2nd edn., Taylor and Francis, New York.

14 Green, D.J. (1998) *An Introduction to the Mechanical Properties of Ceramics*, Cambridge University Press, Cambridge.

15 Lu, Z. (2005) Role of hysteresis in propagating acoustic waves in soils. *Geophys. Res. Lett.*, **32**, L14302.

16 Ishihara, K. (1996) *Soil behavior in Earthquake Geotechnics*, Clarendon Press, Oxford.

2
Microscopic/Macroscopic Formulation of the Traditional Theory of Linear and Nonlinear Elasticity

Following Section 2.1, in which we make a few observations that place the discussion of solids in the context of fluid/solid systems, there are two major sections. Section 2.2 starts with the description of microscopic elasticity and elaborates on the connection between the microscopic description of elasticity and the continuum description of elasticity, while Section 2.3 sets out the essentials of the continuum theory of elasticity, sans microscopic justification. (For those who want to skip over the foundations in Section 2.2, this is the place to start. Of course, one will have to be content to learn μ, λ, A, B, ..., β, ... from experiments.) Many analytic details, Section 2.4, and some useful numbers, Section 2.5, are found at the end of the chapter.

In Section 2.2.1 we develop a description of the energy of a well-ordered solid, in terms of small displacements from equilibrium sites, which is the basis of the microscopic theory; in addition, we introduce the microscopic strains, etc. (Section 2.2.1.1). The dynamics of small displacements, due to forces caused by microscopic strains, leads to the phonon picture, the interacting phonon picture, etc. (Section 2.2.1.2). Some simple numerical estimates that tie microscopic numbers to macroscopic numbers are illustrated, for example, a linear elastic constant or a measure of the cubic anharmonicity. In Section 2.2.2, this mechanical (or quantum mechanical) description is married to an approximate but practical description of a solid in equilibrium with a temperature reservoir. In Section 2.2.2.1 we sketch the principle of the Gruneisen approximation, and in Section 2.2.2.2 we examine the resulting equations at reasonable temperatures, $T \approx 300\,\text{K}$, and find the microscopic basis of other numbers, for example, the thermal expansion. We close Section 2.2 with a formal treatment of the microscopic description that results in the equations of continuum elasticity. Consequently, there is a microscopic link to the parameters of linear and nonlinear continuum elasticity, for example, μ, λ, A, B, ..., β, ...

In Section 2.3 we sketch the theory of linear and nonlinear continuum elasticity without recourse to a microscopic picture. The displacement field, strain, and stress are introduced, as is the elastic energy density, an analytic function of the strain field (Section 2.3.1). The dynamics of the displacement field are treated in Section 2.3.2. The coupling of the displacement field to auxiliary fields, temperature, saturation, ... is described in Section 2.3.3. The generalization to inhomo-

Nonlinear Mesoscopic Elasticity: The Complex Behaviour of Granular Media including Rocks and Soil. Robert A. Guyer and Paul A. Johnson
Copyright © 2009 WILEY-VCH Verlag GmbH & Co. KGaA, Weinheim
ISBN: 978-3-527-40703-3

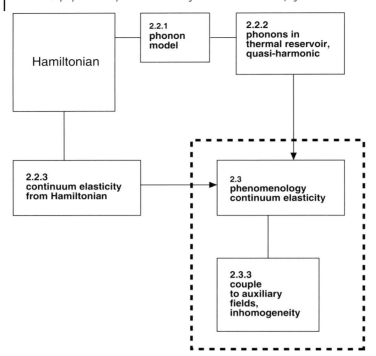

Fig. 2.1 Outline. The discussion in this chapter, from Hamiltonian to continuum elasticity, follows two routes, Sections 2.2.1 and Section 2.2.2, to the phenomenological model of elasticity. These routes supply some of the quantitative underpinnings of the phenomenological theory, which is able to stand on its own.

geneous elastic systems is made in Section 2.3.4 (see Figure 2.1). In Sections 2.4 and 2.5 details used in the chapter are provided.

2.1
Prefatory Remarks

First we step back from our immediate goal to look around. What distinguishes liquids and solids from gases at the atomic level is that in liquids and solids the particles (atoms or molecules) are **self-bound**. This means that the attractive forces between particles are sufficiently strong that they hold the particles near one another while the kinetic energy of the particles (their thermal motion, characterized by the temperature) causes them to move around, to attempt to fly apart. The particles in a gas are not self-bound; you have to put a gas of particles in a container with a lid to keep them together. To remove a particle from a liquid/solid you must reach in and pull with a force strong enough to liberate it from its neighbors. The basic physical state of a collection of particles is determined by the ratio of the strength of the attractive forces, stated as an energy, and the energy of thermal motion, set by the temperature. Sometimes the thermal motions of the particles in a liquid/solid will conspire to deliver a large amount of kinetic energy to one particle and allow it to spontaneously leave the system, that is, evaporate.

And the difference between a liquid and a solid? It is one of degree and structure [1, 2]. In a solid the attractive forces between particles are sufficiently strong, compared to the disordering effect of the thermal motion, that a particular spatial arrangement of particles, each particle sitting advantageously in the attractive potential well of a regular array of neighbors, is the lowest energy state. The energy of a solid arrangement of particles differs from the energy of a liquid arrangement of particles by an amount that is small compared to the energy of either; the heat of fusion (roughly a measure of the energy difference between solid and liquid) is small compared to the heat of vaporization (by, say, a factor of 10 or so, the familiar 80 cal/g and 540 cal/g of freshman physics). The particles in a solid sit at well-defined places relative to their neighbors, and this local arrangement of particles is repeated again and again throughout space, that is, the solid, if it is a single crystal, has translational symmetry [3, 4]. Thus in a solid, where a particle should be is well defined; the departure of a particle from where it should be is also well defined. When you reach into a solid and pull a particle away from where it should be, its neighbors pull back. A set of internal forces arises in reaction to your pull with an accompanying set of displacements. The particle on which you are pulling is displaced and so are the particles that contribute the force trying to hold it in place. These are the manifestations of stress (the forces) and strain (the displacements) at the microscopic level. A description of what is happening at this level, a job for a chemist or a band structure physicist, involves looking at a material electron by electron, chemical bond by chemical bond.

2.2
From Microscopic to Continuum

2.2.1
A Microscopic Description

2.2.1.1 Microscopic Energy and Microscopic Strain
A crystal is an assembly of particles that to good approximation can be taken to reside near a set of lattice sites that are regularly arrayed in space. The symmetry of the crystal, for example, cubic, hexagonal, ..., describes the geometry of this regular array. Since the crystal is self-bound, it is characterized by atomic scale energies, forces, and lengths, ε_0, ε_0/a, and a, respectively, where a is the interparticle spacing. The typical particle is at a distance of a few Angstroms, tenths of a nanometer, from its neighbors and involved in an interparticle interaction of strength $\varepsilon_0 \approx 0.5\,\text{eV}$.

The forces between particles have strength ε_0/a of order 0.5 eV/0.1 nm or 0.1 nN (nanonewton) or 1000 K/Å or 0.1 (GPa)m^2. (The many units displayed here are a reminder that the measure of the importance of any energy/force is its size relative to another, for example, a photon energy in eV, a particle kinetic energy in Kelvin, an applied pressure in Pa, etc.)

The motion of particle R, at x_R, near the lattice site with which it is associated, R, is described by displacement u_R, $x_R = R + u_R$, and the corresponding momentum is $p_R = m\dot{u}_R$. The motion of particles away from their lattice sites is small. Typically at melting one has $|u_R| \approx (0.20 - 0.25)a$ [5]. Thus particle motions are a small fraction of the intersite distance, and the energy of interaction among the particles can be developed as a series in the displacements, u_R. For the energy in the assembly of particles we have

$$\mathcal{E} = \sum_R \frac{p_R^2}{2m} + \frac{1}{2}\sum_R \sum_{R'} V(R - R' - u_R + u_{R'}) = \mathcal{K} + \mathcal{U}, \qquad (2.1)$$

where $V(x_{RR'})$ is the interaction energy between particles separated by $x_{RR'} = x_R - x_{R'} \approx R - R'$, the equilibrium spacing between the lattice sites associated with the particles, Figure 2.2. Using $\Delta^\alpha = u_R^\alpha - u_{R'}^\alpha$ $(\alpha = x, y, z)$ we can write

$$\begin{aligned} V(x_R - x_{R'}) = \Phi_0(R - R') &+ \frac{1}{2!}\Phi_{\alpha\beta}(R - R')\,\Delta^\alpha\,\Delta^\beta \\ &+ \frac{1}{3!}\Phi_{\alpha\beta\gamma}(R - R')\,\Delta^\alpha\,\Delta^\beta\Delta^\gamma \\ &+ \frac{1}{4!}\Phi_{\alpha\beta\gamma\delta}(R - R')\,\Delta^\alpha\,\Delta^\beta\Delta^\gamma\Delta^\delta \\ &+ \ldots, \end{aligned} \qquad (2.2)$$

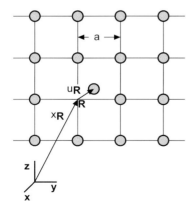

Fig. 2.2 Lattice. The set of vectors R and the displacements u_R allow one to track the particle at $x_R = R + u_R$.

where we use the sum convention on repeated indices and the coefficients Φ, with m subscripts, involve m derivatives of V with respect to R, for example,

$$\Phi_{\alpha\beta}(R) \rightarrow \frac{\partial^2 V(R)}{\partial R_\alpha \partial R_\beta}. \tag{2.3}$$

Since each derivative of V with respect to R brings down a factor of order a, we scale the displacements Δ by a to form the microscopic strain, $e^\alpha = \Delta^\alpha/a$. Then Eq. (2.2) becomes a power series in the microscopic strains involving coefficients that are energies on the order of the energy of interaction, ε_0. Take as an example V given by the Lennard–Jones interaction [6], written here in terms of the near-neighbor distance $a = 2^{1/6}\sigma$ (σ is the hard core radius):

$$V(r) = \varepsilon_0 \left(\left(\frac{a}{r}\right)^{12} - 2\left(\frac{a}{r}\right)^6 \right), \tag{2.4}$$

for which

$$\frac{\partial^2}{\partial x \partial y} \varepsilon_0 \left(\frac{a}{r}\right)^n \bigg|_{r=a\hat{x}} = \varepsilon_0 n(n+1)\hat{x}\hat{y} \frac{1}{a^2} \sim \frac{\varepsilon_0}{a^2}, \tag{2.5}$$

where $\hat{x}$ is the unit vector in direction x.

In terms of the microscopic strains the interaction is

$$\begin{aligned}
V(x_R - x_{R'}) = \overline{\Phi}_0(R - R') &+ \frac{1}{2!}\overline{\Phi}_{\alpha\beta}(R - R')\, e^\alpha e^\beta \\
&+ \frac{1}{3!}\overline{\Phi}_{\alpha\beta\gamma}(R - R')\, e^\alpha e^\beta e^\gamma \\
&+ \frac{1}{4!}\overline{\Phi}_{\alpha\beta\gamma\delta}(R - R')\, e^\alpha e^\beta e^\gamma e^\delta \\
&+ \dots,
\end{aligned} \tag{2.6}$$

where $\overline{\Phi}_{1,\dots,m} = a^m \Phi_{1,\dots,m}$, a power series in the microscopic strain having coefficients with the magnitude set by the strength of interaction, ε_0. (Sometimes there are nontrivial numbers involved, e.g., $\overline{\Phi}_{1,\dots,m+1} \approx n\overline{\Phi}_{1,\dots,m}$. For $n = 12$ this can be significant. See below.)

For a cubic centimeter of material supporting 1 microstrain, the $N_L = 1\,\mathrm{cm}/a \approx 3 \times 10^{+7}$ planes of atoms in the crystal ($a \approx 0.3\,\mathrm{nm}$) support the strain equally and move closer to (further from) one another by $\Delta \approx 10^{-6}\,\mathrm{cm}/N_L \approx 10^{-13}\,\mathrm{cm}$, a distance on the order of the size of the atomic nucleus. Thus $e = \Delta/a \approx 10^{-6}$. What we have written is really

$$e = \frac{\Delta}{a} = \frac{N_L \Delta}{N_L a} = \frac{\Delta L}{L}, \tag{2.7}$$

where $\varepsilon = \Delta L/L$ is the macroscopic strain, that is, the relative motion of particles on the microscopic scale is essentially the same as the relative motion on the macroscopic scale. In a homogeneous sample, where each part of the sample participates equally in taking up the strain field, all strains are equal. With a number like $e \approx 10^{-6}$ Eq. (2.6) is a rapidly converging power series in e. It makes sense to view Eq. (2.1) in the form

$$\mathcal{E} = \mathcal{E}_2 + \mathcal{V}_3 + \mathcal{V}_4 + \dots , \tag{2.8}$$

where

$$\mathcal{E}_2 = \sum_R \frac{p_R^2}{2m} + \frac{1}{2}\frac{1}{2!} \sum_R \sum_{R'} \overline{\Phi}_{\alpha\beta}(R - R') \, e_{RR'}^{\alpha} e_{RR'}^{\beta} = \mathcal{K} + \mathcal{V}_2 ,$$

$$\mathcal{V}_3 = \frac{1}{2}\frac{1}{3!} \sum_R \sum_{R'} \overline{\Phi}_{\alpha\beta\gamma}(R - R') \, e_{RR'}^{\alpha} e_{RR'}^{\beta} e_{RR'}^{\gamma} , \tag{2.9}$$

$$\mathcal{V}_4 = \frac{1}{2}\frac{1}{4!} \sum_R \sum_{R'} \overline{\Phi}_{\alpha\beta\gamma\delta}(R - R') \, e_{RR'}^{\alpha} e_{RR'}^{\beta} e_{RR'}^{\gamma} e_{RR'}^{\delta} ,$$

$$\vdots$$

and the terms in $\overline{\Phi}_0$, a constant energy, have been dropped. The problem posed by this equation is solved systematically taking $\mathcal{E} = \mathcal{E}_2$ as a leading approximation. The $\mathcal{E} = \mathcal{E}_2$ problem is referred to as the **harmonic crystal** problem. The remaining terms, $\mathcal{V}_3$, $\mathcal{V}_4$, etc., in the energy are the **cubic**, **quartic**, etc. **anharmonicities**, to be dealt with using perturbation theory. This perturbation theory philosophy, justified here by the size of $|e^{\alpha}|$, is carried over to continuum elasticity with the continuum strain field playing the role of e^{α}. It is not our intention to *solve* this problem in detail but rather to identify in it those features that work their way into the continuum theory and into the continuum description of phenomena.

2.2.1.2 Phonons

The $\mathcal{E}_2$ problem is usually formulated in terms of the equations of motion for the displacements [3, 4]. We have

$$m\ddot{u}_R^{\gamma} = -\frac{\partial \mathcal{V}_2}{\partial u_R^{\gamma}} = -\sum_{S\alpha} D_{\gamma\alpha}(R - S) u_S^{\alpha} , \tag{2.10}$$

where $D_{\gamma\alpha}(R - S)$, the dynamical matrix, is constructed from $\Phi_{\gamma\alpha}(R - S)$, $D_{\gamma\alpha}(R - S) = \delta_{RS}\overline{\Phi}_{\gamma\alpha} - \Phi_{\gamma\alpha}(R - S)$, and $\overline{\Phi}_{\gamma\alpha} = \sum_S \Phi_{\gamma\alpha}(R - S)$. When one looks for a solution to the equation of motion for u_R^{γ} with time dependence in the form $u_R^{\gamma} \propto \exp(-i\omega t)$, one sees that Eq. (2.10) is a set of homogeneous equations

$$m\omega^2 u_R^{\gamma} = \sum_{S\alpha} D_{\gamma\alpha}(R - S) u_S^{\alpha} , \tag{2.11}$$

involving the *3N* displacements u_R^{γ}, an eigenvalue problem for the frequency and structure of the normal modes. The displacements $\boldsymbol{u}_R$ are taken to have plane-wave-like spatial dependence, $\boldsymbol{u}_R = U \exp(i\boldsymbol{k} \cdot \boldsymbol{R})$, with the result that

$$m\omega^2 U = D(\boldsymbol{k}) \cdot U, \tag{2.12}$$

where $D(k)$ is the Fourier transform of $D(R)$

$$D(k) = \sum_R D(R)e^{-ik\cdot R} .\tag{2.13}$$

As U is a displacement vector, there are three vector amplitudes (polarizations) for each of N wavevectors k, that is, *3N* normal modes.

We have proceeded to this point with some generality. (See [4] for careful delineation of the properties of the solid that allow getting to this point.) Let us look at Eq. (2.12) for a simple model system in which (a) the interaction in Eq. (2.1) is a function of the magnitude of the separation between particles, $V(r) = V(r)$, for example, Eq. (2.4), (b) the interaction is sufficiently short ranged that near neighbors make the only important contribution to $D(R)$ and (c) the particles are on a simple cubic lattice. We have

$$\Phi_{\alpha\beta}(\Delta_{RR'}) = V''(a)e^{\alpha}_{RR'}e^{\beta}_{RR'} = \Gamma e^{\alpha}_{RR'}e^{\beta}_{RR'} ,\tag{2.14}$$

with $\Delta_{RR'}$ being the vector between near neighbors R, R' having magnitude a and components $e^{\alpha}_{RR'}$. Then it follows that $\overline{\Phi}_{\alpha\beta} = 6\Gamma$, where 6 is the number of near neighbors, and

$$D_{\alpha\beta}(\Delta_{RR'}) = \Gamma\delta_{\alpha\beta}(6\delta_{RR'} - 1) .\tag{2.15}$$

Using this result in Eq. (2.12) leads to the frequency/wave vector relation, the phonon dispersion relation,

$$\omega^2 = \frac{\Gamma}{m}\left[6 - 2\cos(k_x a) - 2\cos(k_y a) - 2\cos(k_y a)\right] ,\tag{2.16}$$

with $k = (k_x, k_y, k_z)$. In the long wavelength limit, $|k|a \ll 1$, this equation is transformed into

$$\omega^2 = \frac{\Gamma a^2}{m}k^2 = c^2 k^2 ,\tag{2.17}$$

where c is the phonon velocity. If we estimate Γ from a Lennard–Jones potential, Eq. (2.4), $\Gamma = 72\varepsilon_0/a^2$, we have

$$c^2 = 72\frac{\varepsilon_0}{m} \approx 6 \times 10^{11} \text{ (cm/s)}^2\tag{2.18}$$

upon making the choice $\varepsilon_0 \approx 0.5\,\text{eV}$, $m = 60\,\text{amu}$, and $a \approx 0.4\,\text{nm}$. Using $\varrho = m/a^3 \approx 2\,\text{g/cm}^3$ we have an elastic constant, $K \approx \varrho c^2$, of order $10^{12}\,\text{erg/cm}^3$. Both the estimate of c and the estimate of K are sensible. The series of steps from Eq. (2.10) to Eq. (2.18) serves as a demonstration of the microscopic source of the numbers that characterize the behavior of elastic systems.

When the motion of the displacement is quantized, it is described by phonons of wave vector k, frequency ω, and polarization ε that carry momentum $\hbar k$ and energy $\hbar\omega$. In the harmonic approximation, that is, harmonic crystal problem, the phonon excitations are exact eigenstates of the energy. The $\mathcal{E}_2$ problem, the harmonic crystal problem, is the microscopic analog of the normal mode problem in resonant ultrasound spectroscopy (RUS) [7] that we will encounter in Chapters 8 and 11. The phonons of the harmonic crystal and the normal modes of RUS do not decay; they have infinite lifetimes. The terms $\mathcal{V}_3$, $\mathcal{V}_4$, ... in Eq. (2.8) cause interactions among the phonons, one phonon turning into another, that give the phonons a finite lifetime [8]. Let us sketch the rudiments of what happens. [There are other mechanisms, representing a departure of the physical system from the mathematical ideal, that contribute to phonon (normal mode) lifetime; see Chapter 8.]

Consider the cubic anharmonic term in Eq. (2.9), $\mathcal{V}_3$. This term is of order $\Phi'''(\delta u)^3$, where Φ''' stands for the third derivative of Φ and $\delta u = u_R - u_{R'}$. Using for Φ''' the near-neighbor result for a Lennard–Jones potential, $|\Phi'''| = 21 \cdot 72\varepsilon_0/a^3$ and

$$\mathcal{V}_3 \approx 21 \cdot 72\varepsilon_0 \frac{1}{3!} \left(\frac{\delta u}{a}\right)^3 . \tag{2.19}$$

A similar treatment of $\mathcal{V}_2$ leads to

$$\mathcal{V}_2 \approx 72\varepsilon_0 \frac{1}{2!} \left(\frac{\delta u}{a}\right)^2 , \tag{2.20}$$

so that we can write

$$\mathcal{V}_3 \approx 7\mathcal{V}_2 \left(\frac{\delta u}{a}\right) . \tag{2.21}$$

The energy scales of $\mathcal{V}_2$ and $\mathcal{V}_3$ differ by one factor of the strain and a numerical factor, $7 = 21/3$. A nonlinear parameter β, usually defined in the equation of motion (see below), is essentially this numerical factor. For the dimensionless measure of the leading **atomic nonlinearity** (the cubic anharmonicity) we have β of order 10, Figure 2.3.

When the quantized phonon excitations are used in the description of the displacement field, the cubic anharmonicity brings about the interaction of three phonons. The diagram in Figure 2.4 illustrates a typical process that is allowed by $\mathcal{V}_3$. Two phonons with polarization, wave vector, and frequency, $(\varepsilon_1, k_1, \omega_1)$ and $(\varepsilon_2, k_2, \omega_2)$, interact with strength proportional to β to yield a third phonon (ε, k, ω). In this process energy is conserved, that is, $\omega = \omega_1 + \omega_2$, and momentum is conserved, that is, $k = k_1 + k_2$. This three-phonon process gives the infinitely long-lived phonons of the harmonic crystal a finite lifetime. In continuum elasticity we will encounter nonlinear interactions between strain fields that are the continuum manifestation of this process. Some of the details will differ from those here in important ways because we will have the interaction of three classical fields.

Examination of the microscopic description of a solid, Eq. (2.1), leads to

1. a description of the motion of the displacement field in terms of $3N$ quantized phonons, N pairs of (k, ω) for each of the three polarizations;

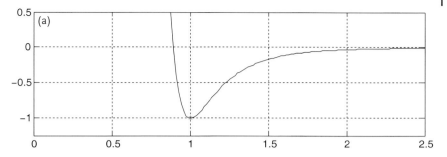

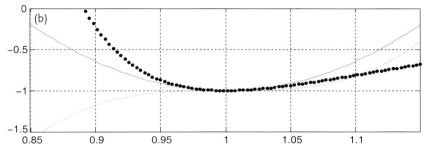

Fig. 2.3 Lennard–Jones Interaction. (a) The Lennard–Jones interaction, scaled by the energy ε, as a function of interparticle separation, scaled by the length a. Eq. (2.4). (b) The Lennard–Jones interaction (dots), the harmonic approximation (dark gray), and the cubic anharmonicity (light gray). These rudimentary energies are the source of the elastic constants in the harmonic approximation and the nonlinear coupling to the third order in displacement (strain).

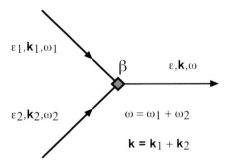

Fig. 2.4 Three-phonon process.

2. a relationship between the velocity of the long-wavelength phonons, the elastic constants, and the microscopic interparticle interaction;

3. a nonlinear mechanism for coupling the phonons, giving them a finite lifetime, that reveals the importance of $k - \omega$ selection rules;

4. an estimate of the strength of the nonlinear interaction between displacement fields based on the microscopic interaction.

2.2.2
Microscopic Description and Thermodynamics

The discussion to this point has been on the microscopic description of the displacement field in a solid in isolation. We encounter the physical realization of solids in particular pressure/temperature/saturation/... circumstances. A thermodynamic description of the solid would let us understand the effect of pressure/temperature/saturation/... on the measured macroscopic quantities. To formulate such a description we use the Helmholtz free energy, $F(T, V, W, N)$, in which the state of the system is specified in terms of the temperatue, T, volume, V, number of particles, N, and an auxiliary field W that can be set by control of the variable μ_W [9, 10]. In one circumstance the field W is the saturation field and μ_W is the chemical potential [11]. The W field is present in this description so that we can see how a generic auxiliary field would couple to temperature, volume, etc. We take W to be extensive, $W \propto V$, and call it saturation for convenience.

At a fiducial point (T_0, V_0, W_0) the free energy associated with the displacement field is made up of three pieces:

$$F(T, V, W, N) = E_S + E_Z + F_T = E_0 + F_T, \tag{2.22}$$

where E_S is the static energy associated with the equilibrium configuration of the particles, the terms in $\overline{\Phi}_0$ in Eq. (2.6), E_Z is the zero-point energy in the phonons, and F_T is the thermal free energy in the phonons. The phonon zero-point energy is [12]

$$E_Z = \frac{1}{2} \sum_\alpha \hbar \omega_\alpha, \tag{2.23}$$

where α is the index specifying the $3N$ modes, N wave vectors with three polarizations each, found from the solution to Eqs. (2.11)–(2.13). The two energies E_S and E_Z are independent of T. For the thermal free energy we have [9]

$$F_T = k_B T \sum_\alpha \ln \left(1 - e^{-\beta \hbar \omega_\alpha}\right) \tag{2.24}$$

$\beta^{-1} = k_B T$. As the temperature, volume, and saturation are changed from (T_0, V_0, W_0) the physical system is assumed to be qualitatively unchanged; there is no chemistry, change in crystal symmetry, particle rearrangement, etc. Thus there are three sources of change in the free energy: (1) a change in F_T due to T and a change in the normal mode frequencies, (2) a change in E_Z due to changes in the normal mode frequencies, and (3) and a change in E_0 due to changes in V and W. The simple Gruneisen model, developed carefully and extensively by Anderson [13], allows us to go quite far in exploring the effects of T, V, and W on the behavior of the system.

2.2.2.1   The Quasiharmonic Approximation, in Principle
In the Gruneisen model the effect of a change in volume, $dV = V - V_0$, is to shift the frequency of the phonon modes by an amount proportional to dV. We assume

that W also changes the phonon frequencies as here

$$\omega_\alpha(V, W) = \omega_\alpha^0 \left(1 - \gamma_\alpha \frac{dV}{V_0} - \Gamma_\alpha \frac{dW}{W_0}\right) = \omega_\alpha^0 (1 - da - db) , \tag{2.25}$$

where $\omega_\alpha^0 = \omega_\alpha(V_0, W_0)$ and

$$\gamma_\alpha = -\frac{V_0}{\omega_\alpha^0} \left(\frac{\partial \omega_\alpha^0}{\partial V}\right) ,$$

$$\Gamma_\alpha = \frac{W_0}{\omega_\alpha^0} \left(\frac{\partial \omega_\alpha^0}{\partial W}\right) \tag{2.26}$$

are the Gruneisen constants of the mode α. The sign of γ is chosen so that a decrease in volume, $dV < 0$, causes an increase in frequency. Quite possibly $W_0 = 0$ so that some care in using the definition of Γ is called for.

The program of manipulations is slightly lengthy. In outline:

1. Develop a generic series representation for F in powers of dT, dV, and dW. This series will have coefficients that are thermodynamic derivatives, for example, $P = -\partial F/\partial V$, $K = -V(\partial F/\partial V)$, $\alpha_0 K = \partial P/\partial T$, where P is the pressure, K is the bulk modulus, α_0 is the thermal expansion, etc.
2. Develop a series representation of F using E_S, E_Z, F_T, and Eq. (2.25); see Eq. (2.29) below.
3. Comparing the two series representations for F results in termwise equations that relate the thermodynamic derivatives to the changes in the phonon frequencies.

We begin by developing a representation of F about (T_0, V_0, W_0) in the form of a generic series in dV, dT, and dW to third-order

$$F = E_S + E_Z + F_T = E_0 + F_T ,$$

$$= \sum_{l+m+n \leq 3} \frac{F^{lmn}}{l!m!n!} dT^l dV^m dW^n , \tag{2.27}$$

where $F^{lmn} = \partial^l \partial^m \partial^n F/\partial T^l \partial V^m \partial W^n$. Many of the coefficients in this equation, F^{lmn}, are defined thermodynamic derivatives. See Eqs. (2.83...) in Section 2.4. These definitions can be used when we write F in the following way:

$$F = F_0 + F_1 + F_2 + F_3 + \dots ,$$

$$F_0 = F^{000} = F(T_0, V_0, W_0) ,$$

$$F_1 = -P dV - S dT + \mu_W dW ,$$

$$F_2 = \frac{K}{V_0} \frac{(dV)^2}{2!} - \frac{C_V}{T_0} \frac{(dT)^2}{2!} + F^{002} \frac{(W)^2}{2!} + \tag{2.28}$$

$$- \alpha_0 K_0(dT dV) + \frac{K_W}{W_0}(dV dW) + F^{101}(dT dW) ,$$

$$F_3 = +\beta \frac{K}{V_0^2} \frac{(dV)^3}{3!} + \alpha_1 \frac{K}{V_0} \frac{(dV^2 dT)}{2!} + \dots ,$$

where we have not written out all of the third-order terms. In the last line β and α_1 have *ad hoc* definitions that are a dimensionless measure of cubic anharmonicity and a nonlinear thermal expansion, respectively. Of particular interest are the second-order terms that couple temperature to strain, $dTdV$, and saturation to strain, $dVdW$ (K_W is defined by $F^{011} = K_W/W_0$), and the third-order terms that correspond to cubic anharmonicity, $(dV)^3$, and couple strain to temperature, $dT(dV)^2$ (the Luxemberg–Gorky effect) [14]. How are the coefficients in Eq. (2.28) related to the microscopic description above?

2.2.2.2 The Quasiharmonic Approximation to *F*

A microscopic description of the quantities on the RHS of Eq. (2.28) is found from a Taylor series expansion of the equation for F. For example, the contribution of F_T is found by putting Eq. (2.24) in the form

$$F_T = k_B T_0 (1 + dc) \ln \left(1 - \exp \left[-x^0 (1 - da - db)/(1 + dc) \right] \right) , \qquad (2.29)$$

where $x_\alpha^0 = \hbar \omega_\alpha^0 / k_B T_0$, $dc = dT/T_0$, da, and db are as in Eq. (2.25), and carrying out a Taylor series expansion to order 3 in da, db, dc. Similarly, series expansions for E_S and E_Z are developed; the details are in Section 2.4.

The resulting series for F is compared to the series in Eq. (2.28) and corresponding terms identified. The outcome of doing this algebra is an equation that gives each thermodynamic quantity a microscopic description. Let us look at a few examples to see what is involved.

1. $P = -\partial F/\partial V = -F^{010}$. There are two terms, one from each of E_Z and F_T, Eqs. (2.97) and (2.101). We have

$$P = \frac{-E_Z^{10}}{V_0} + P_0^T , \qquad (2.30)$$

where E_Z^{10} is the term proportional to dV in E_Z, Eq. (2.97), and P_0^T, the term proportional to $da \propto dV$ in Eq. (2.101), is given by Eq. (2.105):

$$P = \frac{E_Z^{10}}{V_0} + P_0^T = \frac{E_Z^{10}}{V_0} + \frac{k_B T_0}{V_0} \sum_\alpha \gamma_\alpha \frac{x_\alpha}{e^{x_\alpha} - 1} \sim \frac{E_Z^{10}}{V_0} + k_B T_0 \frac{N}{V_0} \langle \gamma_\alpha \rangle . \qquad (2.31)$$

The last term on the right-hand side of this equation is the high-temperature approximation. The pressure is made up of two contributions, one from the zero-point motion and one from the thermal motion of the particles.

2. $\alpha_0 = K^{-1} \partial P/\partial T = -K^{-1} F^{110}$. As this quantity depends on the temperature, it has a contribution only from F_T. Since P comes from dV, we look for the $dadc \propto dVdT$ term in F_T. This is on the second line of Eq. (2.102). From Eq. (2.110)

$$\alpha_0 K = \frac{k_B}{V_0} \sum_\alpha \gamma_\alpha \frac{x_\alpha^2}{4 \sinh^2(x_\alpha/2)} \sim \frac{N}{V_0} \langle \gamma_\alpha \rangle . \qquad (2.32)$$

Again, the last term on the right-hand side is the high-temperature approximation.

What can we learn from the microscopic equations for the thermodynamic quantites? Some remarks and observations:

1. All of the sums in the expressions for the thermal contributions to thermodynamic quantities, that is, contributions from F_T, increase with increasing temperature. Thus the sign and qualitative behavior of these contributions can be read off from Eqs. (2.104)–(2.112). For example:

   a. K^T, proportional to γ^2, is negative and causes an increasing decrease in the bulk modulus as T increases.

   b. The thermal pressure is positive (this depends on the sign of γ, which is assumed/expected to be positive).

2. The thermal expansion is positive (this depends on the sign of γ, which is assumed/expected to be positive).

3. The equations for C_V and α_0 can be combined to give

$$\alpha_0 K = \overline{\gamma} \, \frac{C_V}{V_0} \, , \tag{2.33}$$

where $\overline{\gamma}$ is a weighted average of γ_α.

4. From Eq. (2.28) we have

$$P + dP = -\frac{\partial}{\partial(dV)}(F_1 + F_2) = P - K\frac{dV}{V_0} + \alpha_0 T_0 K\frac{dT}{T_0} \, , \tag{2.34}$$

$$K + dK = V\frac{\partial^2}{\partial(dV)^2}(F_2 + F_3) = K + \beta K\frac{dV}{V_0} + \alpha_1 T_0 K\frac{dT}{T_0} \, . \tag{2.35}$$

   a. For $dT = 0$ and $dP \neq 0$, $dV/V_0 = -dP/K$ and

$$dK = \beta K\frac{dV}{V_0} = -\beta d dP \, . \tag{2.36}$$

   Since we expect K to increase with P, we have $\beta < 0$. The cubic nonlinearity of Eq. (2.19) is the counterpart at the atomic level of description to the β term here.

   b. For $dT \neq 0$ and $dP = 0$, $dV/V_0 = \alpha_0 dT$ and

$$dK = K T_0(\beta\alpha_0 + \alpha_1)\frac{dT}{T_0} \, . \tag{2.37}$$

   Both contributions to the change in K are negative since $\beta < 0$, and from Eqs. (2.110) $\alpha_0 > 0$ and from Eqs. (2.111) $\alpha_1 < 0$. Thus $dT > 0$ produces a reduction in K, softening. The factor β, a measure of the cubic anharmonicity, is often quite large.

5. Finally, let us make an assessment of the numerical size of some of the quantities of interest. There are three energy scales involved, a potential energy scale, a quantum energy scale, and a thermal energy scale, Section 2.5.

The potential energy scale is $e_S = z\varepsilon_0 \approx 10^4$ K, the quantum energy scale is $e_Z = \hbar\omega_E \approx 300$ K, and the thermal energy scale is $e_T = k_B T \approx 300$ K. We have

$$K = K_S + K_T \approx K_S \,, \tag{2.38}$$

$$\alpha_0 = \alpha_T \,, \tag{2.39}$$

$$\beta = \beta_S + \beta_T \approx \beta_S \,, \tag{2.40}$$

where the approximations on the right-hand side are justified because all thermodynamic quantities involving E_ν scale as e_ν, $\nu = S, Z, T$, that is, $K_S \gg K_T$, etc. We have

$$K \approx 8e_S n \approx 10^{12} \text{ erg/cm}^3 = 10^5 \text{ MPa} \,, \tag{2.41}$$

$$\alpha_0 = \alpha_T = e_T n \frac{1}{K} \langle \gamma \rangle \approx \frac{e_T}{8e_S} < 10^{-3} \,, \tag{2.42}$$

$$\beta \approx 56 e_S n \frac{1}{K} \approx 7 \,, \tag{2.43}$$

where the inequality in the equation for α_0 arises because we use an estimate based on the high-temperature limit, an upper limit. These numbers are sensible.

The thermodynamics of a solid in the Gruneisen approximation provides evidence for the type of thermal/mechanical coupling terms we may have in modeling and it provides an estimate of the numerical value of important parameters that is founded in the microscopic description.

2.2.3
From Microscopic Model to Continuum Elasticity

In Section 2.2.1 we carried the microscopic picture forward from the basic energy in Eq. (2.1) to a description in terms of phonons that interact weakly through the cubic anharmonicity. We related numbers in the microscopic picture to the parameters of the phonon picture. In Section 2.2.2, using the Gruneisen treatment of the coupling between phonon frequency and volume change, we married the microscopic picture to thermodynamics and established a connection between microscopic quantities and thermodynamic quantities. It is the thermodynamic quantities that are most closely connected to the parameters of continuum elasticity. A numerical estimate of the bulk modulus, thermal expansion, and coefficient of cubic anharmonicity was possible. Finally, we want to return to the beginning, Eq. (2.1), and find continuum elasticity directly from the microscopic model. We do this following the scheme in Ashcroft and Mermin [4].

Consider a solid, a simple cubic lattice of atoms located at x_R near sites R, having energy of interaction, Eq. (2.1)

$$\mathcal{U} = \frac{1}{2} \sum_R \sum_{R'} V\left(R + u_R - R' - u_{R'}\right) , \tag{2.44}$$

where the atom at $x_R = R + u_R$ near lattice site R is denoted by the site label, R, and $V(r)$ is the energy of interaction between atoms separated by distance $r = |x_R - x_{R'}|$. The microscopic (phonon) description of this solid follows from developing $V(r)$ as a Taylor series in the displacements u_R. The continuum mechanics description of this solid is developed from the same starting point by regarding the displacements u_R as slowly varying functions of R. We will carry through the rudiments of this development for a special case. This will let us illustrate the type of treatment that one employs to do the most general problem and it will provide a recipe for finding the linear and nonlinear parameters of continuum elasticity from the microscopic interaction.

A systematic development of the potential energy in Eq. (2.44) makes use of the translation operator, T:

$$T(a) f(x) = e^{a \cdot \nabla} f(x) = f(x + a) . \tag{2.45}$$

We write

$$V\left(R + u_R - R' - u_{R'}\right) = T\left(u_R - u_{R'}\right) V\left(R - R'\right)$$
$$= e^{\left(u_R - u_{R'}\right) \cdot \nabla_R} V\left(R - R'\right) . \tag{2.46}$$

The idea is to insert this representation V into Eq. (2.44) and to use the Taylor series expansion of the translation operator. What results is a power series in u_R having coefficients that depend on the potential V and derivatives of V. If we carry through this program unmodified, it is simply an alternative way to get to Eq. (2.2), et seq. We carry through this program, but only after treatment of the translation operator that connects it to the continuum mechanics description of the interaction energy. The translation operator works on the atom/atom potential. This potential is typically a function of interatomic separation characterized by a microscopic length scale, a, say the Lennard–Jones interatomic potential in Eq. (2.4). Consider V a function of $(R - R')/a$, define $S = R/a$, and write

$$\left(u_R - u_{R'}\right) \cdot \nabla_R = \frac{\left(u_R - u_{R'}\right)}{a} \cdot \nabla_S . \tag{2.47}$$

For R and R' near one another, within the range of the atomic scale interaction, we assume the displacements u_R to be slowly varying in space, let $u_R \rightarrow u(x)$, and write

$$u_{R'} - u_R = \left[\left(R' - R\right) \cdot \nabla_x\right] u(x)\big|_{x=R} . \tag{2.48}$$

Thus we have

$$\left(u_R - u_{R'}\right) \cdot \nabla_R = \sum_\alpha \frac{\partial u_R^\alpha}{\partial x} \left(X - X'\right) \frac{\partial}{\partial S_\alpha} \equiv \sum_\alpha \Delta_\alpha \frac{\partial}{\partial S_\alpha} , \tag{2.49}$$

where $S = (X, Y, Z)$ and for economy of notation and clarity of presentation we have specialized to the case in which the displacements u_R depend only on x. (To have the complete theory of continuum elasticity, one would need to keep the Y and Z terms in this equation.) Here $\partial u_R^\alpha / \partial x$ means $\partial u^\alpha(x)/\partial x$ evaluated at $x = R$, that is, the strain at R. Going forward we will understand this R to be present and drop it from the notation. The quantities $(X - X')$ are of order 1 so the coefficient of $\partial / \partial S_\alpha$ is of the order of the strain and much less than 1. A series expansion of Eq. (2.46) is a power series in the strain field. Upon using Eqs. (2.46) and (2.49) in Eq. (2.44) we have

$$\mathcal{U} = U_0 + U_1 + U_2 + U_3 + \ldots, \tag{2.50}$$

where

$$U_0 = \frac{1}{2} \sum_R \sum_{R'} V\left(R - R'\right),$$

$$U_1 = \frac{1}{2} \sum_R \sum_\alpha \frac{\partial u^\alpha}{\partial x} C_\alpha,$$

$$U_2 = \frac{1}{2} \frac{1}{2!} \sum_R \sum_{\alpha\beta} \frac{\partial u^\alpha}{\partial x} \frac{\partial u^\beta}{\partial x} C_{\alpha\beta}, \tag{2.51}$$

$$U_3 = \frac{1}{2} \frac{1}{3!} \sum_R \sum_{\alpha\beta\gamma} \frac{\partial u^\alpha}{\partial x} \frac{\partial u^\beta}{\partial x} \frac{\partial u^\gamma}{\partial x} C_{\alpha\beta\gamma},$$

$$\vdots$$

and

$$C_\alpha = \sum_{S'} \left(X - X'\right) \frac{\partial V\left(S - S'\right)}{\partial S_\alpha},$$

$$C_{\alpha\beta} = \sum_{S'} \left(X - X'\right)^2 \frac{\partial^2 V\left(S - S'\right)}{\partial S_\alpha \partial S_\beta}, \tag{2.52}$$

$$C_{\alpha\beta\gamma} = \sum_{S'} \left(X - X'\right)^3 \frac{\partial V^3\left(S - S'\right)}{\partial S_\alpha \partial S_\beta \partial S_\gamma}.$$

The set of elastic constants, C_α, $C_{\alpha\beta}$, and $C_{\alpha\beta\gamma}, \ldots$, is independent of R. We take the strain to be a function of x and replace the sum on R with an integral over x. We have as an example

$$U_2 = \frac{1}{2} \int \frac{dx}{V} \sum_{\alpha\beta} \frac{\partial u^\alpha(x)}{\partial x} \frac{\partial u^\beta(x)}{\partial x} C_{\alpha\beta}, \tag{2.53}$$

where V is the volume of the sample.

To acquire some sense of the magnitude and character of the elastic constants, we will work out the value of C_α, $C_{\alpha\beta}$, and $C_{\alpha\beta\gamma}$ for the case of a central potential. For $V(\boldsymbol{R} - \boldsymbol{R'}) = V(|\boldsymbol{S} - \boldsymbol{S'}|)$ we have

$$C_\alpha = \sum_S X \frac{S_\alpha}{S} V' = 0 \,,$$

$$C_{\alpha\beta} = \sum_S X^2 \frac{S_\alpha S_\beta}{S^2} V'' \,,$$

$$C_{\alpha\beta\gamma} = \sum_S X^3 \left[\frac{S_\alpha S_\beta S_\gamma}{S^3} \left(V''' - 3\frac{V''}{S} \right) + \frac{S_\alpha \delta_{\beta\gamma} + S_\beta \delta_{\alpha\gamma} + S_\gamma \delta_{\beta\alpha}}{S^2} V'' \right] \,.$$

$$(2.54)$$

Again to reduce complexity we have chosen the sum on $\boldsymbol{S'}$ to go over the near neighbors only, replaced $\boldsymbol{S} - \boldsymbol{S'}$ by $\boldsymbol{S}$, and placed the near-neighbor distance at the minimum of the interaction potential. Thus $V' = 0$, $C_\alpha = 0$ and V' does not appear in the equations for $C_{\alpha\beta}$ and $C_{\alpha\beta\gamma}$. To evaluate the sum over $\boldsymbol{S}$ we replace it with an integral over a solid angle (all of the particles contributing to the elastic constants in this model are equidistant from $\boldsymbol{R}$, the first shell of neighbors of $\boldsymbol{R}$, at $|\boldsymbol{S}| = S = 1$). For $C_{\alpha\beta}$ we find

$$C_{\alpha\beta} = C_{xx} \, \delta_{\alpha x}\delta_{\beta x} + C_{yy} \, \delta_{\alpha\beta} \left(\delta_{\alpha y} + \delta_{\alpha z} \right) \,,$$

$$C_{xx} = V'' \sum_S X^4 = \frac{1}{5} z V'' \,,$$

$$(2.55)$$

$$C_{yy} = V'' \sum_S X^2 Y^2 = \frac{2}{15} z V'' \,,$$

with

$$\sum_S = z \int \frac{d\Omega}{4\pi} \,,$$

$$(2.56)$$

where z is the number of near neighbors of site $\boldsymbol{R}$. For $C_{\alpha\beta\gamma}$ we find

$$C_{\alpha\beta\gamma} = C_{xxx}\delta_{\alpha x}\delta_{\beta x}\delta_{\gamma x} + C_{xyy} P\delta_{\alpha x}[\delta_{\beta y}\delta_{\gamma y} + \delta_{\beta z}\delta_{\gamma z}] \,,$$

$$C_{xxx} = \frac{1}{35} z \left(5V''' + 6V'' \right) \,,$$

$$(2.57)$$

$$C_{xyy} = \frac{1}{35} z \left(V''' + V'' \right) \,,$$

where P is the permutation operator on the indices α, β, and γ.

Finally, to see some numbers we need a model for the pair interaction V. Once again we take the Lennard–Jones potential, Eq. (2.4), which can be put in the form

$$V(S) = \varepsilon_0 \left(\frac{1}{S^{12}} - \frac{2}{S^6} \right) \,.$$

$$(2.58)$$

Then

$$V(S = 1) = -\varepsilon_0 ,$$
$$V'(S = 1) = 0 ,$$
$$V''(S = 1) = 72\varepsilon_0 ,$$
$$V'''(S = 1) = -21 \cdot 72\varepsilon_0 ,$$
$$B_{xx} = \frac{72}{5} ,$$
$$B_{yy} = \frac{48}{5} ,$$ (2.59)
$$B_{xxx} = \frac{-7128}{35} ,$$
$$B_{xyy} = \frac{-1440}{35} ,$$

where $B_\nu = C_\nu/(z\varepsilon_0)$. With B_ν we measure the strength of the linear and nonlinear elastic constants in units of the microscopic energy, $e_S = z\varepsilon_0$, Eq. (2.119). Of most interest are the ratios; we have $C_{yy}/C_{xx} = 0.6667$, $C_{xxx}/C_{xx} \approx 14.14$, and $C_{xyy}/C_{xxx} \approx 0.20$. Compare to Eqs. (2.21).

Using the expressions for $C_{\alpha\beta}$ and $C_{\alpha\beta\gamma}$ from Eqs. (2.55) and (2.57) leads to the energy as a function of the strain field given by

$$U_0 = -\frac{z}{2}\varepsilon_0 ,$$
$$U_1 = 0 ,$$
$$U_2 = \frac{1}{2}\frac{1}{2!}\int \frac{dx}{V}\left[C_{xx}\left(\frac{\partial u}{\partial x}\right)^2 + C_{yy}\left(\frac{\partial v}{\partial x}\right)^2 + C_{yy}\left(\frac{\partial w}{\partial x}\right)^2\right] ,$$
$$U_3 = \frac{1}{2}\frac{1}{3!}\int \frac{dx}{V}\left[C_{xxx}\left(\frac{\partial u}{\partial x}\right)^3 + 3C_{xyy}\frac{\partial u}{\partial x}\left[\left(\frac{\partial v}{\partial x}\right)^2 + \left(\frac{\partial w}{\partial x}\right)^2\right]\right] ,$$

(2.60)

where we have used $\boldsymbol{u} = (u, v, w)$. We find the stress σ_{xx} and σ_{yx} from U to be, Eq. (2.75),

$$\sigma_{xx} = \frac{\delta U}{\delta u_x} = \frac{1}{2}C_{xx}\frac{\partial u}{\partial x} + \frac{1}{4}C_{xxx}\left(\frac{\partial u}{\partial x}\right)^2 + \frac{1}{4}C_{xyy}\left[\left(\frac{\partial v}{\partial x}\right)^2 + \left(\frac{\partial w}{\partial x}\right)^2\right] ,$$
$$\sigma_{yx} = \frac{\delta U}{\delta v_x} = \frac{1}{2}C_{yy}\frac{\partial v}{\partial x} + \frac{1}{2}C_{xyy}\frac{\partial u}{\partial x}\frac{\partial v}{\partial x} .$$

(2.61)

From these stresses we find the equation of motion for the x and y strains [15], Eq. (2.76),

$$\varrho \frac{\partial^2 u}{\partial t^2} = \frac{\partial \sigma_{xx}}{\partial x} = \frac{1}{2} C_{xx} \frac{\partial^2 u}{\partial x^2} + \frac{1}{2} C_{xxx} \frac{\partial u}{\partial x} \frac{\partial^2 u}{\partial x^2} + \frac{1}{4} C_{xyy} \frac{\partial}{\partial x} \left[\left(\frac{\partial v}{\partial x} \right)^2 + \left(\frac{\partial w}{\partial x} \right)^2 \right],$$

$$\varrho \frac{\partial^2 v}{\partial t^2} = \frac{\partial \sigma_{yx}}{\partial x} = \frac{1}{2} C_{yy} \frac{\partial^2 v}{\partial x^2} + \frac{1}{2} C_{xyy} \frac{\partial}{\partial x} \left(\frac{\partial u}{\partial x} \frac{\partial v}{\partial x} \right).$$

$$\tag{2.62}$$

These equations have the same form as the equations treated by Goldberg [16], Polyakova [17], McCall [18], and others. We will come back to them below.

So far we have employed the microscopic description of a solid to develop a microscopic description of the displacement field, to establish the proper setting for the coupling between the displacement field and thermodynamics, to see the source of continuum elasticity, and to find numbers. But there are circumstances in which the nature of a physical system is such that stepping back to the microscopics is inconvenient and may not even be a good idea. The theory of elasticity covers these cases.

2.3
Continuum Elasticity and Macroscopic Phenomenology

2.3.1
Displacement, Strain, and Stress

The macroscopic theory of elasticity has no microscopic underpinnings. It follows from several simple assertions. A physical system is taken to be such that all of the pieces of material in the system have a well-defined position when the system is in mechanical and thermal equilibrium. Displacement of a piece of the material away from equilibrium is described by the strain tensor [15],

$$u_{ij} = \frac{1}{2} \left(\frac{\partial u_i}{\partial x_j} + \frac{\partial u_j}{\partial x_i} + \sum_k \frac{\partial u_k}{\partial x_j} \frac{\partial u_k}{\partial x_i} \right) = \hat{u}_{ij} + \frac{1}{2} \left(\sum_k \frac{\partial u_k}{\partial x_j} \frac{\partial u_k}{\partial x_i} \right),$$

$$u_{ij} = u_{ji}, \quad i = 1, 2, 3, \quad j = 1, 2, 3,$$

$$\tag{2.63}$$

where $(x_1, x_2, x_3) = (x, y, z)$ and $(u_1, u_2, u_3) = (u, v, w)$ are the displacements in directions (x, y, z). The quantities $\hat{u}_{ij}$ are termed the rudimentary strains. The energy density, e_V, that results from a strain can be developed as a power series in the strain. The leading terms in this series, correct to the second order in the rudimentary strains, are [15]

$$e_V = \frac{1}{2} \sum_{ijkl} \hat{u}_{ij} c_{ij,kl} \hat{u}_{kl}, \tag{2.64}$$

where $c_{ij,kl}$ is the elastic tensor. The form of the elastic tensor as well as the form of the most general power series for e_V is limited only by the spatial symmetry of the physical system. As we are concerned with physical systems that are a random assembly of mesoscopic pieces, we may assume they are spatially isotropic. In this circumstance, $c_{ij,kl} = (K - 2\mu/3)\delta_{ij}\delta_{kl} + \mu(\delta_{ik}\delta_{jl} + \delta_{il}\delta_{jk})$, where K and μ are the linear elastic parameters, the *bulk* and *shear* modulus, respectively [15]. We then have the energy per unit volume in the form

$$
e_V = \mu S(\boldsymbol{u} \cdot {*}\boldsymbol{u}) + \left(\frac{1}{2}K - \frac{1}{3}\mu\right) T(\boldsymbol{u}) T(\boldsymbol{u})
$$
$$
+ \frac{A}{3} T(\boldsymbol{u} * \boldsymbol{u} * \boldsymbol{u}) + B S(\boldsymbol{u} \cdot {*}\boldsymbol{u}) T(\boldsymbol{u}) + \frac{C}{3} T(\boldsymbol{u}) T(\boldsymbol{u}) T(\boldsymbol{u}) + \ldots \;,
$$

(2.65)

where A, B, and C are the leading nonlinear elastic parameters (the analog of cubic anharmonicity) allowed by isotropic symmetry [15]. The strain field is described by the tensor

$$
\boldsymbol{u} = \begin{bmatrix} u_{xx} & u_{xy} & u_{xz} \\ u_{yx} & u_{yy} & u_{yz} \\ u_{zx} & u_{zy} & u_{zz} \end{bmatrix} ,
$$

(2.66)

where $*$ is standard matrix multiplication, $\cdot{*}$ is element-by-element matrix multiplication, T is the trace of the matrix and S is the sum over all elements of the matrix. For example, $T(\boldsymbol{u}) = \sum_k u_{kk}$, $S(\boldsymbol{u} \cdot {*}\boldsymbol{u}) = \sum_i \sum_j u_{ij} u_{ij} = \sum_i \sum_j u_{ij}^2$, etc. Because of the second-order term in $\partial u/\partial x$ in the definition of u_{ij}, the energy e_V is not in the form of a power series in the rudimentary strain, that is, $\partial u/\partial x$. Putting it in that form we find

$$
e_V = \frac{\mu}{2} \left[S(\boldsymbol{u} \cdot {*}\boldsymbol{u}) + T(\boldsymbol{u} * \boldsymbol{u}) \right] + \left(\frac{K}{2} - \frac{\mu}{3}\right) T(\boldsymbol{u}) T(\boldsymbol{u})
$$
$$
+ \left(\mu + \frac{A}{4}\right) S(\boldsymbol{u} * \boldsymbol{u} \cdot {*}\boldsymbol{u}) + \left(\frac{B}{2} + \frac{K}{2} - \frac{\mu}{3}\right) S(\boldsymbol{u} \cdot {*}\boldsymbol{u}) T((\boldsymbol{u})
$$
$$
+ \frac{A}{12} T(\boldsymbol{u} * \boldsymbol{u} * \boldsymbol{u}) + \frac{B}{2} T(\boldsymbol{u} * \boldsymbol{u}) T(\boldsymbol{u}) + \frac{C}{3} T(\boldsymbol{u}) T(\boldsymbol{u}) T(\boldsymbol{u}) + \ldots \sim \;,
$$

(2.67)

where henceforth $\boldsymbol{u}$ has the simple form

$$
\boldsymbol{u} = \begin{bmatrix} u_x & u_y & u_z \\ v_x & v_y & v_z \\ w_x & w_y & w_z, \end{bmatrix}
$$

(2.68)

with $u_x = \partial u/\partial x$, etc. This representation of the elastic energy is very convenient for examining a series of problems related to the practical use of nonlinear elasticity as well as formulating a description of schemes for learning the values of nonlinear coefficients.

Let us make contact between macroscopic elasticity theory and the results, Eq. (2.60) in Section 2.2.3, from the microscopic description. The situation developed in Section 2.2.3 is limited to variation of the displacements in the *x*-direction. That would mean taking

$$
\boldsymbol{u} = \begin{bmatrix} u_x & 0 & 0 \\ v_x & 0 & 0 \\ w_x & 0 & 0 \end{bmatrix}
$$

(2.69)

in Eq. (2.68) with the result

$$e_V = \left(\frac{2\mu}{3} + \frac{K}{2}\right)\left(\frac{\partial u}{\partial x}\right)^2 + \frac{\mu}{2}\left[\left(\frac{\partial v}{\partial x}\right)^2 + \left(\frac{\partial w}{\partial x}\right)^2\right]$$

$$+ \left(\frac{2\mu}{3} + \frac{K}{2} + \frac{A}{3} + B + \frac{C}{3}\right)\left(\frac{\partial u}{\partial x}\right)^3 \tag{2.70}$$

$$+ \left(\frac{2\mu}{3} + \frac{K}{2} + \frac{A}{4} + \frac{B}{2}\right)\frac{\partial u}{\partial x}\left[\left(\frac{\partial v}{\partial x}\right)^2 + \left(\frac{\partial w}{\partial x}\right)^2\right].$$

The involvement of the strain field in this equation is the same as that found in Eq. (2.60). The macroscopic theory of elasticity has no qualitative phenomena that are not also in the microscopic theory of elasticity. But the quantitative connection is less exact. Since for a uniform system $\mathcal{E} = \int d\mathbf{x} e_V$, comparison of this equation with Eq. (2.60) leads to

$$nC_{xx}/4 = \frac{2\mu}{3} + \frac{K}{2} \tag{2.71}$$

$$nC_{yy}/4 = \frac{\mu}{2}, \tag{2.72}$$

$$nC_{xxx}/12 = \frac{2\mu}{3} + \frac{K}{2} + \frac{A}{3} + B + \frac{C}{3} \tag{2.73}$$

$$nC_{xyy}/12 = \frac{2\mu}{3} + \frac{K}{2} + \frac{A}{4} + \frac{B}{2}, \tag{2.74}$$

where $n = N/V$ is the volume per particle. The generality of an energy based solely on symmetry allows greater flexibility in the behavior of the elastic constants than does the Hamiltonian model. For elastic systems built up from mesoscopic elastic elements we have no reason to reject any of the possibilities in the macroscopic theory.

2.3.2
Dynamics of the Displacement Field

The dynamics of the macroscopic elastic field is given by an equation for finding the stress associated with a strain,

$$\sigma_{ij} = \frac{\partial e_V}{\partial(\partial u_i/\partial x_j)}, \tag{2.75}$$

and the analog of $\mathbf{F} = m\mathbf{a}$,

$$\varrho\ddot{u}_i = \sum_k \frac{\partial \sigma_{ik}}{\partial x_k}, \tag{2.76}$$

where ϱ is the mass density. For example, from these equations and the energy density in Eq. (2.70) (or Eqs. (2.59) and (2.60)) the equation of motion for a y displacement, v, propagating in the x-direction is

$$\varrho\ddot{v} = \frac{\partial \sigma_{yx}}{\partial x} = \frac{1}{2}C_{yy}\frac{\partial^2 v}{\partial x^2} + \frac{1}{2}C_{xyy}\frac{\partial}{\partial x}\left(\frac{\partial u}{\partial x}\frac{\partial v}{\partial x}\right), \tag{2.77}$$

where we have used the definition of linear and nonlinear coefficients from Eq. (2.60) for economy of writing.

2.3.3
Coupling Continuum Elasticity to Auxiliary Fields

The macroscopic elastic field can couple to other fields in the system. The nature of this coupling may be complicated by system-specific details. Let us ignore these for the moment and adopt the isotropic model above for both the temperature and the fluid configurations in the pore space. We couple changes in temperature $dT = T - T_0 \ll T_0$ to the strain field as in Eq. (2.28) with the role of dV/V_0 taken by $\nabla \cdot \boldsymbol{u} = T(\boldsymbol{u})$,

$$e_V^T = -\alpha_0 K T(\boldsymbol{u}) dT + \alpha_1 K \, T(\boldsymbol{u}) T(\boldsymbol{u}) dT \,, \tag{2.78}$$

where α_0 is the linear thermal expansion. The second term here is the analog of the last term in Eq. (2.28). Under certain circumstances such a term can produce the elastic equivalent of the Luxemburg–Gorky effect. We couple the saturation S_W to the strain field in a similar manner, Eq. (2.28),

$$e_V^W = K_W T(\boldsymbol{u}) dS_W \,, \tag{2.79}$$

where K_W is a coefficient to be learned from experiment and/or from examination of the forces exerted between a liquid arrangement in a pore space and the walls of the pore space.

2.3.4
Inhomogeneous Elastic Systems

There are circumstances when the linear and nonlinear elastic constants depend on position, $\boldsymbol{x}$. This might also be true of the constants characterizing the coupling of the elastic field to the auxiliary fields. The range of possibilities is too extensive to attempt to write equations of suitable generality, but a few examples serve to suggest what is possible. For a linear elastic system that is layered in the z-direction (say bedding planes) one would write the energy

$$e_V = \frac{\mu(z)}{2} \left[S(\boldsymbol{u} \cdot \ast \boldsymbol{u}) + T(\boldsymbol{u} \ast \boldsymbol{u}) \right] + \left(\frac{K(z)}{2} - \frac{\mu(z)}{3} \right) T(\boldsymbol{u}) T(\boldsymbol{u}) \,. \tag{2.80}$$

A spatially local nonlinearity might lead to a term in the elastic energy like, see Eq. (2.67),

$$e_V^{(C)} = \frac{C(\boldsymbol{x})}{3} T(\boldsymbol{u}) T(\boldsymbol{u}) T(\boldsymbol{u}) \,. \tag{2.81}$$

A spatially local source of coupling a nonlinear strain field and the temperature field would arise from

$$e_V^T = \alpha_1(\boldsymbol{x}) K T(\boldsymbol{u}) T(\boldsymbol{u}) dT \,. \tag{2.82}$$

In some cases it might be necessary to have equations describing the dynamics of the auxiliary fields, for example, a diffusion equation for the time evolution of temperature. We will encounter some of these more complicated situations below and deal with them appropriately. These few examples are primarily a forewarning of what is possible in a place where general principles are under discussion.

Our next task is to examine some consequences of the theory of continuum elasticity. We must know what it says in order to know when we encounter phenomena beyond its purview.

2.4
Thermodynamics

2.4.1
Thermodynamic Derivatives

The thermodynamic derivatives of interest are

$$S = -\frac{\partial F}{\partial T} = -F^{100}, \tag{2.83}$$

$$P = -\frac{\partial F}{\partial V} = -F^{010}, \tag{2.84}$$

$$\mu_W = \frac{\partial F}{\partial W} = F^{001}, \tag{2.85}$$

$$C_V = T\frac{\partial S}{\partial T} = -T\frac{\partial^2 F}{\partial T^2} = -TF^{200}, \tag{2.86}$$

$$K = -V\frac{\partial P}{\partial V} = V\frac{\partial^2 F}{\partial V^2} = VF^{020}, \tag{2.87}$$

$$K_W = -W\frac{\partial P}{\partial W} = W\frac{\partial^2 F}{\partial W \partial V} = WF^{011}, \tag{2.88}$$

$$\alpha K = \frac{\partial P}{\partial T} = -\frac{\partial^2 F}{\partial V \partial T} = -F^{110}, \tag{2.89}$$

$$\beta = \frac{V^2}{K}\frac{\partial^3 F}{\partial V^3} = \frac{V^2}{K}F^{030} \tag{2.90}$$

where S = entropy, P = pressure, C_V = specific heat at constant volume, K = bulk modulus, α = thermal expansion, and μ_W is the thermodynamic conjugate of W.

2.4.2
Series Expansion for E_S

To have a tractable form for E_S we create the simple model

$$E_S = NzV(r(V, W)), \tag{2.91}$$

where z is a coordination number, $V(r)$ is from Eq. (2.4), and

$$r(V, W) = a \left(1 + \frac{1}{3} \frac{dV}{V_0} + \frac{\lambda}{3} \frac{dW}{W_0} \right) . \tag{2.92}$$

The coefficient λ describes the influence of W on the equilibrium separation between particles. Then

$$\frac{E_S}{Nz\varepsilon_0} = -1 + 4 \left(\frac{dV}{V_0} \right)^2 + 8\lambda \frac{dV}{V_0} \frac{dW}{W_0}$$
$$+ \frac{28}{3} \left(\frac{dV}{V_0} \right)^3 + \dots, \tag{2.93}$$

where we have kept terms out to dV^3 but only the leading term in dW. There is no first-order term in dV or dW. The contribution of E_S to the thermodynamic derivatives of interest is

$$K^S = 8z\varepsilon_0 \frac{N}{V_0} , \tag{2.94}$$

$$K_W^S = 8\lambda z\varepsilon_0 \frac{N}{V_0} , \tag{2.95}$$

$$\beta^S = 56z\varepsilon_0 \frac{N}{KV_0} . \tag{2.96}$$

2.4.3
Series Expansion for E_Z

Using Eq. (2.25) in Eq. (2.23) we have

$$E_Z = E_Z^{00} + E_Z^{10} \frac{dV}{V_0} + E_Z^{01} \frac{dW}{W_0} , \tag{2.97}$$

where

$$E_Z^{00} = -\frac{1}{2} \sum_\alpha \hbar\omega_\alpha^0 ,$$

$$E_Z^{10} = -\frac{1}{2} \sum_\alpha \hbar\omega_\alpha^0 \gamma_\alpha , \tag{2.98}$$

$$E_Z^{01} = -\frac{1}{2} \sum_\alpha \hbar\omega_\alpha^0 \Gamma_\alpha .$$

The contribution of E_Z to the thermodynamic derivatives of interest is

$$P^Z = -\frac{E_Z^{10}}{V_0} , \tag{2.99}$$

$$\mu_W^Z = \frac{E_Z^{01}}{V_0} . \tag{2.100}$$

2.4.4
Series Expansion for F_T

1. Terms in dT^0

$$\frac{F_T^0}{k_B T_0} = \sum_\alpha \ln(1 - e^{-x_\alpha})$$

$$- \sum_\alpha \frac{x_\alpha}{e^{x_\alpha} - 1} da - \sum_\alpha \frac{x_\alpha}{e^{x_\alpha} - 1} db$$

$$- \sum_\alpha \frac{x_\alpha^2}{4 \sinh^2(x_\alpha/2)} \frac{(da)^2}{2!} - \sum_\alpha \frac{x_\alpha^2}{4 \sinh^2(x_\alpha/2)} da\, db$$

$$- \sum_\alpha \frac{x_\alpha^2}{4 \sinh^2(x_\alpha/2)} x_\alpha \coth(x_\alpha/2) \frac{(da)^3}{3!}$$

$$+ \cdots .$$

$$\text{(2.101)}$$

in
2. Terms in dT^1

$$\frac{F_T^1}{k_B T_0} = \sum_\alpha \left(-\frac{x_\alpha}{e^{x_\alpha} - 1} + \ln(1 - e^{-x_\alpha}) \right) dc$$

$$- \sum_\alpha \frac{x_\alpha^2}{4 \sinh^2(x_\alpha/2)} da\, dc - \sum_\alpha \frac{x_\alpha^2}{4 \sinh^2(x_\alpha/2)} db\, dc$$

$$- \sum_\alpha \frac{x_\alpha^2}{4 \sinh^2(x_\alpha/2)} \left(x_\alpha \coth(x_\alpha/2) - 1 \right) \frac{(da)^2}{2!} dc \qquad \text{(2.102)}$$

$$+ \cdots .$$

3. Terms in dT^2

$$\frac{F_T^2}{k_B T_0} = - \sum_\alpha \frac{x_\alpha^2}{4 \sinh^2(x_\alpha/2)} \frac{dc^2}{2!}$$

$$- \sum_\alpha \frac{x_\alpha^2}{4 \sinh^2(x_\alpha/2)} \left(x_\alpha \coth(x_\alpha/2) - 2 \right) da \frac{(dc)^2}{2!} \qquad \text{(2.103)}$$

$$+ \cdots .$$

From these results find the thermal contribution to the following thermodynamic derivatives:

$$S = k_B \sum_\alpha \left(\frac{x_\alpha}{e^{x_\alpha} - 1} - \ln(1 - e^{-x_\alpha}) \right) \sim k_B \sum_\alpha \ln(x_\alpha), \qquad \text{(2.104)}$$

$$P^T = \frac{k_B T_0}{V_0} \sum_\alpha \gamma_\alpha \frac{x_\alpha}{e^{x_\alpha} - 1} \sim \frac{k_B T_0}{V_0} \sum_\alpha \gamma_\alpha, \qquad \text{(2.105)}$$

$$\mu_W^T = -\frac{k_B T_0}{V_0} \sum_\alpha \Gamma_\alpha \frac{x_\alpha}{e^{x_\alpha} - 1} \sim \frac{k_B T_0}{V_0} \sum_\alpha \Gamma_\alpha, \tag{2.106}$$

$$C_V = k_B \sum_\alpha \frac{x_\alpha^2}{4 \sinh^2(x_\alpha/2)} \sim k_B \sum_\alpha 1, \tag{2.107}$$

$$K^T = -\frac{k_B T_0}{V_0} \sum_\alpha \gamma_\alpha^2 \frac{x_\alpha^2}{4 \sinh^2(x_\alpha/2)} \sim -\frac{k_B T_0}{V_0} \sum_\alpha \gamma_\alpha^2, \tag{2.108}$$

$$K_W^T = -\frac{k_B T_0}{V_0} \sum_\alpha \gamma_\alpha \Gamma_\alpha \frac{x_\alpha^2}{4 \sinh^2(x_\alpha/2)} \sim -\frac{k_B T_0}{V_0} \sum_\alpha \gamma_\alpha \Gamma_\alpha, \tag{2.109}$$

$$\alpha_0 K = \frac{k_B}{V_0} \sum_\alpha \gamma_\alpha \frac{x_\alpha^2}{4 \sinh^2(x_\alpha/2)} \sim \frac{k_B}{V_0} \sum_\alpha \gamma_\alpha, \tag{2.110}$$

$$\alpha_1 K = -\frac{k_B}{V_0} \sum_\alpha \gamma_\alpha^2 \frac{x_\alpha^2}{4 \sinh^2(x_\alpha/2)} (x_\alpha \coth(x_\alpha/2) - 1) \sim -\frac{k_B}{V_0} \sum_\alpha \gamma_\alpha^2 \frac{x_\alpha^2}{3}, \tag{2.111}$$

$$\beta^T = -\frac{k_B T_0}{K V_0} \sum_\alpha \gamma_\alpha^3 \frac{x_\alpha^2}{4 \sinh^2(x_\alpha/2)} x_\alpha \coth(x_\alpha/2) \sim -2 \frac{k_B T_0}{K V_0} \sum_\alpha \gamma_\alpha^3. \tag{2.112}$$

In these equations the second formula on the right-hand side comes from a high-temperature treatment of the integrand in each sum.

In those cases where the quantity of interest is related to a temperature change, the only contribution comes from F_T and there is no T superscript, for example, S, C, ... from Eqs. (2.83), (2.86), ... For quantites like P and K, Eqs. (2.84), (2.87), ..., there are nonthermal contributions from E_S and E_Z.

2.4.5
Assemble the Pieces

For the quantities that have nonthermal contributions we have:
P:

$$P = \frac{E_Z^{10}}{V_0} + P_0^T = \frac{E_Z^{10}}{V_0} + \frac{k_B T_0}{V_0} \sum_\alpha \gamma_\alpha \frac{x_\alpha}{e^{x_\alpha} - 1} \sim \frac{E_Z^{10}}{V_0} + k_B T_0 \frac{N}{V_0} \langle \gamma_\alpha \rangle; \tag{2.113}$$

μ_W:

$$\mu_W = \frac{E_Z^{01}}{V_0} + \mu_W^T = \frac{E_Z^{01}}{V_0} - \frac{k_B T_0}{V_0} \sum_\alpha \Gamma_\alpha \frac{x_\alpha}{e^{x_\alpha} - 1} \sim \frac{E_Z^{01}}{V_0} - k_B T_0 \frac{N}{V_0} \langle \Gamma_\alpha \rangle; \tag{2.114}$$

K:

$$K = 8 z \varepsilon_0 \frac{N}{V_0} + K_0^T = 8 z \varepsilon_0 \frac{N}{V_0} - \frac{k_B T_0}{V_0} \sum_\alpha \gamma_\alpha^2 \frac{x_\alpha^2}{4 \sinh^2(x_\alpha/2)} \sim 8 z \varepsilon_0 \frac{N}{V_0}$$
$$- k_B T_0 \frac{N}{V_0} \langle \gamma_\alpha^2 \rangle; \tag{2.115}$$

K_W:

$$
\begin{aligned}
K_W &= W_0 \frac{\partial^2 F}{\partial W \partial V} = 8\lambda z \varepsilon_0 \frac{N}{V_0} - \frac{k_B T_0}{V_0} \sum_\alpha \gamma_\alpha \Gamma_\alpha \frac{x_\alpha^2}{4\sinh^2(x_\alpha/2)} \\
&\sim 8\lambda z \varepsilon_0 \frac{N}{V_0} - k_B T_0 \frac{N}{V_0} \langle \gamma_\alpha \Gamma_\alpha \rangle ;
\end{aligned}
\tag{2.116}
$$

β:

$$
\begin{aligned}
\beta = \beta^S + \beta^T &= 56z \frac{\varepsilon_0}{K} \frac{N}{V_0} - \frac{k_B T_0}{K V_0} \sum_\alpha \gamma_\alpha^3 \frac{x_\alpha^2}{4\sinh^2(x_\alpha/2)} x_\alpha \coth(x_\alpha/2) , \\
&= 56z \frac{\varepsilon_0}{K} \frac{N}{V_0} - 2\frac{k_B T_0}{K} \frac{N}{V_0} \langle \gamma_\alpha^3 \rangle .
\end{aligned}
\tag{2.117}
$$

In these equations we use the definition

$$
\langle X \rangle = \frac{1}{N} \sum_\alpha X_\alpha .
\tag{2.118}
$$

Typically γ is of order 1, so $\langle \gamma \rangle$ and $\langle \gamma^2 \rangle$ are numbers of order 3 since α has *3N* values.

2.5
Energy Scales

The quantities that are assembled to form the thermodynamic variables have values that can be estimated as here. There are three energy scales, the potential energy that scales as $z\varepsilon_0$, the zero-point energy (a quantum mechanical energy), and the thermal energy, $k_B T$. We consider the case in which the mass of each particle is $m = 60$ amu, the strength of the interparticle interaction is $\varepsilon_0 = 0.5$ eV, and the interparticle spacing is $a = 0.4$ nm.

1. Potential energy:

$$
e_S = z\varepsilon_0 \approx 60\,000\ \mathrm{K}, \quad z = 12 .
\tag{2.119}
$$

2. Quantum energy: We use the Einstein oscillator approximation

$$
E_Z = \frac{1}{2} \sum_\alpha \hbar \omega_\alpha \approx \frac{3}{2} N \hbar \omega_E ,
\tag{2.120}
$$

where $\omega_E^2 = \Gamma_2/m$ and Γ_2 from Eq. (2.20) is

$$
\Gamma_2 = 36z \frac{\varepsilon_0}{a^2} .
\tag{2.121}
$$

Thus the basic quantum energy is to within a numerical factor the geometric mean of the $z\varepsilon_0$ and $\hbar^2/ma^2$:

$$e_Z = \hbar\omega_E = 6\sqrt{z\varepsilon_0 \frac{\hbar^2}{ma^2}} \approx 300\,\text{K}. \tag{2.122}$$

3. Thermal energy:

$$e_T = k_\text{B}T \approx 300\,\text{K} \tag{2.123}$$

at room temperature.

4. We have

$$e_S \gg e_Z \approx e_T. \tag{2.124}$$

The high-temperature approximations in Eqs. (2.104)–(2.112) are for $e_T \gg e_Z$.

References

1 Guyer, R.A. (1969) *Solid State Physics*, vol. 23 (eds Seitz, F., Turnbull, D. and Ehrenreich, H.) John Wiley & Sons, Inc., New York.

2 Barranco, M., Dalfovo, F., and Navarro, J. (2002) *Quantum Fluid Clusters*, 5th Workshop, Trento.

3 Kittel, C. (1996) *Introduction to Solid State Physics*, 7th edn, John Wiley & Sons, Inc, New York.

4 Ashcroft, N.W. and Mermin, N.D. (1976) *Solid State Physics*, Brooks/Cole, United States.

5 Ross, M. (1969), Generalized Lindemann Melting Law. *Phys. Rev.*, **184**, 233–242.

6 Hirschfelder, J.O., Curtiss, C.F., and Bird, R.B., *Molecular theory of gases and liquids*, John Wiley & Sons, Inc., New York, (1964).

7 Migliori, A. and Sarrao, J.L. (1997) *Resonant Ultrasound Spectroscopy*, John Wiley & Sons, Inc., New York.

8 Carruthers, P. (1961) Theory of thermal conductivity of solids at low temperatures. *Rev. Mod. Phys.*, **33**, 92–138.

9 Pathria, R.K. (1972) *Statistical Mechanics*, Pergamon Press, Oxford.

10 Huang, K. (1987) *Statistical Mechanics*, 2nd edn, John Wiley & Sons, Inc., New York.

11 Guyer, R.A. (2005) *The Science of Hysteresis III*, (eds Bertotti, G., Mayergoyz, I.) Academic Press, San Diego.

12 Ziman, J.M. (1972) *Principles of the Theory of Solids*, Cambridge University Press, Cambridge.

13 Anderson, O.L. (1995) *Equations of State of Solids for Geophysics and Ceramic Science*, Oxford University Press, New York, (1995).

14 Zaitsev, V., Gusev, V., and Castagnede, B. (1999) The Luxembourg–Gorky effect retooled for elastic waves: a mechanism and experimental evidence. *Phys. Rev. Lett.*, **89**, 105502–105505.

15 Landau, L.D. and Lifshitz, E.M. (1999) *Theory of Elasticity*, Butterworth and Heinemann, Oxford.

16 Goldberg, Z.A. (1960) Interaction of plane longitudinal and transverse elastic waves. *Sov. Phys.-Acoust.*, **6**, 306–310.

17 Polyakova, A.L. (1964) Nonlinear effects in a solid. *Sov. Phys.-Solid State*, **6**, 50–53.

18 McCall, K.R. (1994) Theoretical study of nonlinear elastic wave propagation. *J. Geophys. Res.*, **99**, 2591–2600.

3
Traditional Theory of Nonlinear Elasticity, Results

In this chapter we sketch some of the consequences of the traditional theory of nonlinear elasticity described in Chapter 2. In so doing our idea is not to produce the contents of a text on elasticity theory but rather to look at those phenomena that will receive treatment later on, using nontraditional nonlinear elasticity. To have a standard for comparison we want to see what things look like in the traditional theory. Further, much of the analytic apparatus we employ is introduced as we go. The chapter is organized in five sections that go from *quasistatic to dynamic* and from *linear to nonlinear*. See Figure 3.1, which serves as a rough outline. In Sections 3.1.1 and 3.1.2 we examine the linear and nonlinear quasistatic stress-strain response. The essential consequence of nonlinearity is the coalesence of strain fields. Thus the leading nonlinear quasistatic response is the joining of two quasistatic strain fields to form a third. In Section 3.2 we discuss the linear dynamic stress-strain response, that is, the linear wave equation. Because of nonlinear coupling, a quasistatic strain and a dynamic strain interact to give a dynamic strain at shifted velocity. This is discussed in detail in Section 3.3.

The bulk of the chapter is devoted to nonlinear elasticity in dynamics, Section 3.4. The basic equations, nonlinear coupled wave equations, are set out in Section 3.4.1. Wave propagation and resonance bar phenomena are described in Sections 3.4.2.1 and 3.4.2.2, respectively. Two particular processes, $l + l \rightarrow l$ and $l + t \rightarrow t$, are decribed for both wave propagation and in a resonance bar. Higher-order processes and selection rules for interacting waves are discussed in Section 3.4.2.3.

In Section 3.5 we examine the Luxemberg–Gorky effect, an example of the nonlinear coupling of a strain field to an auxiliary field, the temperature. In this section and throughout we emphasize the diagrammatic description of interaction processes. This description allows us to identify the important physical variables in a process from a picture of the process. The Luxemberg–Gorky effect provides a nice example of this way of thinking.

A final section, Section 3.6, has some details of the Green functions that are used in the description of displacement field propagation.

Nonlinear Mesoscopic Elasticity: The Complex Behaviour of Granular Media including Rocks and Soil. Robert A. Guyer and Paul A. Johnson
Copyright © 2009 WILEY-VCH Verlag GmbH & Co. KGaA, Weinheim
ISBN: 978-3-527-40703-3

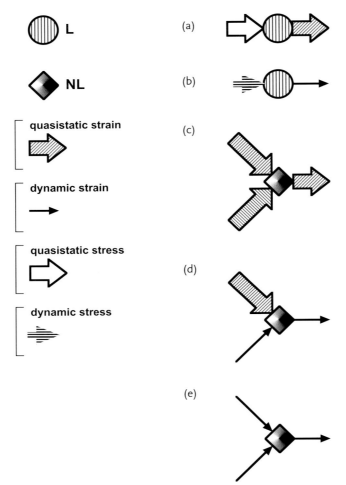

Fig. 3.1 Outline. The phenomenological theory of nonlinear elasticity is in the leading approximation a linear theory. A quasistatic stress produces a quasistatic strain; a dynamic stress produces a dynamic strain. Nonlinear coupling produces coalesence of these strains. (a) Linear quasistatic: a quasistatic stress produces a quasistatic strain through the agency of a linear elastic constant. (b) Linear dynamic: a dynamic stress produces a dynamic strain through the agency of a linear elastic constant. Cubic nonlinearity couples pairs of strain fields in three qualitatively different ways: (c) two quasistatic strain fields produce a nonlinear quasistatic strain, (d) a quasistatic strain distorts a material and causes a shift in the velocity of propagation of a dynamic strain field, and (e) two dynamic strain fields coalesce to produce a third dynamic strain field.

3.1
Quasistatic Response; Linear and Nonlinear

3.1.1
Quasistatic Response; Linear

The linear elasticity of an isotropic system is described by the elastic energy in the first line of Eq. (2.67):

$$e_V = \frac{\mu}{2} \left[S(\boldsymbol{u} \cdot *\boldsymbol{u}) + T(\boldsymbol{u} * \boldsymbol{u}) \right] + \left(\frac{K}{2} - \frac{\mu}{3} \right) T(\boldsymbol{u}) T(\boldsymbol{u}) , \tag{3.1}$$

in conjunction with the equation for the stress, Eq. (2.75). The recipe is to (a) set the strain and (b) find the stress. The elastic constants K and μ are associated with simple compression and simple shear. We show this by considering two choices for $\boldsymbol{u}$ in Eq. (2.68), Figure 3.2.

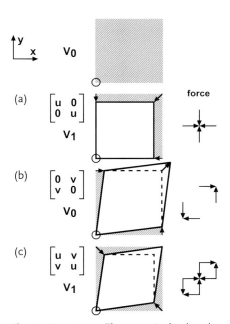

Fig. 3.2 Notation, $\boldsymbol{u}$. The energy is developed using Eq. (2.67) and the matrix of strains $\boldsymbol{u}$ in Eq. (2.68). By convention the displacements are (u, v, w), so the diagonal elements of $\boldsymbol{u}$ correspond to compressional strains. When the strains are uniform, the strain from the first column is used for all; (a)–(c) are two-dimensional examples of the notation. When it is necessary to have a single strain for the purposes of calculating the stress, the appropriate displacement is denoted (U, V, W).

For the case of *uniform* compression choose

$$\boldsymbol{u} = \boldsymbol{u}_K = \begin{pmatrix} U_x & 0 & 0 \\ 0 & u_x & 0 \\ 0 & 0 & u_x \end{pmatrix}, \tag{3.2}$$

where $v_y = w_z = u_x$ and U_x is denoted specially so that we can implement the stress equation. We have

$$e_V = \frac{\mu}{2} \left(2U_x^2 + 4u_x^2 \right) + \left(\frac{K}{2} - \frac{\mu}{3} \right) \left(U_x + 2u_x \right)^2, \tag{3.3}$$

$$\sigma_{xx} = \frac{\partial e_V}{\partial U_x} = 3K \frac{\partial u}{\partial x}. \tag{3.4}$$

From the symmetry of $\boldsymbol{u}_K$ we have $\sigma_{yy} = \sigma_{zz} = \sigma_{xx}$.

For the case of *uniform* shear choose

$$\boldsymbol{u} = \boldsymbol{u}_\mu = \begin{pmatrix} 0 & v_x & v_x \\ V_x & 0 & v_x \\ v_x & v_x & 0 \end{pmatrix}, \tag{3.5}$$

where $v_x = V_x$, and we have

$$e_V = \frac{\mu}{2} \left(V_x^2 + 2V_x v_x + 9v_x^2 \right), \tag{3.6}$$

$$\sigma_{xy} = \frac{\partial e_V}{\partial V_x} = 2\mu \frac{\partial v}{\partial x}. \tag{3.7}$$

From the symmetry of $\boldsymbol{u}_\mu$, $\sigma_{xy} = \sigma_{yz} = \sigma_{zx}$, and $\sigma_{ij} = \sigma_{ji}$, $i \neq j$. We read causality in Eqs. (3.3) and (3.7) from right to left. In response to a uniform compressive strain the system carries a uniform compressional stress proportional to the strain. In response to a uniform shear strain the system carries a uniform shear stress proportional to the strain. These results provide a recipe for learning K and μ from suitable experiments. What could be learned from similar experiments on nonlinear materials?

3.1.2
Quasistatic Response; Nonlinear

Using $\boldsymbol{u}_K$ from Eq. (3.2) in the full equation for e_V, Eq. (2.67), we find

$$\begin{aligned}
e_V = {}& 2 \left(K - \frac{2\mu}{3} \right) U_x u_x + \left(\frac{K}{2} + \frac{2\mu}{3} \right) U_x^2 \\
& + \left(K - \frac{2\mu}{3} + 2B + 4C \right) U_x u_x^2 + \left(K - \frac{2\mu}{3} + 2B + 2C \right) U_x^2 u_x \\
& + \left(\frac{K}{2} + \frac{2\mu}{3} + \frac{A}{2} + B + \frac{C}{3} \right) U_x^3 \\
& + \left(\frac{K}{2} + \frac{2\mu}{3} \right) u_x^2 + \left(2K + \frac{2\mu}{3} + \frac{2A}{3} + 4B + \frac{8C}{3} \right) u_x^3,
\end{aligned} \tag{3.8}$$

from which

$$\sigma_{xx} = \frac{\partial e_V}{\partial U_x} = 3K \frac{\partial u}{\partial x} + \left(\frac{9K}{2} + A + 9B + 9C\right) \left(\frac{\partial u}{\partial x}\right)^2 , \tag{3.9}$$

cf. Eq. (3.4).

Using u_μ from Eq. (3.5) in the full equation for e_V, Eq. (2.67), we find

$$
\begin{aligned}
e_V &= \frac{\mu}{2} \left(2V_x v_x + V_x^2\right) \\
&+ (3\mu + A) V_x v_x^2 \\
&+ \frac{9\mu}{2} v_x^2 + (3\mu + A) v_x^3 ,
\end{aligned}
\tag{3.10}
$$

from which

$$\sigma_{yx} = \frac{\partial e_V}{\partial V_x} = 2\mu \left(\frac{\partial v}{\partial x}\right) + (3\mu + A) \left(\frac{\partial v}{\partial x}\right)^2 , \tag{3.11}$$

cf. Eq. (3.7).

Using the simple mixed strain

$$\boldsymbol{u} = \boldsymbol{u}_{K\mu} = \begin{pmatrix} U_x & 0 & 0 \\ V_x & 0 & 0 \\ 0 & 0 & 0 \end{pmatrix} \tag{3.12}$$

in the full equation for e_V, Eq. (2.67), we find

$$
\begin{aligned}
\sigma_{xx} &= \frac{\partial e_V}{\partial U_x} = \left(K + \frac{4\mu}{3}\right) \frac{\partial u}{\partial x} \\
&+ \left(\frac{3K}{2} + 2\mu + A + 3B + C\right) \left(\frac{\partial u}{\partial x}\right)^2 + \left(\frac{K}{2} + \frac{2\mu}{3} + \frac{A}{4} + \frac{B}{2}\right) \left(\frac{\partial v}{\partial x}\right)^2 ,
\end{aligned}
\tag{3.13}
$$

$$\sigma_{yx} = \frac{\partial e_V}{\partial V_x} = \mu \left(\frac{\partial v}{\partial x}\right) + \left(K + \frac{4\mu}{3} + \frac{A}{2} + + \frac{B}{2}\right) \left(\frac{\partial u}{\partial x}\right) \left(\frac{\partial v}{\partial x}\right) , \tag{3.14}$$

cf. Eqs. (3.4) and (3.7).

The three strain choices $\boldsymbol{u}_K$, $\boldsymbol{u}_{K\mu}$, and $\boldsymbol{u}_\mu$ have three different nonlinear responses that afford the means to measure the phenomenological coefficients, A, B, and C. The recipe is to (a) set the strain and (b) find the stress, just as above. This is the theoretical view. One could equally well take the point of view that stress is applied and strain results. As an example, rearrange Eq. (3.14) using the notion that the strains are small. The sequence of moves is

$$u_x^{(1)} = \frac{\sigma_{xx}}{K + \frac{4\mu}{3}},$$

$$v_x^{(1)} = \frac{\sigma_{yx}}{\mu},$$

$$\frac{\partial v}{\partial x} = \frac{\sigma_{yx}}{\mu} - \frac{K + \frac{4\mu}{3} + \frac{A}{2} + \frac{B}{2}}{\mu} \left(\frac{\partial u}{\partial x} \right) \left(\frac{\partial v}{\partial x} \right),$$

$$\frac{\partial v}{\partial x} = \frac{\sigma_{yx}}{\mu} - \frac{K + \frac{4\mu}{3} + \frac{A}{2} + \frac{B}{2}}{\mu} u_x^{(1)} v_x^{(1)} + \cdots,$$

$$\frac{\partial v}{\partial x} = \frac{\sigma_{yx}}{\mu} - \beta_1 \left(\frac{\sigma_{xx}}{\mu} \right) \left(\frac{\sigma_{yx}}{\mu} \right) + \cdots, \qquad (3.15)$$

where $\beta_1 = 1 + (\frac{A}{2} + \frac{B}{2})/(K + \frac{4\mu}{3})$. We read meaning into the last line. The coefficient β_1 is the strength of the coupling between two quasistatic stress fields, one compressional and one shear. The coupling of these quasistatic stress fields brings about the nonlinear behavior of the quasistatic shear strain. The shear strain $\partial v/\partial x$ is uniform throughout the system. As we expect less (more) strain per unit stress at higher stress, we expect $\beta_1 > 0$ ($\beta_1 < 0$). From a measurement of v_x as a function of σ_{xx} at fixed σ_{yx} one could in principle learn about A and B. Employing other stress fields would allow measurements that get at A alone, Eq. (3.11), or combinations of A, B, and C, Eq. (3.9), and would provide a means to learn these (see Chapter 10).

3.2
Dynamic Response; Linear

The dynamic response of an elastic system is found from the $F = ma$ equation, the equation of motion for the displacement field, Eq. (2.76),

$$\varrho \ddot{u}_i = \sum_k \frac{\partial \sigma_{ik}}{\partial x_k}. \qquad (3.16)$$

The dynamics of concern to us are those of small-amplitude displacement fields, elastic waves, or sound waves. Because the system is isotropic, the propagation properties of a small-amplitude disturbance are independent of direction. Thus we consider waves propagating along the x-axis. These could have a displacement amplitude, $\boldsymbol{d}$, parallel or perpendicular to the direction of propagation.

For the case of a longitudinal wave, $\boldsymbol{d} = (u(x,t), 0, 0)$, $u_x = \partial u/\partial x$, and in the linear approximation,

$$\boldsymbol{u} = \boldsymbol{u}_L = \begin{pmatrix} u_x & 0 & 0 \\ 0 & 0 & 0 \\ 0 & 0 & 0 \end{pmatrix}, \qquad (3.17)$$

$$\sigma_{xx} = \left(\frac{4}{3}\mu + K \right) u_x, \qquad (3.18)$$

and

$$\ddot{u} = \left(\frac{\frac{4}{3}\mu + K}{\varrho} \right) \frac{\partial u_x}{\partial x} = c_L^2 \frac{\partial^2 u}{\partial x^2} \,, \tag{3.19}$$

where c_L is the speed of a longitudinal wave.

In the case of a shear wave, displacement amplitude $\boldsymbol{d} = (0, v(x,t), 0)$, $v_x = \partial v(x,t)/\partial x$, and in the linear approximation,

$$\boldsymbol{u} = \boldsymbol{u}_T = \begin{pmatrix} 0 & 0 & 0 \\ v_x & 0 & 0 \\ 0 & 0 & 0 \end{pmatrix} \,, \tag{3.20}$$

$$\sigma_{yx} = \mu v_x \,, \tag{3.21}$$

and

$$\ddot{v} = \left(\frac{\mu}{\varrho} \right) \frac{\partial v_x}{\partial x} = c_T^2 \frac{\partial^2 v}{\partial x^2} \,, \tag{3.22}$$

where c_T is the speed of a transverse wave.

The elastic-wave excitations, typically thought of as displacement fields or strain fields, are accompanied by a stress field, or stress waves. An equation of motion for the stress field is found by replacing u and v in Eqs. (3.19) and (3.22) with the corresponding stress from Eqs. (3.18) and (3.21). For an elastic wave of amplitude A and polarization $v = L, T$ propagating with wave vector k, the amplitude of the corresponding stress is $\sigma \approx \varrho c_v^2 k A$, where c_v^2 is the appropriate sound speed.

3.3
Quasistatic/Dynamic Response; Nonlinear

Above we described quasistatic stress-strain fields in a nonlinear system. Methods of examining their behavior, learning about nonlinear parameters, etc. involve quasistatic measurements. As the behavior of an elastic wave is sensitive to the elastic environment in which it moves, it is possible to probe the nonlinearity of an elastic system by monitoring the effect of a quasistatic stress field on a dynamic stress field. As an example consider $\boldsymbol{u}$ given by

$$\boldsymbol{u} = \begin{pmatrix} U_x + u_x & 0 & 0 \\ 0 & 0 & 0 \\ 0 & 0 & 0 \end{pmatrix} \,, \tag{3.23}$$

where U_x is a *quasistatic* strain field and $u_x(x,t)$ is a *dynamic* strain field. From the formulation applied above we find

$$\sigma_{xx} = \frac{\partial e_V}{\partial U_x} = \left(\frac{4}{3}\mu + K \right) U_x + \left(\frac{4}{3}\mu + K \right) u_x$$
$$+ (4\mu + 3K + 2A + 6B + 2C) U_x u_x. \tag{3.24}$$

Because formally U_x and u_x enter $\boldsymbol{u}$ and e_V on an equal footing, we can find the stress from the dependence of e_V on either. Confirmation of this comes from the symmetry of σ_{xx} under interchange $U_x \leftrightarrow u_x$. Since U_x is a quasistatic strain field and u_x is a dynamic strain field, we split σ_{xx} into two parts, a quasistatic part, $\sigma_{xx}^S = (4\mu + 3K)\,U_x/3$, and a dynamic part that gives the force in the equation of motion, $\sigma_{xx} = \sigma_{xx}^S + \sigma_{xx}^D$,

$$\sigma_{xx}^D = \left(\frac{4}{3}\mu + K\right) u_x + (4\mu + 3K + 2A + 6B + 2C)\,U_x u_x \,. \tag{3.25}$$

We have

$$\ddot{u} = \frac{\partial \sigma_{xx}^D}{\partial x} = c_L^2\,(1 - \beta_2 U_x)\,\frac{\partial^2 u}{\partial x^2} = c_L^2(U_x)\frac{\partial^2 u}{\partial x^2}\,, \tag{3.26}$$

where $\beta_2 = 3 + 3(2A + 6B + 2C)/(4\mu + 3K)$. So a quasistatic strain field, in the direction of propagation of a longitudinal elastic wave, produces a velocity shift proportional to the strength of the quasistatic field. In the language introduced above, the nonlinearity makes itself known as a coupling between a quasistatic strain field and a dynamic strain field, coupling constant β_2, that modifies the behavior of the dynamic strain field. Since for quasistatic compression $U_x < 0$ we

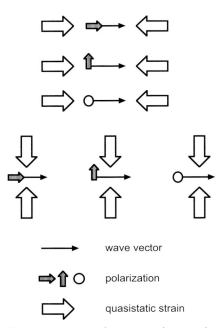

wave vector

polarization

quasistatic strain

Fig. 3.3 Quasistatic/dynamic coupling. For the case of a uniaxial compressive quasistatic stress there are six elastic waves that can be used to probe the consequences of that stress, as illustrated here. Not all of these potential experiments are independent. See Table 3.1.

Table 3.1 Examples of quasistatic/dynamic experiments. Column: (1) the matrix u, Eq. (2.68), (2) coefficient of linear elasticity, (3) coefficient of quasistatic field, Eq. (3.25).

u	linear coefficient	quasistatic coefficient
$\begin{pmatrix} U_x + u_x & 0 & 0 \\ 0 & 0 & 0 \\ 0 & 0 & 0 \end{pmatrix}$	$K + \frac{4}{3}\mu$	$3K + 4\mu + 2A + 6B + 2C$
$\begin{pmatrix} U_x & 0 & 0 \\ v_x & 0 & 0 \\ 0 & 0 & 0 \end{pmatrix}$	μ	$K + \frac{4}{3}\mu + \frac{1}{2}A + B$
$\begin{pmatrix} U_x & 0 & 0 \\ 0 & 0 & 0 \\ w_x & 0 & 0 \end{pmatrix}$	μ	$K + \frac{4}{3}\mu + \frac{1}{2}A + B$
$\begin{pmatrix} u_x & 0 & 0 \\ 0 & V_y & 0 \\ 0 & 0 & 0 \end{pmatrix}$	$K + \frac{4}{3}\mu$	$K - \frac{2}{3}\mu + 2B + 2C$
$\begin{pmatrix} 0 & 0 & 0 \\ v_x & V_y & 0 \\ 0 & 0 & 0 \end{pmatrix}$	μ	$K + \frac{4}{3}\mu + \frac{1}{2}A + B$
$\begin{pmatrix} 0 & 0 & 0 \\ 0 & V_y & 0 \\ w_x & 0 & 0 \end{pmatrix}$	$K + \frac{4}{3}\mu$	$K - \frac{2}{3}\mu + B$

expect $(1 - \beta_2 U_x) > 1$ or $\beta_2 > 0$. Measurement of the velocity shift brought about by U_x can be employed to learn about A, B, and C. This could in principle be done by time of flight or resonant bar methods. By employing various combinations of quasistatic strain fields and dynamic strain fields (direction and polarization) complete determination of A, B, and C is possible, for example, [2]. See Figures 3.1 and 3.3, Table 3.1 and Chapter 10.

3.4
Dynamic Response; Nonlinear

3.4.1
Basic Equations

Finally the nonlinear coupling in Eq. (2.67) brings about the coupling of two dynamic strain fields. Our study of this problem will be quite extensive as experiments involving coupled dynamic strain fields are fundamental to understanding and to many investigations of elastic nonlinearity. We begin with the problem treated by Goldberg [3], Polyakova [4], McCall [5], and others of strain fields propagating in the x-direction with displacements ($\parallel$ or $\perp$ to the x-axis) that depend on x only. To describe this situation we write

$$\boldsymbol{u} = \begin{pmatrix} u_x & 0 & 0 \\ v_x & 0 & 0 \\ 0 & 0 & 0 \end{pmatrix} . \tag{3.27}$$

From this $\boldsymbol{u}$ we find the pair of equations

$$\varrho \ddot{u} = \left(K + \frac{4\mu}{3} \right) \frac{\partial^2 u}{\partial x^2} + \beta \frac{\partial}{\partial x} \left(\frac{\partial u}{\partial x} \right)^2 + \frac{\gamma}{2} \frac{\partial}{\partial x} \left(\frac{\partial v}{\partial x} \right)^2 ,$$

$$\varrho \ddot{v} = \mu \frac{\partial^2 v}{\partial x^2} + \gamma \frac{\partial}{\partial x} \left(\frac{\partial u}{\partial x} \right) \left(\frac{\partial v}{\partial x} \right) , \tag{3.28}$$

where

$$\beta = \frac{3K}{2} + 2\mu + A + 3B + C ,$$

$$\gamma = K + \frac{4\mu}{3} + \frac{A}{2} + B , \tag{3.29}$$

cf. Eqs. (3.9) and (3.14). The coefficient β is a measure of the nonlinear coupling of a longitudinal wave to itself (or another longitudinal wave), $l + l \rightarrow l$, and the coefficient γ is a measure of the nonlinear coupling of a longitudinal wave to a transverse wave, $l + t \rightarrow t$, or of a transverse wave to itself (or another transverse wave), $t + t \rightarrow l$. Let us explore the consequences of Eqs. (3.28) in detail.

3.4.2
Wave Propagation

3.4.2.1 $l + l \rightarrow l$
We begin with the case that there is only an x displacement, u, and suppose there is a source at $x = 0$ with time dependence $f(t)$. For the equation of motion of u we have

$$\frac{\partial^2 u}{\partial t^2} + \frac{1}{\tau_0} \frac{\partial u}{\partial t} = \frac{1}{\varrho} \frac{\partial}{\partial x} K_0 \left(\frac{\partial u}{\partial x} + \beta \left(\frac{\partial u}{\partial x} \right)^2 + \delta \left(\frac{\partial u}{\partial x} \right)^3 + \dots \right) + \frac{f(x, t)}{\varrho} , \tag{3.30}$$

where $K_0 = K + 4\mu/3$ and we have introduced several generalizing features that require comment. (1) On the left there is a simple phenomenological damping. (2) On the right we have added a term in strain cubed, the term in δ, that is in the form appropriate to quartic anharmonicity in e_V. (3) And we have failed to pass ϱ through $\partial/\partial x$ and $\partial/\partial x$ through K_0, β, and δ because, while the systems we are dealing with are isotropic, they may not be homogeneous. In principle we have to be prepared for the case that ϱ, K_0, β, and δ depend on x. We will encounter examples of these finer points below. For now let us proceed with the simplest case

$$\frac{\partial^2 u}{\partial t^2} + \frac{1}{\tau_0}\frac{\partial u}{\partial t} = c_L^2 \frac{\partial^2 u}{\partial x^2} + \beta c_L^2 \frac{\partial}{\partial x}\left(\frac{\partial u}{\partial x}\right)^2 + F(x,t)\,, \tag{3.31}$$

where $c_L^2 = K_0/\varrho$ and $F = f/\varrho$. We use a Fourier representation of $u(x,t)$ and $F(t)$ in the form

$$F(x,t) = \int \frac{d\omega}{2\pi} F_\omega(x) e^{-i\omega t}\,,$$
$$u(x,t) = \int \frac{d\omega}{2\pi} u_\omega(x) e^{-i\omega t}\,. \tag{3.32}$$

Then the amplitude $u_\omega(x)$ obeys

$$\left(-\omega^2 - i\frac{\omega}{\tau_0} - c_L^2\frac{\partial^2}{\partial x^2}\right) u_\omega = c_L^2\beta \int \frac{d\omega'}{2\pi}\frac{\partial u_{\omega'}}{\partial x}\frac{\partial u_{\omega-\omega'}}{\partial x} + F_\omega(x)\,. \tag{3.33}$$

We turn this equation into a hierarchy of equations in powers of the amplitude of F_ω. To do this replace F_ω with λF_ω, write u_ω as a power series in λ,

$$u_\omega = \lambda u_\omega^{(1)} + \lambda^2 u_\omega^{(2)} + \lambda^3 u_\omega^{(3)} + \dots\,, \tag{3.34}$$

and substitute into Eq. (3.33). Grouping terms with like powers of λ leads to the hierarchy

$$D_\omega^{-1} u_\omega^{(1)} = F_\omega(x)\,,$$
$$D_\omega^{-1} u_\omega^{(2)} = c_L^2\beta\frac{\partial}{\partial x}\int \frac{d\omega'}{2\pi}\frac{\partial u_{\omega'}^{(1)}}{\partial x}\frac{\partial u_{\omega-\omega'}^{(1)}}{\partial x}\,,$$
$$D_\omega^{-1} u_\omega^{(3)} = c_L^2\beta\frac{\partial}{\partial x}\int \frac{d\omega'}{2\pi}\left(\frac{\partial u_{\omega'}^{(1)}}{\partial x}\frac{\partial u_{\omega-\omega'}^{(2)}}{\partial x} + \frac{\partial u_{\omega'}^{(2)}}{\partial x}\frac{\partial u_{\omega-\omega'}^{(1)}}{\partial x}\right)\,, \tag{3.35}$$
$$\vdots$$

where

$$D_\omega^{-1}(x) = \left(-\omega^2 - i\frac{\omega}{\tau_0} - c_L^2\frac{\partial^2}{\partial x^2}\right)\,. \tag{3.36}$$

From F_ω you learn $u_\omega^{(1)}$, from $u_\omega^{(1)}$ you learn $u_\omega^{(2)}$, etc. The hierarchy in Eq. (3.35) is solvable by a systematic procedure. We will be content to see how this works with the first two of Eqs. (3.35).

A solution is found using the Green function that is the inverse of D_ω^{-1},

$$D_\omega^{-1}(x) G_\omega(x|x') = \delta(x - x')\,, \tag{3.37}$$

and the identity

$$\int dx' D_\omega^{-1}(x) G_\omega(x|x') A(x') = A(x) . \tag{3.38}$$

Then

$$u_\omega^{(1)} = \int dx' G_\omega(x|x') F_\omega(x') ,$$

$$u_\omega^{(2)} = c_L^2 \beta \int \frac{d\omega'}{2\pi} \int dx' G_\omega(x|x')$$

$$\times \frac{\partial}{\partial x'} \int dx'' \int dx''' \frac{\partial G_{\omega'}(x'|x'')}{\partial x'} \frac{\partial G_{\omega-\omega'}(x'|x''')}{\partial x'} F_{\omega'}(x'') F_{\omega-\omega'}(x''') . \tag{3.39}$$

This formal apparatus takes on some meaning when we consider a particular example. We retreat to the case originally discussed $F_\omega(x) = A_\omega \delta(x)$ and have

$$u_\omega^{(1)}(x) = G_\omega(x|0) A_\omega \tag{3.40}$$

and

$$u_\omega^{(2)}(x) = c_L^2 \int \frac{d\omega'}{2\pi} \int dx' \frac{\partial G_\omega(x|x')}{\partial x} \beta(x') \frac{\partial G_{\omega'}(x'|0)}{\partial x'} \frac{\partial G_{\omega-\omega'}(x'|0)}{\partial x'} A_{\omega'} A_{\omega-\omega'} , \tag{3.41}$$

where we have used one integration by parts and $G(x|x') = G(x - x')$ to move $\partial/\partial x'$ onto x. A source at frequency ω at 0 gives rise to a displacement at frequency ω at $x > 0$ given by $G_\omega(x|0) A_\omega$, Eq. (3.40). If displacement $\propto G$, then strain $\propto \partial G/\partial x$. In Eq. (3.41) two amplitudes, $A_{\omega'}$ and $A_{\omega-\omega'}$, propagate to x' and produce two strains that interact there through the agency of $\beta(x')$. The disturbance at x' propagates on to x, $G_\omega(x|x')$, to produce the second-order contribution to the displacement at frequency ω. We have used $\beta(x')$ to call attention to the working of the nonlinearity pointwise throughout the domain $0 < x' < x$.

To get an explicit answer we require the Green function defined by Eq. (3.37) and a choice of A_ω. For the case at hand we have, Eq. (3.112),

$$G_\omega(x|x') = \frac{i}{2c_L^2 k_\omega} e^{ik_\omega|x-x'|} , \tag{3.42}$$

where we use $c_L^2 k_\omega^2 = \omega^2 + i\omega/\tau_0$. For the amplitude A_ω choose

$$A_\omega = -2\pi i k_\omega c_L^2 A_0 \left[\delta(\omega - \omega_0) + \delta(\omega + \omega_0) \right] . \tag{3.43}$$

Then

$$u_\omega^{(1)}(x) = G_\omega(x|0) A_\omega = \pi A_0 e^{ik_\omega x} \left[\delta(\omega - \omega_0) + \delta(\omega + \omega_0) \right] ,$$

$$u^{(1)}(x,t) = \int \frac{d\omega}{2\pi} u_\omega^{(1)}(x) e^{-i\omega t} = 2A_0 \cos(k_{\omega_0} x - \omega_0 t) , \tag{3.44}$$

where $k_{-\omega} = -k_{\omega}^*$. The source at $x = 0$ has been so chosen that at x, in the absence of nonlinearity, a plane wave signal is detected. Using Eqs. (3.42) and (3.43) in the equation for $u_{\omega}^{(2)}(x)$ leads to terms involving combinations of frequency delta functions, $\delta(\omega' - \omega_0)\delta(\omega - \omega' - \omega_0)$, $\delta(\omega' - \omega_0)\delta(\omega - \omega' + \omega_0)$, $\delta(\omega' + \omega_0)\delta(\omega - \omega' - \omega_0)$, and $\delta(\omega' + \omega_0)\delta(\omega - \omega' + \omega_0)$, that give $\omega = 2\omega_0$, $\omega = 0$ (two terms), and $\omega = -2\omega_0$. For the term involving $\omega = 2\omega_0$ we have

$$u_{\omega}^{(2)}(x) = \frac{1}{2}(k_{\omega_0}A_0)^2 e^{ik_{2\omega_0}x}\delta(\omega - 2\omega_0)\int_0^x dx'\beta(x')e^{i(2k_{\omega_0}-k_{2\omega_0})x'}. \tag{3.45}$$

If the damping due to τ_0 is weak, $\omega_0\tau_0 \gg 1$, $k_{\omega} = \omega/c_L + i\kappa$, $\kappa = 1/c_L\tau_0$, and the phase factor $\exp i(2k_{\omega_0} - k_{2\omega_0})x' = e^{-\kappa x'} \approx 1$ for x within the attenuation length $b = \kappa^{-1}$. Then, for the case $\beta(x') = \beta$ at all x', the integration over the domain of the nonlinear coupling yields a factor $x\beta$. Thus

$$u_{\omega}^{(2)}(x) = \frac{1}{2}\beta(k_{\omega_0}A_0)^2 x e^{ik_{2\omega_0}x}\delta(\omega - 2\omega_0),$$

$$u^{(2)}(x,t) = \frac{1}{2}\beta(k_{\omega_0}A_0)^2 x e^{i(k_{2\omega_0}x - 2\omega_0 t)}. \tag{3.46}$$

This is a well-known result. We derive it in a way that calls attention to the physics that is operating. The source at ω_0 broadcast a strain that couples to itself with strength β at all points beyond the point of initial launch. At each coupling point a signal of strength proportional to $\beta(kA)^2$ is broadcast. As the broadcast signals are in phase they add over the path from 0 to x to give the factor x in Eq. (3.46). It is apparent how a number of variations on this result would go. For example, if the nonlinearity were present at a single point in space, $\beta(x) = \beta_0 a\delta(x - x_0)$, or in a strip of length d, $\beta(x) = \beta_0\theta(x - x_0)\theta(x_0 + d - x)$, we would replace βx in Eq. (3.46) with $\beta_0 a$ and $\beta_0 d$, respectively. The influence of attenuation would make itself known in an obvious way. The two signals that couple at x' are each attenuated by $\exp -\kappa x'$. The signal broadcast from x' to x is subject to attenuation over the path $|x - x'|$.

There is a term in $u^{(2)}$ with the same amplitude as in Eq. (3.46) with frequency $\omega = -2\omega_0$ and a DC term that would be seen in principle by a conventional strain gauge, that is, at $\omega = 0$ [6, 7].

3.4.2.2 $t + t \rightarrow l$ and $l + t \rightarrow t$

Consider the case in which there are two sources at $x = 0$; one launches a longitudinal wave with (frequency,amplitude) = (ω, A_0) and the other launches a shear wave with (frequency,amplitude) = (Ω, B_0). These waves will bring into play the terms in γ in Eqs. (3.28). We can talk through the basic physical idea using the understanding from above. Let us consider the effect on u first.

1. There will be two first-order disturbances $u_{\omega}^{(1)}(x') \propto A_0$ and $v_{\Omega}^{(1)}(x') \propto B_0$ at x' that are carried there by the Green functions G_{ω} and g_{Ω} for u and v, respectively.

2. Because of the coupling $\beta(x')$, the longitudinal wave $u_{\omega}^{(1)}(x')$ will interact with itself and launch a longitudinal wave that will propagate to x. This is described above, $l + l \rightarrow l$.

3. Because of the coupling $\gamma(x')$, the transverse wave $v_\Omega^{(1)}(x')$ will interact with itself at x' and launch a longitudinal wave that will propagate to x, $t + t \rightarrow l$.

   a. The longitudinal wave launched from x' is at frequency 2Ω and has an amplitude proportional to $v^{(1)} \times v^{(1)}$ and phase $\exp(i2q_\Omega x')$.

   b. The longitudinal wave launched from x' to x has a further phase evolution $\exp(ik_{2\Omega}[x - x'])$ so that the phase of the longitudinal wave at x is

   $$e^{ik_{2\Omega}x}e^{i(2q_\Omega - k_{2\Omega})x'} = e^{ik_{2\Omega}x}e^{\phi(x')} \,, \tag{3.47}$$

   where $q_\Omega = \Omega/c_T$, $c_T^2 = \mu/\varrho$ from the the second of Eqs. (3.28). For the case studied above, $l + l \rightarrow l$, the phase factor $\phi(x')$ was essentially zero and the longitudinal waves launched from all x', $0 \le x' \le x$ to give the factor x. Here that is not the case and we have

   $$\int_0^x dx e^{i\phi(x)} = -ia(1 - e^{i(2q_\Omega - k_{2\Omega})x}) = -ia\mathcal{F} \,, \tag{3.48}$$

   an oscillatory function of x with amplitude a, $2\pi/a = 2\Omega(1/c_T - 1/c_L)$.

   c. The amplitude of $u^{(2)}(x)$ scales as $\gamma(q_\Omega B_0)^2 a$ and is multiplied by the oscillatory factor $\mathcal{F}$.

4. Because of the coupling $\gamma(x')$, the transverse wave $v_\Omega^{(1)}(x')$ will interact with the longitudinal wave $u_\omega^{(1)}(x')$ and launch a transverse wave that will propagate to x, $l + t \rightarrow t$.

   a. The transverse wave launched from x' is at frequency $\omega + \Omega$.

   b. The transverse wave launched from x' to x has phase

   $$e^{iq_{\omega+\Omega}x}e^{i(q_\Omega + k_\omega - q_{\omega+\Omega})x'} = e^{iq_{\omega+\Omega}x}e^{\phi(x')} \,. \tag{3.49}$$

   We have

   $$\int_0^x dx e^{i\phi(x)} = -ia(1 - e^{i(k_\omega - q_\omega)x}) = -ia\mathcal{G} \,, \tag{3.50}$$

   an oscillatory function of x with amplitude a, $2\pi/a = 2\omega(1/c_T - 1/c_L)$.

   c. The amplitude of $v^{(2)}(x)$ scales as $\gamma(q_\Omega B_0)(k_\omega A_0)a$ and is multiplied by the oscillatory factor $\mathcal{G}$.

The three processes we have looked at are schematically $l + l \rightarrow l$, $t + t \rightarrow l$, and $l + t \rightarrow t$.

3.4.2.3 $l + l + l \rightarrow l$, $l + 2l \rightarrow l$ and more

There are many possibilities beyond those considered above. In Figure 3.4 we show schematically several interaction processes among dynamical strain fields that are variations on the theme discussed here. For the most general case of cubic anharmonicity in an isotropic solid Jones and Kobett [8] have worked through the details of what is possible and presented the results as a set of selection rules. Their calculations do not use the restricted geometry and particular circumstance used here

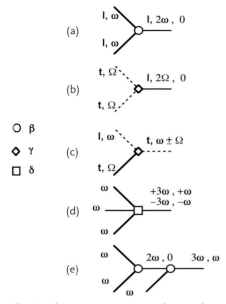

Fig. 3.4 Elastic-wave interactions due to cubic anharmonicity. There are three rudimentary elastic wave interactions due to cubic anharmonicity. These are shown schematically in (a) $l+l \to l$, (b) $t+t \to l$, and (c) $l+t \to t$. Quartic anharmonicity leads to the process shown in (d) in which three longitudinal elastic waves coalesce to form a fourth, $l+l+l \to l$. A process similar to the quartic process can occur as a second-order cubic process (e).

and as a consequence contain results of great generality. What one wants to take away is not just the results, selection rules, but additionally the set of physical conditions that are necessary to have two elastic waves coalesce in a volume of space to form a distinctive third wave. The physical content of the calculation of Jones and Kobett is illustrated in Figure 3.5.

1. Plane waves (A_1, k_1, ω_1) and (A_2, k_2, ω_2) are launched from transducers 1 and 2 and carried into the interior of the system by the appropriate Green functions.
2. In the interaction volume V, where they encounter one another, these plane waves interact through the cubic anharmonicity producing a system of sources with strength proportional to $(k_1 A_1)(k_2 A_2)$, spatial structure determined by k_1 and k_2, and time dependence determined by ω_1 and ω_2.
3. Broadcast from the system of sources is carried out of the interaction volume by the appropriate Green function. The broadcast amplitude is large in those directions in which the phase factors,

$$e^{i\phi(x)} = e^{ik_1 \cdot x} e^{ik_2 \cdot x} e^{-ik_3 \cdot x} , \qquad (3.51)$$

add over the interaction volume. Here k_3 is the wave vector of the excitation leaving the interaction volume at a frequency related to ω_1 and ω_2.

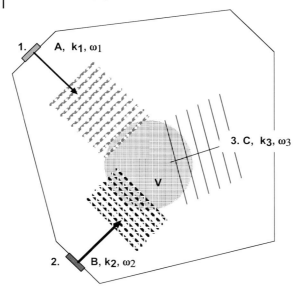

Fig. 3.5 Three-wave processes. Waves (A, k_1, ω_1) and (B, k_2, ω_2) interact in volume V, via a cubic anharmonic interaction, from which elastic energy is broadcast. Under certain conditions the broadcast from V is sufficiently coherent that a third elastic wave is created, (C, k_3, ω_3), $C \propto AB$, and wavevector and frequency related by the selection rules, Eq. (3.52).

4. When specific choices are made for the characteristics of the two incident waves, one can assess the characteristics of the outgoing waves. For example, when the two incident waves are transverse, a coherent outgoing wave that is longitudinal can be formed.

5. For $t + t \rightarrow l$ the wavevector-frequency selection rules are

$$
\begin{aligned}
k_1 + k_2 &= k_3 , \\
\omega_1 + \omega_2 &= \omega_3 ,
\end{aligned}
\tag{3.52}
$$

essentially momentum and energy conservation. The amplitude of the outgoing wave scales as

$$
A_3 \sim \Gamma (k_1 A_1)(k_2 A_2) \frac{kV}{r} ,
\tag{3.53}
$$

where Γ is a measure of the strength of the cubic anharmonicity, cf. the discussion below Eq. (3.48). The amplitude of the outgoing wave scales as the product of the strain fields in the interaction volume and as V, the analog of x in Eq. (3.46). An extensive body of careful further work extended these theoretical developments and a large body of experimental work has confirmed them in all details for appropriate materials. See Chapter 10.

3.4.3
Resonant Bar

We look at two cases: (1) the analog of $l + l + l \rightarrow l$ and (2) the analog of $l + t \rightarrow t$.

3.4.3.1 $l + l + l \rightarrow l$

Consider the situation shown in Figure 3.6, a resonant bar system driven at the left. To describe this, we write the equation of motion, Eq. (3.30),

$$\ddot{u} + \frac{1}{\tau}\dot{u} = c_L^2 \frac{\partial^2 u}{\partial x^2} + c_L^2 \delta \frac{\partial}{\partial x} \left(\frac{\partial u}{\partial x}\right)^3 + F(t)\frac{a}{m}\delta(x), \qquad (3.54)$$

where a is a constant that carries the dimensions lost in $\delta(x)$ and in place of the terms from cubic anharmonicity that are in the equations above, proportional to β, γ, we have a term from the quartic anharmonicity, proportional to δ. The reason for this choice will be clear as we proceed. The method of analysis starts much the same as above. Using Fourier analysis of u and F and the development of u_ω in the form $u_\omega^{(1)} + u_\omega^{(3)} + \ldots$ we have the set of equations

$$D_\omega^{-1} u_\omega^{(1)} = \frac{a}{m} F_\omega \delta(x), \qquad (3.55)$$

$$D_\omega^{-1} u_\omega^{(3)} = \delta c_L^2 \frac{\partial}{\partial x} \int \frac{d\omega'}{2\pi} \int \frac{d\omega''}{2\pi} \frac{\partial u_{\omega'}^{(1)}}{\partial x} \frac{\partial u_{\omega''}^{(1)}}{\partial x} \frac{\partial u_{\omega-\omega'-\omega''}^{(1)}}{\partial x}, \qquad (3.56)$$

$$\vdots \qquad (3.57)$$

Using the Green function properties in Eqs. (3.37) and (3.38) we find

$$u_\omega^{(1)}(x) = G_\omega(x|0)\frac{a}{m}F_\omega, \qquad (3.58)$$

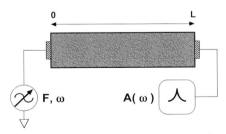

Fig. 3.6 Resonant bar experiment. In a typical resonant bar experiment, the bar is driven at one end and the strain field or its functional equivalent is detected at the other end. It is usual to fix the drive amplitude, F_n, $n = 1 \cdots N$, and sweep the frequency over a resonance. The location and shape of the resonance as a function of drive amplitude is studied. Since $A_n(\omega)$ is simply proportional to F_n for a linear system, one often studies $A_n(\omega)/F_n$.

$$u_\omega^{(3)}(x) = \delta c_L^2 \frac{a^3}{m^3} \int dx' \, G_\omega(x|x') \int \frac{d\omega'}{2\pi} \int \frac{d\omega''}{2\pi}$$

$$\times \frac{\partial}{\partial x'} \left(\frac{\partial G_{\omega'}(x'|0)}{\partial x'} \frac{\partial G_{\omega''}(x'|0)}{\partial x'} \frac{\partial G_{\omega-\omega'-\omega''}(x'|0)}{\partial x'} \right)$$

$$\times F_{\omega'} F_{\omega''} F_{\omega-\omega'-\omega''} . \tag{3.59}$$

For F_ω we take $F_\omega = A[\delta(\omega - \omega_0) + \delta(\omega + \omega_0)]$. Let us focus on the amplitude at $x = L$, the detection point, at frequency ω_0. We have one term in $u_\omega^{(1)}$ proportional to $\delta(\omega - \omega_0)$. In $u_\omega^{(3)}$ we can get $\delta(\omega - \omega_0)$ from the three combinations of ω', ω'', and $\omega - \omega' - \omega''$ that yield ω_0. (The cubic nonlinearity, treated above in wave propagation, does not appear prominently in the description of the resonant bar because the interaction does not return the fundamental frequency.) We have

$$u_\omega^{(1)}(L) = \delta(\omega - \omega_0) \frac{aA}{m} G_\omega(L|0) , \tag{3.60}$$

$$u_\omega^{(3)}(L) = \delta(\omega - \omega_0) 3\delta c_L^2 \frac{a^3 A^3}{m^3} \int dx' \, G_{\omega_0}(L|x')$$

$$\times \frac{\partial}{\partial x'} \left(\frac{\partial G_{\omega_0}(x'|0)}{\partial x'} \frac{\partial G_{\omega_0}(x'|0)}{\partial x'} \frac{\partial G_{-\omega_0}(x'|0)}{\partial x'} \right) . \tag{3.61}$$

To develop these equations further we use the Green function for the resonant bar

$$G_\omega(x|x') = \frac{1}{kc_L^2} \frac{\cos kx' \cos k(L-x)}{\sin kL} , \quad x > x' , \tag{3.62}$$

where $\omega^2 + i\omega/\tau = c_L^2 k^2$ and $k(-\omega) = -k(\omega)^*$. Then

$$u_\omega^{(1)} = -\delta(\omega - \omega_0) \frac{1}{kc_L^2} \frac{aA}{m} \frac{1}{\sin kL} \tag{3.63}$$

and

$$u_\omega = u_\omega^{(1)} + u_\omega^{(3)} = -\delta(\omega - \omega_0) \frac{1}{kc_L^2} \frac{aA}{m} \left(\frac{1}{\sin kL} + 3\frac{\delta}{c_L^4} \left(\frac{aA}{m} \right)^2 \frac{1}{\sin^4 kL} J \right) , \tag{3.64}$$

where J, from the integral over x', is of the form $J = J_s \sin kL + J_c \cos kL$. This equation undergoes a series of rearrangements, trying to bring it into the form of Eq. (3.63), with the result

$$u_\omega = \delta(\omega - \omega_0) B \frac{1}{\sin \left(kL \left[1 - \frac{9}{8} \delta (ku_{\omega_0})^2 \right] \right)} , \tag{3.65}$$

where B is a complicated constant. This equation is an implicit equation for u_ω. The resonance of the bar at $k_0 L \approx n\pi$ for $\delta = 0$ is shifted for $\delta \neq 0$ approximately as here

$$kL \approx k_0 L \left(1 + \frac{9}{8} \delta \varepsilon_\omega^2 \right) , \tag{3.66}$$

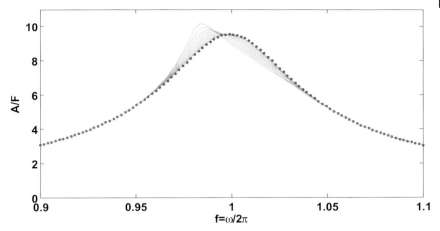

Fig. 3.7 Resonant bar example. The amplitude $A_n(\omega)/F_n$ as a function of ω, for the quartic problem in Eq. (3.54), from Eq. (3.65). Resonance curves like this put simple definitions of measures of attenuation to a severe test.

where $\varepsilon_\omega = k u_\omega$. Thus there is a frequency shift, proportional to the strain field squared, that respects the sign of δ, Figure 3.7. In addition to the frequency shift there is a nonlinear attenuation that causes a reduction (increase) in the amplitude on the side of the resonance peak from (to) which the frequency shifts. This result makes it clear that a nonlinear contribution to the elasticity need not necessarily produce an addition to the attenuation. It is a matter of phase. The attenuation depends on the phase relationship between the drive and the velocity field. For $F = F_0 \sin \omega t$ and $v(t) = A \cos \omega t + B \sin \omega t$

$$W = \int F(t) dx = \frac{1}{2} F_0 B = F_0 R \sin \phi \,, \tag{3.67}$$

$R = \sqrt{A^2 + B^2}$ and $\tan \phi = B/A$. In the resonant bar case the phase of the strain fields in the bar affect the outcome markedly. For examples of sensitivity to phase in the case of wave propagation, see below, Eq. (3.47), and the discussion of RTMF in Chapter 8.

3.4.3.2 $l + t \rightarrow t$

Consider the truncated form of Eq. (3.28) in which we retain only one of the nonlinear terms, the uv term in the equation for v,

$$\ddot{u} = c_L^2 \frac{\partial^2 u}{\partial x^2} + F(t) \frac{a}{m} \delta(x) \,, \tag{3.68}$$

$$\ddot{v} = c_T^2 \frac{\partial^2 v}{\partial x^2} + \gamma c_T^2 \frac{\partial}{\partial x} \left(\frac{\partial u}{\partial x} \right) \left(\frac{\partial v}{\partial x} \right) + H(t) \frac{a}{m} \delta(x) \,, \tag{3.69}$$

where $F(t)$ and $H(t)$ are forces due to a compressional and shear transducer, respectively, and $\partial u/\partial x = \partial v/\partial x = 0$ at $x = 0$ and $x = L$. We want to look at the case where F drives the compressional displacement near a normal mode and H drives the shear displacement near a normal mode. We then have $u = u^{(1)}$ and $v = v^{(1)} + v^{(2)} + \ldots$, where

$$u^{(1)}(x, \omega) = G_L(x|0, \omega) F(\omega),$$

$$v^{(1)}(x, \Omega) = G_T(x|0, \Omega) H(\Omega)$$

$$v^{(2)}(x, \omega) = -\gamma c_T^2 \int_0^L dx' \frac{\partial G_T(x|x'; \omega)}{\partial x'} \tag{3.70}$$

$$\times \int \frac{d\omega'}{2\pi} \frac{\partial u^{(1)}(x', \omega')}{\partial x'} \frac{\partial v^{(1)}(x', \omega - \omega')}{\partial x'},$$

and the factors of a/m have been absorbed into the definitions of $F(\omega)$ and $H(\Omega)$. Using the equations for $u^{(1)}$ and $v^{(1)}$ leads to

$$v^{(2)}(x, \omega) = -\gamma c_T^2 \int_0^L dx' \frac{\partial G_T(x|x'; \omega)}{\partial x'}$$

$$\times \int \frac{d\omega'}{2\pi} \frac{\partial G_L(x'|0; \omega')}{\partial x'} \frac{\partial G_T(x'|0; \omega - \omega')}{\partial x'} \tag{3.71}$$

$$\times F(\omega') H(\omega - \omega').$$

Choose the forces to be $F(\omega) = F_0 \delta(\omega - \omega_1)$ and $H(\omega) = F_0 \delta(\omega - \omega_2)$. Then

$$v^{(2)}(L, \omega) = -\delta(\omega - \omega_3) \gamma c_T^2 F_0 H_0$$

$$\times \int_0^L dx' \frac{\partial G_T(L|x'; \omega_3)}{\partial x'} \frac{\partial G_L(x'|0; \omega_1)}{\partial x'} \frac{\partial G_T(x'|0; \omega_2)}{\partial x'}, \tag{3.72}$$

where $\omega_3 = \omega_1 + \omega_2$. Use the Green functions

$$G_L(x|x'; \omega) = \frac{1}{c_L^2 k_\omega} \frac{\cos k_\omega x' \cos k_\omega (L - x)}{\sin k_\omega L}, \quad x > x', \tag{3.73}$$

$$G_T(x|x'; \omega) = \frac{1}{c_T^2 q_\omega} \frac{\cos q_\omega x' \cos q_\omega (L - x)}{\sin q_\omega L}, \quad x > x', \tag{3.74}$$

with $k_\omega = \omega/c_L$ and $q_\omega = \omega/c_T$. Then

$$\frac{v^{(2)}(L, \omega)}{L} = \delta(\omega - \omega_3) \gamma \frac{F_0}{c_L^2 \sin(k_{\omega_1} L)} \frac{H_0}{c_T^2 \sin(q_{\omega_2} L)} \mathcal{F}, \tag{3.75}$$

where

$$\mathcal{F} = \frac{1}{L} \int_0^L dx \frac{\sin(q_{\omega_3} x)}{\sin(q_{\omega_3} L)} \sin(k_{\omega_1} x) \sin(q_{\omega_2} x). \tag{3.76}$$

The factor $\mathcal{F}$ contains detailed information about how the modes couple. It has no sharp features as a function of frequency unless there is a shear mode with frequency ω_3. Further interpretation of Eq. (3.75) uses the equations for $u^{(1)}$ and $v^{(1)}$ in the form

$$k_\omega u^{(1)}(L, \omega) = \frac{F_0}{c_L^2 \sin(k_{\omega_1} L)} \delta(\omega - \omega_1) = \varepsilon_L(\omega_1)\delta(\omega - \omega_1),$$ (3.77)

$$q_\omega v^{(1)}(L, \omega) = \frac{H_0}{c_T^2 \sin(q_{\omega_2} L)} \delta(\omega - \omega_2) = \varepsilon_T(\omega_2)\delta(\omega - \omega_2).$$ (3.78)

Thus the amplitude $v^{(2)}$ is proportional to the product of the strain fields created by $F(t)$ and $H(t)$, the dimensionless nonlinear parameter γ, and the mode-dependent form factor $\mathcal{F}$,

$$\frac{v^{(2)}(L, \omega)}{L} = \delta(\omega - \omega_1 - \omega_2)\gamma\varepsilon_L(\omega_1)\varepsilon_T(\omega_2)\mathcal{F}.$$ (3.79)

3.5
Exotic Response; Nonlinear

Let us examine a situation in which the displacement field couples nonlinearly to an auxiliary field. Consider the interaction of a pump wave (displacement field 1) and a probe wave (displacement field 2) due to nonlinear coupling to a dynamic "auxiliary" field, a local temperature field. The physical system is a resonant bar, $0 \leq x \leq L$. We go outside of the domain of traditional nonlinear elasticity to construct a nonlinear coupling. Suppose the coupled elastic/thermal system is described by the free energy

$$F(T) = F_0(T) + \frac{1}{2}K\varepsilon^2 - K\alpha_0\delta T\varepsilon - K\alpha_0\gamma(x)\delta T\frac{\varepsilon^2}{2},$$ (3.80)

where $\alpha_0 = (\partial V/\partial T)/V$ and $\delta T = T - T_0$. The first three terms on the right-hand side are the standard terms from linear elasticity, Section 2.2.2.1. The last term, which we will eventually take to be spatially local, is the nonlinear coupling between strain and the temperature suggested as the analog of the Luxemburg–Gorky effect [9]. We introduced such a term in Eq. (2.28). The equation of motion for the displacement field is

$$\varrho\left(\ddot{u} + \frac{1}{\tau_0}\dot{u}\right) = \frac{\partial\sigma}{\partial x} = \frac{\partial}{\partial x}\frac{\partial F}{\partial\varepsilon}$$ (3.81)

and the equation of motion for the temperature field is

$$T_0\frac{\partial S}{\partial t} = -T_0\frac{\partial}{\partial t}\frac{\partial F}{\partial T} = \kappa_T\frac{\partial^2\delta T}{\partial x^2},$$ (3.82)

where $S = -\partial F/\partial T$ is the entropy and κ_T is the thermal conductivity. Using Eq. (3.80) and $C = T_0\partial S/\partial T$ we have

$$\left(\ddot{u} + \frac{1}{\tau_0} \dot{u} \right) = c_L^2 \frac{\partial}{\partial x} \left[\varepsilon - \alpha \delta T - \alpha \gamma(x) \delta T \varepsilon \right] , \tag{3.83}$$

$$C \delta \dot{T} = \kappa_T \frac{\partial^2 \delta T}{\partial x^2} - K T_0 \alpha_0 \dot{\varepsilon} - K T_0 \alpha_0 \gamma(x) \varepsilon \dot{\varepsilon} . \tag{3.84}$$

Use the definitions $\delta T = \delta T / T_0$, $\Gamma_0 = T_0 \alpha_0$, $\Gamma = \Gamma_0 \gamma$, $D_T = \kappa_T / C$, and $r = K/(C T_0)$. Find

$$\left(\ddot{u} + \frac{1}{\tau_0} \dot{u} \right) = c_L^2 \frac{\partial}{\partial x} \varepsilon - c_L^2 \Gamma_0 \frac{\partial \delta T}{\partial x} - c_L^2 \frac{\partial}{\partial x} \Gamma \delta T \varepsilon + F \delta(x) + f \delta(x) , \tag{3.85}$$

$$\delta \dot{T} = D_T \frac{\partial^2 \delta T}{\partial x^2} - r \Gamma_0 \dot{\varepsilon} - r \Gamma \varepsilon \dot{\varepsilon} , \tag{3.86}$$

where F and f are the pump and probe forces, respectively. We will drop the linear terms, proportional to Γ_0 in both equations (these terms lead to the linear attenuation and adiabatic sound speed) and direct our attention to the two nonlinear terms, involving $\delta T \varepsilon$ and $\varepsilon \dot{\varepsilon}$. In what is by now a familiar procedure, we perform a Fourier analysis on the equations of motion. With the definition

$$g(t) = \int \frac{d\omega}{2\pi} g_\omega e^{-i\omega t} \tag{3.87}$$

and $u(t) \leftrightarrow u_\omega$, $\delta T(t) \leftrightarrow A_\omega$, $F(x,t) \leftrightarrow F_\omega \delta(x)$, $f(x,t) \leftrightarrow f_\omega \delta(x)$ we have

$$\left[-\omega^2 - \frac{i\omega}{\tau_0} - c^2 \frac{\partial^2}{\partial x^2} \right] u_\omega = -c^2 \frac{\partial}{\partial x} \Gamma \int \frac{d\omega'}{2\pi} A_{\omega-\omega'} u_{\omega'} + (F_\omega + f_\omega) \delta(x) , \tag{3.88}$$

$$\left[-i\omega - D_T \frac{\partial^2}{\partial x^2} \right] A_\omega = i r \Gamma \int \frac{d\omega'}{2\pi} \omega' u_{\omega-\omega'} u_{\omega'} . \tag{3.89}$$

We imagine the pump displacement field to be relatively large and the probe displacement field to be relatively small. The temperature field is only nonzero because it is driven by the displacement fields in places where $\Gamma(x)$ is nonzero. To focus on the sidebands of the probe frequency we divide u_ω into two parts, $u_\omega = U_\omega + u_\omega$, where in leading order $U_\omega \propto F_\omega$ and $u_\omega \propto f_\omega$. We are interested in u and A. If Λ and λ are measures of the size of F and f, respectively, a systematic treatment of U, u, and A would use $U_\omega = \Lambda U_\omega^{(10)}$, $u_\omega = \lambda u_\omega^{(01)} + \Lambda^2 \lambda u_\omega^{(21)} + \Lambda \lambda^2 u_\omega^{(12)} + \ldots$, and $A_\omega = \Lambda^2 A_\omega^{(20)} + \Lambda \lambda A_\omega^{(11)} + \ldots$, where the superscripts (μ, ν) correspond to the powers of Λ and λ, respectively. Using these expansions in Eq. (3.89) we have

$$G_\omega^{-1} U_\omega^{(10)} = F_\omega \delta(x) , \tag{3.90}$$

$$G_\omega^{-1} u_\omega^{(01)} = f_\omega \delta(x) , \tag{3.91}$$

$$G_\omega^{-1} u_\omega^{(21)} = -c^2 \frac{\partial}{\partial x} \Gamma \int \frac{d\omega'}{2\pi} A_{\omega-\omega'}^{(20)} u_{\omega'}^{(01)'} - c^2 \frac{\partial}{\partial x} \Gamma \int \frac{d\omega'}{2\pi} A_{\omega-\omega'}^{(11)} U_{\omega'}^{(10)'} , \tag{3.92}$$

$$G_\omega^{-1} u_\omega^{(12)} = -c^2 \frac{\partial}{\partial x} \Gamma \int \frac{d\omega'}{2\pi} A_{\omega-\omega'}^{(11)} u_{\omega'}^{(01)'} \,, \tag{3.93}$$

$$\vdots$$

$$\tag{3.94}$$

$$g_\omega^{-1} A_\omega^{(20)} = i r \Gamma \int \frac{d\omega'}{2\pi} \omega' U_{\omega-\omega'}^{(10)'} U_{\omega'}^{(10)'} \,,$$

$$g_\omega^{-1} A_\omega^{(11)} = i r \Gamma \int \frac{d\omega'}{2\pi} \omega' \left(U_{\omega-\omega'}^{(10)'} u_{\omega'}^{(01)'} + u_{\omega-\omega'}^{(01)'} U_{\omega'}^{(10)'} \right) \,, \tag{3.95}$$

where

$$G_\omega^{-1} = \left[-\omega^2 - \frac{i\omega}{\tau_0} - c_L^2 \frac{\partial^2}{\partial x^2} \right] \,,$$

$$g_\omega^{-1} = \left[-i\omega - D_T \frac{\partial^2}{\partial x^2} \right] \,, \tag{3.96}$$

and $u' = \partial u / \partial x$, etc. Look at the equation for $u_\omega^{(21)}$. Consider the ways in which modulation of the pump U could feed into u. The two diagrams in Figure 3.8 il-

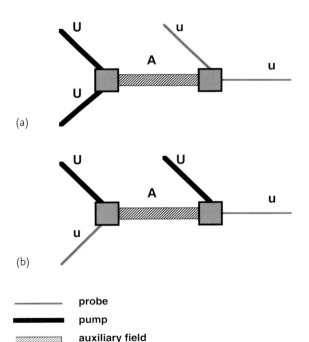

(a)

(b)

⎯⎯⎯⎯ **probe**

▬▬▬▬ **pump**

▨▨▨▨ **auxiliary field**

Fig. 3.8 Pump, probe, and temperature I. Possible schemes for coupling the temperature, pump elastic wave, and probe elastic wave. Elastic waves create a temperature disturbance from which there is further elastic-wave scattering. The temperature carries aspects of the elastic waves that created it and transfers these to the scattered wave.

lustrate the two terms in the equation for $u_\omega^{(21)}$. The diagram at the top involves the pump driving the A field; the product $U U$ transfers modulation of U to A, which is then probed by u, the first term on the right-hand side of the equation for $u_\omega^{(21)}$. The diagram at the bottom involves the pump and probe driving the A field which is then "probed" by U, the second term on the right-hand side of the equation for $u_\omega^{(21)}$. We examine the case corresponding to the top diagram. Simplify notation with $U = U^{(10)}$, $v = u^{(01)}$, $w = u^{(21)}$, and $A = A^{(20)}$ and use Eqs. (3.96) to write

$$U_\omega(x) = G_\omega(x|0) F_\omega \,, \tag{3.97}$$

$$v_\omega(x) = G_\omega(x|0) f_\omega \,, \tag{3.98}$$

$$w_\omega(x) = c^2 \int_0^L dx' \frac{\partial G_\omega(x|x')}{\partial x'} \Gamma(x') \int \frac{d\omega'}{2\pi} A_{\omega-\omega'}(x') v_{\omega'}(x') \,, \tag{3.99}$$

$$A_\omega(x) = ir \int_0^L dx' g_\omega(x|x') \Gamma(x') \int \frac{d\omega'}{2\pi} \omega' U'_{\omega-\omega'}(x') U'_{\omega'}(x') \,, \tag{3.100}$$

where in the equation for w_ω we have done an integration by parts. We are interested in w_ω. Since U_ω and v_ω are simply created by the applied forces, it is straightforward to write the equation for w_ω. We find, taking $\Gamma = \Lambda \delta(x - R)$ and detecting w_ω at the bar end $x = L$,

$$
\begin{aligned}
w_\omega(L) = &\, i r c_L^2 \Lambda^2 \int \frac{d\omega'}{2\pi} \int \frac{d\omega''}{2\pi} \\
&\times \frac{\partial G_\omega(L|R)}{\partial R} g_{\omega-\omega'}(R|R) \frac{\partial G_{\omega-\omega'-\omega''}(R|0)}{\partial R} \frac{\partial G_{\omega'}(R|0)}{\partial R} \frac{\partial G_{\omega''}(R|0)}{\partial R} \\
&\times F_{\omega-\omega'-\omega''} F_{\omega'} f_{\omega''}.
\end{aligned}
\tag{3.101}
$$

Suppose that the time dependence of the pump and probe are such that

$$
\begin{aligned}
F_\omega = F_0 &\big(\delta(\omega - \omega_P) + \delta(\omega + \omega_P) \\
&+ m \big[\delta(\omega - \omega_P + \Omega) + \delta(\omega - \omega_P - \Omega) \\
&+ \delta(\omega + \omega_P + \Omega) + \delta(\omega + \omega_P - \Omega) \big] \big) \,,
\end{aligned}
\tag{3.102}
$$

$$f_\omega = f_0(\delta(\omega - \omega_p) + \delta(\omega + \omega_p)) \,, \tag{3.103}$$

where m is the amplitude of the modulation of ω_P at Ω. There are many possibilities for transfer of the frequencies of the pump to the probe. Consider, for example, the transfer of $+\Omega$ to the probe. This can happen in two ways, Figure 3.9a, $(\omega' = -\omega_P + \Omega, \omega - \omega' - \omega'' = \omega_P)$ and $(\omega' = \omega_P + \Omega, \omega - \omega' - \omega'' = -\omega_P)$, with the result

$$w_\omega(L) \propto \delta(\omega - \omega_p - \Omega) m r c_L^2 \Lambda^2 g_\Omega(R|R) H_p(R) f_0 H_P(R) F_0^2 \,, \tag{3.104}$$

where $H_P(R)$ and $H_p(R)$ are form factors that describe the consequences of the system geometry, for example, the location of the coupling point, R, source point,

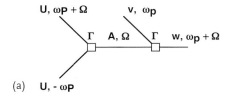

(a)

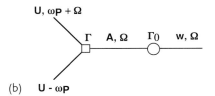

(b)

Fig. 3.9 Pump, probe, and temperature II.
(a) The physical process followed in detail in
Eqs. (3.100)–(3.106) is described by this elaborated version of the diagram in Figure 3.8
in which the amplitudes and frequencies of
the participants are labeled. (b) A process
not described in detail, but of potential interest, is that in which a modulated thermal
disturbance, caused by the pump, "emits"
a modulated elastic wave. The amplitudes and
frequencies involved are shown.

and detection point relative to the structure of the disturbance in the bar at frequencies ω_P and ω_p,

$$H_p(R) = \frac{\sin k_\omega R}{\sin k_\omega L} \frac{\sin k_p(L-R)}{\sin k_p L}, \tag{3.105}$$

$$H_p(R) = \frac{\sin k_P(L-R)}{\sin k_P L} \frac{\sin k_P(L-R)}{\sin k_P L}, \tag{3.106}$$

$k_P = k_{\omega_p}$, and $k_p = k_{\omega_p}$. The pump drives the temperature field at R at frequency Ω with amplitude $H_R F_0^2$. The probe field samples the temperature field at R, picks up the frequency Ω, and carries the consequences of its encounter to L with amplitude $H_p f_0$. The temperature field may propagate through the sample, but in the specific case chosen for study both the pump and probe encounter the temperature field at the single point R, hence the factor $\Lambda^2 g_\Omega(R|R)$.

The diagram in Figure 3.9 exhibits the pieces you have to assemble to carry through the calculation for the model, Eqs. (3.85) and (3.86), and provides a guide to what has to be considered in getting to the final answer. This answer contains the evidence of the pieces assembled in a distinctive way that, if the answer is supposed to describe the outcome of a measurement, provides a test of the physical model, for example, the dependence on F_0^2, m, Ω, etc.

We have looked at one of many possibilities associated with Eqs. (3.85) and (3.86), that associated with the two terms in Γ. There are many other possibilities (phenomena). For example that shown in the lower panel of Figure 3.9 in which the pump produces a local temperature source at R at frequency Ω that, through Γ_0, the linear coupling, broadcasts a strain field at Ω directly into the sample.

The subject in this chapter is wide ranging. However, there are several ingredients present in all aspects of the discussion:

1. a Green function description of the fields that propagate in the system, elastic wave at ω, elastic wave at 3ω, diffusive auxiliary field, ...;
2. nonlinear parameters as coupling constants that bring about the interaction between these fields;
3. a diagramatic picture of the resulting physical processes, from which most of the physical variables involved in a description of the processes can be identified.

3.6
Green Functions

3.6.1
Green Function, Free Space

The equation for the Green function (in the frequency domain) is

$$\left(-\omega^2 - i\frac{\omega}{\tau_0} - c^2 \frac{\partial^2}{\partial x^2} \right) G_\omega(x|x') = \delta(x - x'), \tag{3.107}$$

where $c^2 = K_0/\varrho_0$. The Green function satisfies the boundary conditions

$$G^<(x|x') = G^>(x|x'), \quad x = x',$$

$$\frac{\partial G^<(x|x')}{\partial x} - \frac{\partial G^>(x|x')}{\partial x} = \frac{1}{c^2}, \quad x = x',$$

$$G^<(x|x') \to 0, \quad x \ll x',$$

$$G^>(x|x') \to 0, \quad x \gg x', \tag{3.108}$$

where $G = G^<$ at $x \le x'$ and $G = G^>$ at $x \ge x'$. When there is no ambiguity about the frequency ω associated with the Green function, we suppress the ω subscript. In instances when frequency is an issue, we use $G_\omega^{(<,>)}$. The solution for G is found by choosing

$$G^< = A e^{ikx},$$

$$G^> = B e^{ikx}, \tag{3.109}$$

where

$$k^2 c^2 = \omega^2 + i\frac{\omega}{\tau_0}, \tag{3.110}$$

and subjecting this ansatz to the boundary conditions. The complex wave vector k carries the sign of ω and has an imaginary part that leads to the appropriate behavior for G at $|x - x'| \to \infty$, that is, $k(-\omega) = -k(\omega)^*$. For $\omega\tau_0 \gg 1$ we have

$$k = \sqrt{\frac{\omega^2}{c^2} + i\frac{\omega}{c^2\tau_0}} = \frac{\omega}{c} + i\frac{1}{2c\tau_0} + \ldots \tag{3.111}$$

For G we find

$$G^<(x|x') = \frac{i}{2c^2k} \exp ik\left(x' - x\right),$$
$$G^>(x|x') = \frac{i}{2c^2k} \exp ik\left(x - x'\right).$$

(3.112)

Because of translational invariance, G depends only on $x - x'$; we have

$$\frac{\partial G(x|x')}{\partial x} = -\frac{\partial G(x|x')}{\partial x'}.$$

(3.113)

When the displacement satisfies the equation of motion

$$\left(-\omega^2 - i\frac{\omega}{\tau_0} - c^2\frac{\partial^2}{\partial x^2}\right) u_\omega(x) = S_U(\omega)\delta(x - x_1),$$

(3.114)

we have

$$u_\omega(x) = G_\omega(x|x_1)S_U(\omega).$$

(3.115)

A wave propagating to the right, that is,

$$u(x, t) = \int d\omega u_\omega(x)e^{-i\omega t} = U_0 \cos\omega_0\left(\frac{x}{c} - t\right),$$

(3.116)

results from

$$S_U(\omega) = A_U(\omega)\frac{\delta(\omega - \omega_0) + \delta(\omega + \omega_0)}{2}$$
$$A_U(\omega) = \frac{2c\omega}{i} U_0.$$

(3.117)

A tone burst propagating to the right, that is,

$$u(x, t) = \int d\omega u_\omega(x)e^{-i\omega t} = U_0 \cos\omega_0\left(\frac{x}{c} - t\right)e^{-\frac{a^2}{2}\left(\frac{x}{c} - t\right)^2},$$

(3.118)

results from

$$S_U(\omega) = A_U(\omega)\frac{1}{a\sqrt{2\pi}}\left[\frac{e^{-\frac{(\omega-\omega_0)^2}{2a^2}} + e^{-\frac{(\omega+\omega_0)^2}{2a^2}}}{2}\right]$$
$$A_U(\omega) = \frac{2c\omega}{i} U_0.$$

(3.119)

3.6.2
Green Function, Resonant Bar

The equation for the Green function is

$$\left(-\omega^2 - i\frac{\omega}{\tau_0} - c^2\frac{\partial^2}{\partial x^2}\right) G_\omega(x|x') = \delta(x - x'),$$

(3.120)

where $c^2 = K_0/\varrho_0$ and $0 \leq x \leq L$. The Green function satisfies the boundary conditions

$$G^<(x|x') = G^>(x|x'), \quad x = x',$$

$$\frac{\partial G^<(x|x')}{\partial x} - \frac{\partial G^>(x|x')}{\partial x} = \frac{1}{c^2}, \quad x = x',$$

$$\frac{\partial G^<(x|x')}{\partial x} = 0, \quad x = 0,$$

$$\frac{\partial G^>(x|x')}{\partial x} = 0, \quad x = L,$$

(3.121)

where $G = G^<$ at $0 \leq x \leq x'$ and $G = G^>$ at $x' \leq x \leq L$. When there is no ambiguity about the frequency ω associated with the Green function, we suppress the ω subscript. In instances when frequency is an issue, we use $G_\omega^{(<,>)}$. The solution for G is found by choosing

$$G^< = A\cos(kx),$$
$$G^> = B\cos(k(L-x)),$$

(3.122)

where

$$k^2 c^2 = \omega^2 + i\frac{\omega}{\tau_0},$$

(3.123)

and subjecting this ansatz to the boundary conditions.

For G we find

$$G^<(x|x') = \frac{-1}{c^2 k}\frac{\cos k(L-x')\cos kx}{\sin kL},$$

$$G^>(x|x') = \frac{-1}{c^2 k}\frac{\cos kx' \cos k(L-x)}{\sin kL}.$$

(3.124)

We have

$$G^<(x|x') = G^>(x'|x)$$
$$G_{-\omega} = G_\omega^*$$

(3.125)

but no simple relationship between $\partial/\partial x$ and $\partial/\partial x'$.

When the displacement satisfies the equation of motion

$$\left(-\omega^2 - i\frac{\omega}{\tau_0} - c^2\frac{\partial^2}{\partial x^2}\right) u_\omega(x) = S_U(\omega)\delta(x),$$

(3.126)

there is a source on the left at $x = 0$, the displacement on the right, at $x = L$, is

$$u_\omega(L) = G_\omega^>(L|0)S_U(\omega).$$

(3.127)

To have $u(L) = U_0\cos\omega_0 t$ we require

$$u(L,t) = \int d\omega u_\omega(L)e^{-i\omega t} = U_0\cos\omega_0 t$$

(3.128)

or

$$S_U(\omega) = A_U(\omega) \left(\frac{\delta(\omega - \omega_0) + \delta(\omega + \omega_0)}{2} \right)$$

$$A_U(\omega) = \frac{U_0}{G_{\omega_0}^{>}(L|0)} .$$

(3.129)

References

1 Boitnott, G.N. (1997) Experimental characterization of the nonlinear rheology of rock. *Int. J. Rock Mech. Min. Sci.*, **34**, 379–388.

2 Winkler, K. and Liu, X. (1996) Measurements of third-order elastic constants in rocks. *J. Acoust. Soc. Am.*, **100**, 1392–1398.

3 Goldberg, Z.A. (1960) Interaction of plane longitudinal and transverse elastic waves. *Soviet Physics-Acoustics*, **6**, 306–310.

4 Polyakova, A.L. (1964), Nonlinear effects in a solid. *Soviet Physics-Solid State*, **6**, 50–53.

5 McCall, K.R. (1994) Theoretical study of nonlinear elastic wave propagation. *J. Geophys. Res.*, **99**, 2591–2600.

6 Yost, W.T. and Cantrell, J.H. (1984) Acoustic-radiation stress in solids. II. Experiment. *Phys. Rev. B*, **30**, 3221–3227.

7 Jacob, X., Takatsu, R., Barrire, C., and Royer, D. (2006) Experimental study of the acoustic radiation strain in solids. *Appl. Phys. Lett.*, **88**, 134111.

8 Jones, G.L. and Kobett, D.R. (1963) Interaction of Elastic Waves in an Isotropic Solid. *J. Acoust. Soc. Am.*, **35**, 5–10.

9 Zaitsev, V., Gusev, V., and Castagnede, B. (1999) The Luxembourg–Gorky effect retooled for elastic waves: a mechanism and experimental evidence. *Phys. Rev. Lett.*, **89**, 105502–105505.

4
Mesoscopic Elastic Elements
and Macroscopic Equations of State

In this chapter we begin the transition away from traditional nonlinear elasticity. In Section 4.1 we note where we have been in the previous chapters and sketch what is to come, a description of elastic materials with properties importantly influenced by mesoscopic elastic elements. We begin in Section 4.2 with a description of candidate mesoscopic elastic elements, Hertz–Mindlin contacts (Section 4.2.1), hysteretic Hertzian contacts (Section 4.2.2), Hertzian asperities (Section 4.2.3), van der Waal surfaces (Section 4.2.4), and other (Section 4.2.5). In Section 4.3 we develop a simple effective medium theory that will let us learn the macroscopic elastic constants of a material comprised of mesoscopic elastic elements. In Section 4.4 we take Hertzian contacts, Section 4.4.1, and van der Waals surfaces, Section 4.4.2, through the recipe of Section 4.3. The van der Waals surfaces are hysteretic and the description of an assembly of them requires a *bookkeeping space*, a Preisach space. Generalization and caveats are found in Section 4.4.3.

4.1
Background

We have just completed a description of phenomena found in traditional nonlinear elasticity. The possibility of these phenomena depends on the structure of the theory. The importance of these phenomena depends on the numbers that go into the theory, linear and nonlinear elastic constants, the parameters in the equation of state (EOS) of an elastic material, etc. In Chapter 2 we suggested that for certain materials these numbers have their source in the interatomic potential. For other elastic materials, those of primary concern to us, these numbers arise from the workings within the material of mesoscopic elastic elements, things like cracks, Hertzian contacts, asperities, etc. A mesoscopic elastic element, for example a crack, has an equation of state, resides in an elastic material, and contributes to the elasticity of that material in a way that depends on how it is placed. The elastic element may be dilutely distributed, uniformly distributed, . . . throughout an otherwise unexceptional elastic material. It may be densely distributed in the mortar of a bricks-and-mortar elastic material having bricks of unexceptional elas-

Nonlinear Mesoscopic Elasticity: The Complex Behaviour of Granular Media including Rocks and Soil. Robert A. Guyer and Paul A. Johnson
Copyright © 2009 WILEY-VCH Verlag GmbH & Co. KGaA, Weinheim
ISBN: 978-3-527-40703-3

tic properties. We must understand how the EOS of a mesoscopic elastic element feeds into the EOS of an elastic material.

In this chapter we look at several candidates for mesoscopic elastic elements. We develop an effective medium theory that allows us to gain some understanding of the contributions the EOS of a mesoscopic elastic element makes to the EOS of an elastic material. When we face the question of practical implementation, we will bring to the fore a number of important issues that arise in attempting to turn mesoscopic EOSs into macroscopic EOSs.

4.2
Elastic Elements

4.2.1
Hertz–Mindlin Contacts

The most basic continuum elastic element is a Hertzian contact [1], for example, an isotropic elastic sphere, radius R, that is pushed down against a hard flat surface with external force N as in Figure 4.1. The volume $\delta V \approx a^2 h$ of the sphere, which

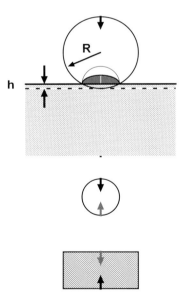

Fig. 4.1 Hertzian contact, compressive forces. The force across a Hertzian contact is due to the pressure at the contact surface that arises from the excess material pushed up into the contact. When the thickness of the indented layer is h, the radius of the contact area is $a \approx \sqrt{Rh}$, and material of volume a^2h is pushed into volume a^3. The resulting strain h/a causes the pressure, Eq. (4.1).

would be below the surface of the hard material if this material were absent ($a^2 \approx Rh$ is the contact surface area), is pushed up into a volume of order a^3. The strain in this volume, $\varepsilon \approx \delta V / a^3 \approx h/a$, produces pressure $P \approx K\varepsilon \approx Kh/a$ that pushes down on the contact surface area. Thus the force carried by the contact surface is $N \approx Pa^2 \approx Kah$, or

$$N \approx K \sqrt{Rh^3}, \tag{4.1}$$

that is, h is set by N.

Suppose the sphere is pushed against the surface with force N and then a shearing force T is applied at, say, the sphere center, as shown in Figure 4.2. If the sphere contact area is "glued" to the surface, then the surface contact area is unchanged. A displacement Δs will build up across the volume δV producing shear strain $\Delta s/a$, traction $\tau \approx \mu(\Delta s/a)$, and shear force

$$T \approx a^2 \tau \approx \mu a \Delta s. \tag{4.2}$$

The shear displacement Δs is set by T and dependent on N through a.

Equations (4.1) and (4.2) describe a Hertz–Mindlin contact [2]. Careful treatment of these contacts in the domain of elasticity theory is found in many places [3–5]. Our back-of-the-envelope description is intended to emphasize that the basic physics resides in the domain of continuum elasticity. In addition, the elastically

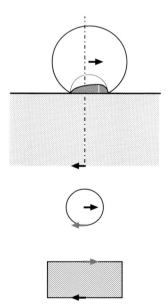

Fig. 4.2 Hertzian contact, shear forces. A shearing force T displaces the center of the sphere by Δs. This displacement occurs over volume a^3, that is, there is a shear strain $\Delta s/a$ and a shear stress on contact area a^2 that balances T, Eq. (4.2).

active part of the elastic element is $a^3/R^3 \approx N/(KR^2)$. At an applied pressure of 10 MPa, $N/R^2 \approx 10$ MPa and $a^3/R^3 \approx 10^{-3}$. The elastically active part of the elastic element is small.

4.2.2
Hysteretic Hertzian Contacts

Hysteresis is not intrinsic to elastic elements made from continuum elastic material, for example, Hertz–Mindlin contacts. It is conferred by the force protocol to which the contact is subjected. Consider a Hertzian contact subject to an (N, T) protocol, for example, the sequence of normal/shear force pairs $(N_1, T_1) \ldots (N_M, T_M)$. The EOS for the contact will depend on the rule that determines the relative motion of adjacent segments of the contact area [6, 7]. In one extreme, pieces of the contact area that first touch one another upon an increase in N to above N_m become/stay glued to one another until N decreases to below N_m. In another extreme, pieces of contact area slip by one another as long as the traction on them exceeds a prescribed limit set by the coefficient of friction. In these and related situations the behavior of a single contact depends importantly on the (N, T) protocol and the rules for material engagement. In some cases the contacts exhibit hysteresis. We discuss several simple results for hysteretic Hertzian contacts in Chapter 6.

4.2.3
Hertzian Asperities

The interaction of two superficial surfaces, each of area wL, is sometimes modeled as the encounter between a set of Hertzian contacts carried by one of the surfaces toward the other [8, 9]. The contacts model a set of asperities (Figure 4.3). Suppose the contacts all have the same radius, R, and that the probability density for their height, z, measured from a fiducial surface, line B, is $\phi(z)$, where

$$\int_R^{b_0} \phi(z)dz = 1 . \tag{4.3}$$

A force N applied to the surfaces, just touching at $N = 0$ and taken to be rigid along lines A and B, causes the indentation of those contacts for which $z > b$. The set of forces from these contacts carries force N. A contact at height $z > b$ carries force

$$\delta N \approx a^2 \delta P \approx KR^{1/2}(z-b)^{3/2} , \tag{4.4}$$

where $a^2 = R(z-b)$, and the total force carried by the set of all contacts is

$$N = N_A wL \int_b^{b_0} KR^{1/2}(z-b)^{3/2}\phi(z)dz , \tag{4.5}$$

where N_A, the number of contacts per unit area, is chosen by some argument such as: when all the contacts are crushed, $b < R$, the force is $N = KwL(b_0 - b)/b_0$. For

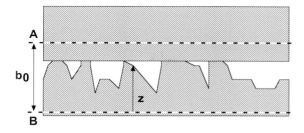

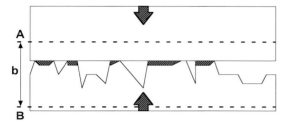

Fig. 4.3 Set of asperities. The interface between two surfaces is modeled as the encounter between a flat surface and a set of independent Hertzian contacts. Displacement of the upper surface toward the lower causes indentation of some of the contacts and a net force that is the sum of the force at each contact, Eq. (4.6).

example, for the case $\phi(z)$ equal to a constant for $R \le z \le b_0$, we would have

$$N = KwL\frac{(1 - b/b_0)^{5/2}}{(1 - R/b_0)^{3/2}} = KwL\frac{1}{(1 - R/b_0)^{3/2}}\left(\frac{h}{b_0}\right)^{5/2}, \tag{4.6}$$

where $b = b_0 - h$, cf. Eq. (4.1). It is apparent that by suitable choice of $\phi(z)$ and of a probability density for the radii, $\chi(R)$, a broad range of $N(h)$ EOSs could be constructed. An aspect of this model is that in both static and dynamic responses the contacts are taken to be independent of one another. This simplification suits computation but may need to be abandoned in certain circumstances, for example, it is possible that the quasistatic response of a contact can modify the response of another or an elastic wave launched by one contact can disturb (trigger) the response of another.

4.2.4
Van der Waals Surfaces

Consider a half-space of material made up of atoms at density ϱ that interact with one another through a Lennard–Jones interaction [10, 11],

$$v(r) = 4\varepsilon_0\left[\left(\frac{\sigma}{r}\right)^{12} - \left(\frac{\sigma}{r}\right)^{6}\right], \tag{4.7}$$

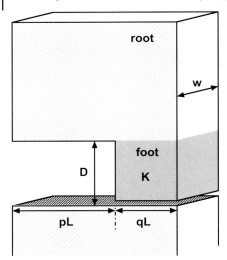

Fig. 4.4 Van der Waals contact. Two surfaces that interact through the van der Waals force are initially separated by D and in parallel with a piece of elastic material, the foot. The root is taken to be relatively rigid so that a force across the whole causes strain in the foot/van der Waals system.

where ε_0 is an atomic energy scale, say 0.5 eV, and σ is an atomic length scale, say 4 Angstroms, Chapter 2. Adjacent to this half-space is a half-space of similar material of size pLw, $0 \le p \le 1$, see Figure 4.4. The potential energy of this system, when the separation between the two half-spaces is D, is

$$V_0(D) = \varepsilon \varrho^2 \, pLw\sigma^4 \left[\left(\frac{\sigma}{D} \right)^8 - \left(\frac{\sigma}{D} \right)^2 \right] , \qquad (4.8)$$

where $\varepsilon \propto \varepsilon_0$ contains a collection of numerical factors and σ here is $4^{1/6}\sigma$ from Eq. (4.7). Parallel to the two half-spaces is a uniform piece of material called the *foot*. When the separation of the half-spaces is D, the foot is stretched uniformly by $D - D_0$, where D_0 is its unstretched length. The foot has energy

$$V_1(D) = \frac{1}{2} D_0 q Lw K \left(\frac{D - D_0}{D_0} \right)^2 , \qquad (4.9)$$

where $q = 1 - p$. The root, Figure 4.4, has negligible elastic energy compared to the energy of the encounter of the half-spaces and the foot. Suppose this system is subjected to force N. The internal force carried by the elastic element, $-\partial(V_0 + V_1)/\partial D$, is equal to N. We can gain some understanding of this mechanical equilibrium problem by looking at the energy

$$\mathcal{E} = V_0 + V_1 = p\varepsilon \frac{Lw}{\sigma^2} \left(\left[\left(\frac{\sigma}{D} \right)^8 - \left(\frac{\sigma}{D} \right)^2 \right] + \frac{\Lambda}{2} \left(\frac{D}{\sigma} - \frac{D_0}{\sigma} \right)^2 \right) , \qquad (4.10)$$

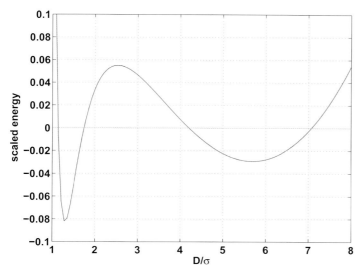

Fig. 4.5 Energy of a van der Waals contact. The energy of a van der Waals contact, Figure 4.4, as a function of D/σ has two minima, one due to the elasticity of the foot and a sharper, deeper one due to the van der Waals interaction between the surfaces. Certain applied forces can be supported at multiple values of the contact separation, D/σ. Thus the van der Waals contact is a hysteretic elastic element.

where we have used the simple approximations $\varrho \approx 1/\sigma^3$, $K \approx \varepsilon/\sigma^3$ to get at the essential structure. The parameter $\Lambda = q\sigma/(D_0 p)$ controls the relative size of the two energy contributions. The energy $\mathcal{E}$ is plotted in Figure 4.5. From the figure we see that a given force, slope of $\mathcal{E}$ vs. D/σ, has the same value at locations separated by points of inflection. Thus there will be a hysteretic solution to the equations of mechanical equilibrium, the EOS. We return to this below where we discuss the van der Waals surfaces in detail.

Note that in contrast to the cases involving Hertzian contacts, where we use only continuum elasticity, the description of van der Waals surfaces has an energy based in microscopic physics, ε and σ, as well as an energy that comes from continuum elasticity.

4.2.5
Other

There are many other possibilities for elastic elements that we will not discuss here in detail. Among these are various types of cracks (simple cracks, cohesive cracks, ...) [12, 13], systems of linked rods, frictional sliders [14], fluid-modified contacts [15], etc. Some of these will come up in later chapters. The point of the discussion here is to introduce some of the basic elastic elements that one might be prepared to consider in building up a material. *An elastic element isn't an elastic material.* We turn to our primary concern: finding the EOS of a material given the set of elastic elements it comprises.

4.3
Effective Medium Theory

The description above is of particular models of elastic elements. If one imagines that such an elastic element is present in a system, how do you go about assessing its consequences. One procedure is to construct models in which the elastic element is the only actor, for example, a system of carefully arranged spheres for Hertzian contacts or cubes for van der Waals surfaces, what might be called

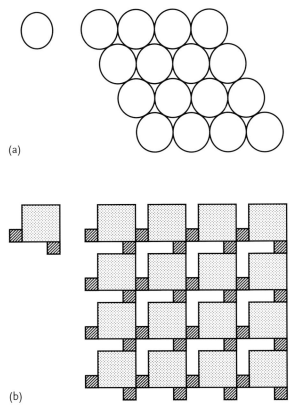

(a)

(b)

Fig. 4.6 Simple elastic systems. When an elastic system is built up of a regular array of identical elastic elements, (a) a close-packed set of spheres or (b) a uniform tiling of van der Waals contacts, the elastic properties of the system are essentially those of the elastic element.

bathroom-tile models, Figure 4.6. In these cases the EOS of the elastic element basically becomes the EOS of the material. But one expects that there would be different arrangements of a particular elastic element in a material, for example, van der Waals surfaces arranged at random among conventional pieces of material as illustrated in Figure 4.7. Or there might be a topology that places a particular set of elastic elements in a featured location, for example, a bricks-and-mortar arrange-

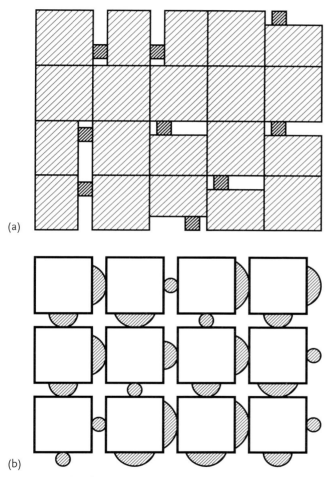

(a)

(b)

Fig. 4.7 Complex elastic systems. (a) A system built up of a random arrangement of five types of elastic elements, van der Waals contacts with four possible orientations and a *normal* elastic element. (b) A system built up of a uniform array of *normal* elastic elements that see one another through a random arrangement of Hertzian contacts. Elastic constants for (a) can be found from a simple effective medium theory. Elastic constants for (b) come from an extension of the simple EMT that allows one to preserve the bricks-and-mortar geometry.

ment, with a system of Hertzian contacts (mortar) that mediate the interaction between conventional pieces of material (bricks), Figure 4.7. Location matters.

In order to assess the participation of a system of elastic elements in the elastic properties of a piece of material, we develop a simple effective medium theory (EMT) for the elastic EOS of materials like those pictured in Figure 4.7, that is, materials that are built up out of elastic elements with known properties. In addition to providing a recipe for doing EMT, these calculations serve a pedagogic purpose. They allow us to call attention to a number of points we think important. The EMT we develop here can be placed in the context of the highly developed subject of EMT for elastic systems, on which there are several excellent references [16, 17].

The idea behind EMT, as we will employ it, is sketched in Figure 4.8. To find the effective elastic constants of a material, made up, say, of a random mixture of elastic elements having van der Waals surfaces and elastic elements of normal material (both the van der Waals elastic elements and the normal elastic elements could have a spectrum of properties), proceed as here.

1. Have in hand the EOSs for all instances of the elastic elements from which the material is built, 1EE, 2EE, ... NEE; denote these elastic elements by iEE. Also, have in hand the EOS of an elastic element for a uniform isotropic material, called the medium; denote this elastic element by mEE. The elastic parameters of the mEE are found in step 5.

2. Solve the problem of a system made up of many mEE with one iEE, called the *representative elastic element*, embedded in the interior. This problem is solved with applied forces set by 3.

3. If one is interested in the compressive response of the material, put a set of compressive forces at the boundaries, etc.

4. If, for example, compressive forces have been placed on the boundaries, find the compressive response, that is, compressive displacement, of the representative elastic element. This displacement will depend on the particular instance of the iEE and on the parameters assigned to the mEE.

5. Average the compressive displacement of the representative elastic element over all instances. Ask that this average be the same as that which would be found if the entire system were made of mEE. The resulting equation connects the parameters assigned to the mEE to an average over the iEE. Solve for the parameters of the mEE.

6. Approximate the properties of the elastic system by those of the mEE.

7. This procedure results in a set of elastic constants for the material. These can be manipulated to find the EOS of the material, nonlinear elastic constants, etc.

Let us work through this program. (Numerous simplifications will be made without sacrificing the essentials. This is so we can be explicit while at the same time have a manageable set of equations. We work in two dimensions and call the elastic elements *tiles*. Thus it is the EOS of instances of tiles that is input to these calculations.) Consider a uniform system of tiles (mEE from above), size $a \times a$, that have displacement in only one direction, y (Figure 4.9). The lower left corner of each tile

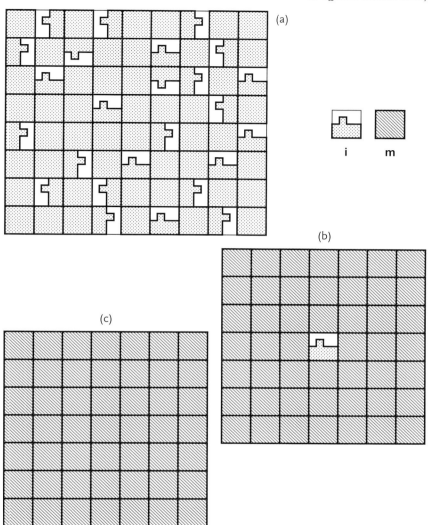

Fig. 4.8 EMT, the principle. The principle of effective medium theory is to relate the properties of three elastic systems, (a) an elastic systems built from a disordered arrangement of elastic elements, (b) an elastic system built as an ordered arrangement of elastic elements (having *to be found* elastic parameters) into which an elastic element from (a) is embedded, and (c) an ordered elastic system having elastic parameters found from an average of (b) over all instances of the embedded elastic element.

is tracked with the displacement $y_\alpha = y_{mn}$. The tiles are taken to be "glued" to their neighbors along the common borders [18]. When a particular tile is being referred to, it is labeled y_α and the three neighbors that locate its other corners are $y_{\alpha 1}$, $y_{\alpha 7}$, and $y_{\alpha 8}$, as shown in Figure 4.9. There are three rudimentary strains associ-

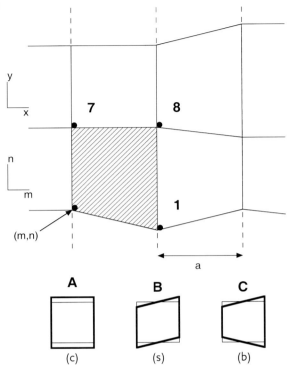

Fig. 4.9 Tile space. The elastic elements, tiles, are in a $D = 2$ arrangement, glued along common edges to their neighbors so that each is described by a single displacement (lower left corner). There are three distortions of each tile, compression (c), shear (s), and bending (b) associated with elastic constants A, B, and C, respectively. These distortions involve the displacement coordinate of a tile and the displacement coordinate of its 7,8,1 neighbors, Eqs. (4.11)–(4.13).

ated with each tile and an elastic energy associated with each strain. The strains are compressive, c, shear, s, and bending, b, illustrated at the bottom of Figure 4.9, given by

$$\Delta y_\alpha^c = [-y_\alpha - y_{\alpha 1} + y_{\alpha 7} + y_{\alpha 8}], \tag{4.11}$$

$$\Delta y_\alpha^s = [-y_\alpha + y_{\alpha 1} - y_{\alpha 7} + y_{\alpha 8}], \tag{4.12}$$

$$\Delta y_\alpha^b = [+y_\alpha - y_{\alpha 1} - y_{\alpha 7} + y_{\alpha 8}], \tag{4.13}$$

and the corresponding energies are

$$E_\alpha = \frac{A}{2}\left(\Delta y_\alpha^c\right)^2 + \frac{B}{2}\left(\Delta y_\alpha^s\right)^2 + \frac{C}{2}\left(\Delta y_\alpha^b\right)^2, \tag{4.14}$$

where numerical factors and factors of a have been lumped into the elastic constants, A, B, and C, the parameters of the effective medium. The total energy is

$\vec{E}_0 = \sum E_\alpha$ and the equation of motion for γ_α is given by

$$-\varrho \ddot{\gamma}_\alpha = \frac{\partial E_0}{\partial \gamma_\alpha} - F_\alpha , \tag{4.15}$$

where F_α is the applied force on tile α (applied forces are typically on the edges of the system; it is not necessary to specify them in detail).

The first step in developing EMT is to consider the system that differs from the uniform system (all mEE tiles) with energy E_0 by having one tile that is an instance of the elastic elements, iEE. Assume that the tile at $(0,0)$ has elastic constants A_i, B_i, C_i; i is the index for instances. The energy of this new system can be written as

$$E = E_0 + \left(\frac{\delta A_i}{2} \left(\Delta \gamma_{00}^c \right)^2 + \frac{\delta B_i}{2} \left(\Delta \gamma_{00}^s \right)^2 + \frac{\delta C_i}{2} \left(\Delta \gamma_{00}^b \right)^2 \right) , \tag{4.16}$$

where $\delta A_i = A_i - A$, $\delta B_i = B_i - B$, and $\delta C_i = C_i - C$. Let us follow one of the changes in elastic energy, choose δA, through the equations of motion. From Eq. (4.15) $(E_0 \rightarrow E)$ we have

$$-\varrho \ddot{\gamma}_\alpha = \frac{\partial E_0}{\partial \gamma_\alpha} + \delta A_i \Delta \gamma_{00}^c D_{\alpha,00}^c - F_\alpha , \tag{4.17}$$

where

$$D_{\alpha,00}^c = [-\delta_{\alpha,00} - \delta_{\alpha,10} + \delta_{\alpha,01} + \delta_{\alpha,11}] . \tag{4.18}$$

For the coupling between the displacements, $\partial E_0 / \partial \gamma_\alpha$, we have

$$\begin{aligned} \frac{\partial E_0}{\partial \gamma_{mn}} &= 4(A+B)\gamma_{mn} \\ &\quad - (A+B)(\gamma_{m+1n-1} + \gamma_{m-1n-1} + \gamma_{m-1n+1} + \gamma_{m+1n+1}) \\ &\quad + 2(A-B)(\gamma_{mn} - \gamma_{mn-1} + \gamma_{m-1n} - \gamma_{mn+1}) . \end{aligned} \tag{4.19}$$

(For simplicity we keep only two of the elasticities in E_0 as this is enough to illustrate the interplay of the different displacements.) Let us agree to look at the quasistatic response, that is, drop the inertial term. To find γ perform a Fourier analysis of Eq. (4.19) as here:

$$\delta_{\alpha,\alpha'} = \frac{1}{N^2} \sum_\mu \sum_\nu e^{i\mu(m-m')} e^{i\nu(n-n')} , \tag{4.20}$$

$$\gamma_\alpha = \frac{1}{N^2} \sum_\mu \sum_\nu e^{i\mu m} e^{i\nu n} U_{\mu\nu} , \tag{4.21}$$

$$D_{\alpha,00}^c = \frac{1}{N^2} \sum_\mu \sum_\nu e^{i\mu m} e^{i\nu n} D_{\mu\nu}^c , \tag{4.22}$$

$$F_\alpha = \frac{1}{N^2} \sum_\mu \sum_\nu e^{i\mu m} e^{i\nu n} F_{\mu\nu} , \tag{4.23}$$

with

$$D_{\mu\nu}^c = [-1 - e^{-i\mu} + e^{-i\nu} + e^{-i\mu}e^{-i\nu}]. \tag{4.24}$$

From Eq. (4.17) find

$$U_{\mu\nu} = \frac{1}{M_{\mu\nu}} F_{\mu\nu} - \delta A_i \Delta\gamma_{00}^c \frac{1}{M_{\mu\nu}} D_{\mu\nu}^c, \tag{4.25}$$

where

$$M_{\mu\nu} = 4(A + B)(1 - \cos\mu\cos\nu) + 4(A - B)(\cos\mu - \cos\nu). \tag{4.26}$$

Equation (4.25) is a mixed representation of γ, involving $U_{\mu\nu}$ and $\Delta\gamma_{00}^c$. Use $U_{\mu\nu}$ from Eq. (4.25) to find $\Delta\gamma_{00}^c$, that is, from Eq. (4.13), $\Delta\gamma_{00}^c = -\gamma_{00} - \gamma_{10} + \gamma_{01} + \gamma_{11}$, and Eq. (4.23) obtain

$$
\begin{aligned}
\Delta\gamma_{00}^c &= \frac{1}{N^2} \sum_{\mu}\sum_{\nu} e^{i\mu m} e^{i\nu n} [-\delta_{mn,00} - \delta_{mn,10} + \delta_{mn,01} + \delta_{mn,11}] U_{\mu\nu}, \\
&= \frac{1}{N^2} \sum_{\mu}\sum_{\nu} D_{\mu\nu}^{c*} U_{\mu\nu}, \tag{4.27} \\
&= \mathcal{S}\left(D_{\mu\nu}^{c*} \frac{1}{M_{\mu\nu}} F_{\mu\nu} \right) - \delta A_i \Delta\gamma_{00}^c \mathcal{S}\left(D_{\mu\nu}^{c*} \frac{1}{M_{\mu\nu}} D_{\mu\nu}^c \right),
\end{aligned}
$$

where $\mathcal{S}$ stands for $N^{-2}\sum_{\mu}\sum_{\nu}$. Solve for $\Delta\gamma_{00}^c$ and rearrange the resulting equation in the form

$$\Delta\gamma_{00}^c = \mathcal{S}\left(D_{\mu\nu}^{c*} \frac{1}{M_{\mu\nu}} F_{\mu\nu} \right) - \frac{\delta A_i \Lambda_{cc}}{1 + \delta A_i \Lambda_{cc}}, \tag{4.28}$$

with

$$\Lambda_{cc} = \mathcal{S}\left(D_{\mu\nu}^{c*} \frac{1}{M_{\mu\nu}} D_{\mu\nu}^c \right) = \frac{1}{N^2} \sum_{\mu}\sum_{\nu}\left(D_{\mu\nu}^{c*} \frac{1}{M_{\mu\nu}} D_{\mu\nu}^c \right). \tag{4.29}$$

Central Dogma

The basic equation of effective medium theory comes from the equation for $\Delta\gamma_{00}^c$. The first term on the RHS of this equation gives the compressive response at tile $(0, 0)$ of uniform mEE material under compressive stress. If we require that the average of $\Delta\gamma_{00}^c$, over the iEE, be that of uniform mEE material, we must have

$$\left\langle \frac{\delta A_i \Lambda_{cc}}{1 + \delta A_i \Lambda_{cc}} \right\rangle = 0, \tag{4.30}$$

or, since Λ_{cc} is simply a number,

$$\left\langle \frac{A_i - A}{1 + (A_i - A)\Lambda_{cc}} \right\rangle = \sum_{i=1}^{M} \frac{A_i - A}{1 + (A_i - A)\Lambda_{cc}(A, B, C)} = 0. \tag{4.31}$$

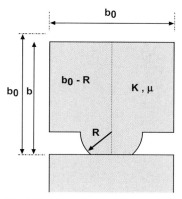

Fig. 4.10 Hertz tile. A tile appropriate to a Hertzian contact consists of a root and attached half-sphere having elastic constants K, μ. The root is treated as rigid so that the elasticity of the tile is that of the half-sphere, Hertzian contact. The size of the tile is b_0 with the contact just touching the lower surface. When the tile has size $b < b_0$, the contact is indented by $h = b_0 - b$.

With B and C fixed, A is chosen so that the average over the A_i yields zero. The elastic constant characterizing a particular strain depends on the geometry of that strain through the factors $D^t_{\mu\nu}$, $t = c, s, b$, different for the three elastic constants, and the elasticity of the A, B, C system through $M_{\mu\nu}$. It may be that Λ has to be calculated numerically for each A. Various strategies for finding A, B, and C are possible.

4.4
Equations of State; Examples

Let us look in detail at two examples of the use of EMT (1) a system of Hertzian contacts and (2) a system of van der Waals surfaces.

4.4.1
Hertzian Contacts

In Figure 4.10 we show a tile appropriate to a Hertzian contact. The equation of state for a tile of size $b_0 \times b_0$, all of whose elasticity is taken to be that of a contact of radius R_i, is from Eq. (4.1)

$$b_0 - b = \left(\frac{F}{K \sqrt{R_i}} \right)^{\frac{2}{3}} ,$$

(4.32)

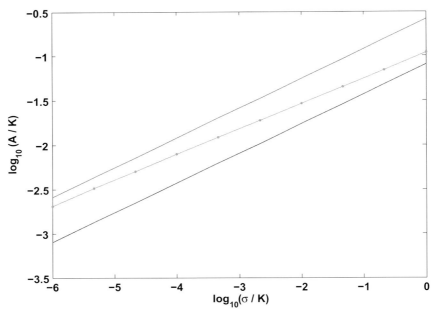

Fig. 4.11 EMT elastic constants for Hertzian contacts I. The compressional elastic constant *A* from solution to Eq. (4.31) is plotted as a function of σ/K, the middle curve. The value of *B* is fixed at 0.02*K*. The upper and lower curves, each with slope 1/3, are $\langle A \rangle$ and $\langle A^{-1} \rangle^{-1}$, respectively. The middle curve, *A*, trends from $\langle A \rangle$ at low σ to $\langle A^{-1} \rangle^{-1}$ at high σ. This evolution from parallel-like to serieslike behavior is due to the shear support of compressional distortion.

where $h \rightarrow b_0 - b$ and *K* is the elastic constant of the contact material. For the compressive elastic constant of the tile, A_i, we have

$$\frac{A_i}{K} = \left(\frac{\sigma}{K}\right)^{\frac{1}{3}} \left(\frac{R_i}{b_0}\right)^{\frac{1}{3}}, \tag{4.33}$$

where $\sigma = F/b_0^2$ is the compressive stress on the tile. We take the shear elastic constants of all contact tiles to be the same, $B_i = \mu \; \forall \; i$. Thus the set of instances of elastic constants for the contact tiles is (A_i, μ) with the spectrum of A_i determined by the probability density of $r_i = R_i/b_0$. These tiles may be mixed with tiles of uniform material, having elastic constants (A_0, B_0), at concentration $x = N_c/(N_0 + N_c)$, where N_c and N_0 are the number of contact and normal tiles, respectively. It is apparent from Eq. (4.33) that a convenient dimensionless description of this system is given upon scaling all elastic constants by *K* and all lengths by b_0. We do this.

In Figure 4.11 we show A/K, the effective compressional elastic constant, as a function of σ/K from the solution to Eq. (4.31) for the case $x = 1.0$, $B = \mu = 0.02K$, $10^{-6} \leq \sigma/K \leq 1$, and the probability density for *r*

$$p(r) \propto \frac{1}{r^{1.001}}, \quad 10^{-5} \leq r \leq 1.0. \tag{4.34}$$

The upper and lower curves in the figure, both with slope 1/3, are $A = \langle A_i \rangle$ and $A = 1/\langle A_i^{-1} \rangle$, respectively, and the middle curve, with open circles, is from the solution to Eq. (4.31). (Here $\langle \ldots \rangle$ is the average over the probability density $p(r)$.) In these calculations the value of the shear modulus was chosen to be in the middle of the range of compressional moduli to illustrate the way in which the two moduli affect one another. As background for this discussion we note:

1. When a set of springs, with a spectrum of elastic constants, is arranged to carry a force in series, with each spring supporting the same force, the effective elastic constant of the assembly is $\Gamma = 1/\langle \gamma_i^{-1} \rangle$, where the individual spring elastic constants are γ_i and $\langle \ldots \rangle$ is the average over these. The effective spring constant notices the weakest spring as this must carry the same force as the strongest.

2. When a set of springs, with a spectrum of elastic constants, is arranged in parallel, with each spring having the same displacement, the effective elastic constant of the assembly is $\Gamma = \langle \gamma_i \rangle$, where the individual spring elastic constants are γ_i and $\langle \ldots \rangle$ is the average over these. The effective spring constant notices the strongest spring as this must be displaced by the same amount as the weakest.

3. It can be established by formal rearrangments of Eq. (4.31) alone that [19]
   a. For $Q = B/A \ll 1$

$$A = \frac{1}{\langle A^{-1} \rangle} \left(1 + Q \frac{\langle (\delta A^{-1})^2 \rangle}{\langle A^{-1} \rangle^2} \right) + \ldots, \tag{4.35}$$

   where $\delta A_i^{-1} = (1/A_i) - 1/A$;
   b. For $Q = A/B \ll 1$

$$A = \langle A \rangle \left(1 - Q \frac{\langle (\delta A)^2 \rangle}{\langle A \rangle^2} \right) + \ldots, \tag{4.36}$$

   where $\delta A_i = A_i - A$.
   Thus $\langle A^{-1} \rangle^{-1} \leq A \leq \langle A \rangle$.

When a set of elastic elements is embedded in a medium, the compressional force the elastic element must support, say from top to bottom, is shared with its neighbors, to the left and right, in a way that depends on the shear modulus. If the shear coupling is zero, there is no way to transfer force to the left/right neighbors and the elastic element responds to force as if in a series arrangement. If the shear coupling to the neighbors is very strong, then much of the force the elastic element must support can be shared by the left/right neighbors and the elastic element responds as if in a parallel arrangement. Thus the geometry of an elastic system has an important effect on the sensitivity of the system to extremes in the behavior of the elastic elements. The ability to shunt force around an unusual elastic element typically makes systems of elastic elements much less exciting than individual elastic elements. Sociology is much less exciting than psychology.

Consider a set of Hertzian contact tiles mixed at concentration x with tiles of normal material. In Figure 4.12 we show A/K, the effective compressional elastic constant, as a function of x from the solution to Eq. (4.31) for the case $0.01 \leq x \leq 1$, $B_i/K = (1/3)A_i/K \;\; \forall i$, the probability density for r given by Eq. (4.34), and the constituent with concentration $1 - x$ having $A_i/K = 1$ and $B_i/K = (1/3)A_i/K$. The contacts dominate the behavior of the whole for $x > 0.40$, and in the regime where they dominate, their behavior is rather closely *serieslike*, that is, the shear coupling, $B_i = A_i/3$, does not allow much force to be shunted.

From results like those in Figure 4.11 one can construct an EOS for the material, for example, for the light gray $A = d\sigma/d\varepsilon \sim (\sigma/K)^{0.29}$. Thus

$$\varepsilon(\sigma) - \varepsilon(0) = \int_0^\sigma \frac{d\sigma'}{A(\sigma')} \sim \sigma^{0.71} . \tag{4.37}$$

We are able to proceed in this very naive way because by constuction the shear modulus is constant as A varies. So this calculation suggests a principle, but the

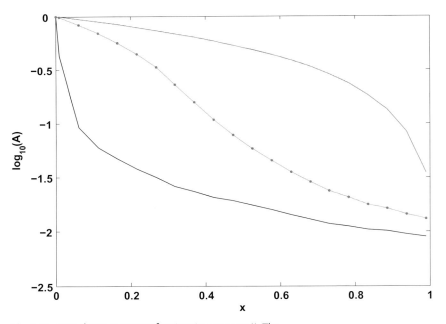

Fig. 4.12 EMT elastic constants for Hertzian contacts II. The compressional elastic constant A from solution to Eq. (4.31) is plotted as a function of the concentration of Hertzian contacts, x, the middle curve, for fixed $\sigma/K = 0.001$. The upper and lower curves are $\langle A \rangle$ and $\langle A^{-1} \rangle^{-1}$, respectively. A 40% concentration of contacts reduces the elastic constant by an order of magnitude.

proper venue for such calculations is the surfaces of A and B in (σ, τ) space. We will have more to say about this below.

It is also possible from results like these to deduce nonlinear elastic constants. We have

$$\frac{A}{K} \sim \left(\frac{\sigma}{K}\right)^{\mu}, \quad \varepsilon \sim \left(\frac{\sigma}{K}\right)^{\nu}, \quad \frac{A}{K} \sim \varepsilon^{\mu/\nu},$$
$$\beta \sim \frac{\partial}{\partial \varepsilon} \log(A) \sim \frac{1}{\varepsilon} \sim \left(\frac{K}{\sigma}\right)^{\nu},$$

(4.38)

where $\mu = 0.29$ and $\nu = 0.71$. The nonlinear coefficient is singular at $\sigma \to 0$ and decreases from large values as σ increases. It is this qualitative property, already present in Eq. (4.1), that makes the Hertzian contact a popular model for the description of nonlinear elastic systems.

4.4.2
Van der Waals Surfaces

In Figure 4.13 we show a tile appropriate to an encounter between van der Waals surfaces. This case is considerably more complicated than that of Hertzian contacts. Each instance of a tile has two geometrical parameters, p, which gives the fraction of the surface that involves the van der Waals interaction, and D_0, the inset

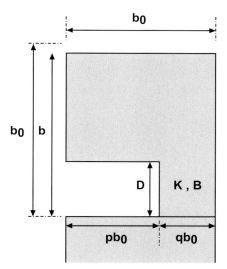

Fig. 4.13 Van der Waals tile. A tile appropriate to a van der Waals contact consists of a root and foot having elastic constants K, B and a pair of interfacing van der Waals surfaces. The root is treated as rigid so that the elasticity of the tile is that of the foot in parallel with the van der Waals contact. The size of the tile is $b_0 \times b_0$ when the van der Waals surfaces are separated by D_0.

of the van der Waals surface, (D_0, p), $q = 1 - p$. The foot of the tile is very soft. Thus the root is taken to be infinitely stiff as is the half-space on which the foot rests. As a consequence, a force is carried by the elastic element through the adjustment of the length D. This adjustment involves the elasticity of the foot in parallel with the elasticity of the van der Waals encounter. For the energy of the tile in the presence of an applied force F we have

$$\mathcal{E} = V_0 \left[\left(\frac{\sigma}{D} \right)^8 - \left(\frac{\sigma}{D} \right)^2 \right] + \frac{1}{2} V_1 \left[\frac{D - D_0}{D_0} \right]^2 + FD, \tag{4.39}$$

where $V_0 = \varepsilon\varrho^2 pb_0^2\sigma^4$, $V_1 = Kqb_0^2$, and $D - D_0 = b - b_0$. For purposes of illustration we use $K = \varepsilon/\sigma^3$, scale all lengths by σ, scale $\mathcal{E}$ by ε, scale F by ε/σ, and write

$$\mathcal{E} = pG \left[\left(\frac{1}{D} \right)^8 - \left(\frac{1}{D} \right)^2 \right] + \frac{1}{2} qG \frac{1}{D_0} (D - D_0)^2 + FD,$$

$$\mathcal{E} = p \left[\left(\frac{1}{D} \right)^8 - \left(\frac{1}{D} \right)^2 \right] + \frac{1}{2} q \frac{1}{D_0} (D - D_0)^2 + \Sigma D = \mathcal{E}_0 + \mathcal{E}_1 + \Sigma D, \tag{4.40}$$

where $G = b_0^2/\sigma^2$ and the second line follows from dividing by G and defining the effective stress $\Sigma = F/G$. For each instance of the tile, (D_0, p), we want the EOS of the tile, b, as a function of Σ, and the elastic constant of the tile, A, as a function of Σ. For the EOS we solve the problem $-\partial\mathcal{E}/\partial D = 0$. As illustrated in Figure 4.5, this equation has multiple solutions. Thus we specify a Σ protocol and follow the state of the tile through that protocol. Both the state, b or D, and the elastic constant, A, will be a hysteretic function of Σ. We will go through this case in some detail to illustrate what is called for. To make it possible to write out most of what is involved, we approximate the problem set by Eq. (4.40) as here.

1. Replace the van der Waals interaction, $\mathcal{E}_0$, by a Taylor series expansion to second order in $D - r$ about its minimum at $r = 2^{\frac{1}{3}}$:

$$\mathcal{E}_0 \approx p \left[-\frac{3}{4r^2} + \frac{1}{2}\frac{12}{r^4}(D - r)^2 \right] \theta(D_c - D),$$

$$\mathcal{E}_1 = \frac{1}{2}q\frac{1}{D_0}(D - D_0)^2\theta(D - D_c), \tag{4.41}$$

   where $\theta(z)$ is the Heaviside function and D_c is the value of D at which $\mathcal{E}_0 = \mathcal{E}_1$ (Figure 4.14).

2. For $D > D_c$ the tile is said to be in the elastic state $\eta = -1$, and for $D < D_c$ the tile is said to be in the elastic state $\eta = +1$. We adopt the convention that stresses that tend to compress the tile are positive and those that expand it are negative.

3. a. If the tile is in the elastic state -1 and Σ becomes more positive, the state of the tile makes the transition $-1 \rightarrow +1$ at $\Sigma = \Sigma_c > 0$. The new equilibrium of the tile is at D in $\mathcal{E}_0$, where the stress is Σ_c. A further increase in Σ drives D up the left wall of $\mathcal{E}_0$.

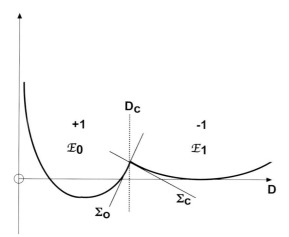

Fig. 4.14 Energy of the van der Waals tile. The energy of the van der Waals tile is approximated by two adjacent parabolic contributions, one for the foot, $\mathcal{E}_1$, and the second for the interfacing van der Waals surfaces, $\mathcal{E}_0$. The stresses (Σ_c, Σ_o) are determined by the slope of $\mathcal{E}$ vs D at D_c. The state variable η is +1 for $D < D_c$ and −1 for $D > D_c$.

b. If the tile is in the elastic state −1 and Σ is reduced, the tile will pass through $\Sigma = 0$ and become negative while remaining in the state $\eta = +1$ until $\Sigma = \Sigma_o < 0$, when $\eta = +1 \rightarrow -1$. A further decrease in Σ drives D up the right wall of $\mathcal{E}_1$.

In summary

$$\Sigma_c = (1-p)(D_0 - D_c)/D_0, \quad \eta = -1 \rightarrow +1,$$
$$\Sigma_o = 12\,p(r - D_c)/r^4, \quad \eta = +1 \rightarrow -1,$$

(4.42)

where $r < D_c < D_0$.

4. The elastic constant of a tile depends on its state

$$A(p, D_0) = \frac{1-p}{D_0}, \quad \eta = -1,$$
$$A(p, D_0) = p\,\frac{12}{r^4}, \quad \eta = +1.$$

(4.43)

To carry through an EMT calculation for van der Waals surfaces
1. choose a set of instances,
2. choose a stress protocol and set the initial state of each instance according to the initial stress,
3. follow the state of each instance through the stress protocol and at each stress value determine the elastic constant of each instance, and
4. at each value of the stress in the protocol use the elastic constants from the set of instances to determine an effective elastic constant from Eq. (4.31).

We show the results of a calculation of A from Eq. (4.31) in a series of figures.

1. The instances of van der Waals tiles were chosen with the probability density for D_0 proportional to D_0^{-1} for $D_< \leq D_0 \leq D_>$ ($D_< = 5r(1 + 1/\sqrt{8})/4$, $D_> = 8D_<$) and for p uniformly distributed between 0.2 and 0.8.

2. The spectrum of elastic constants associated with these instances is shown in Figure 4.15, a histogram of $\log_{10}(A(D_0, p))$. There are two parts to the probability density: (a) a set of weak elastic constants, $A \approx 0.01$, associated with the foot and due to the factor D_0 in Eq. (4.39) and (b) a set of strong elastic constants that are associated with the van der Waals surface. When an instance is in the state $\eta = -1$, its elastic constant is from those in (a), and when it is in the state $\eta = +1$, its elastic constant is from those in (b).

3. The stress pairs (Σ_c, Σ_o) associated with the instances are shown in Figure 4.16. The space of control variable pairs for two state units is called a Preisach space [20, 21]. Here the control variable is the stress and the Preisach space is (Σ_c, Σ_o) space, Figure 4.16. Such a space is a bookkeeping

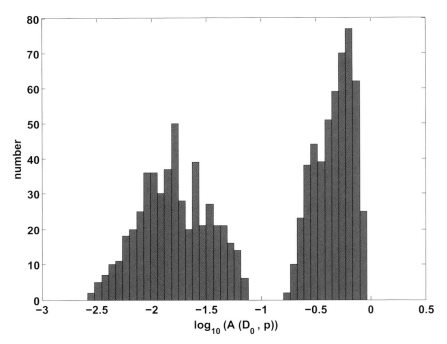

Fig. 4.15 Spectrum of instances of A. The compressive elastic constants for the instances of the van der Waals tile is approximately bimodal, Eq. (4.43). The small elastic constants occur for $\eta = -1$ and the large elastic constants occur for $\eta = +1$. The spectrum arises from choosing a spectrum of D_0 and a spectrum of p, $8.53 \leq D_0 \leq 68.21$ and $0.20 \leq p \leq 0.80$. See Figure 4.14.

device that can be used very effectively to provide qualitative understanding and sometimes even quantitative understanding. We will see further examples of Preisach bookkeeping spaces as we proceed. Note that the closing stresses are positive and are weaker than the opening stresses, which are negative.

4. We choose a stress protocol, based on the range of (Σ_c, Σ_o) in the Preisach space, that begins at $\Sigma = -1.5$, goes to $\Sigma = +0.5$, and returns to $\Sigma = -1.5$. Thus the initial state for all tiles is $\eta = -1$.

5. In Figure 4.17 we show the average state of the tiles, $\langle \eta \rangle$, as a function of Σ. Initially $\eta = -1$ for all instances and $\langle \eta \rangle = -1$. Since all of the (Σ_c, Σ_o) pairs lie in the lower right quadrant, transitions $\eta = -1 \rightarrow +1$ begin at $\Sigma = 0^+$. Similarly, as Σ decreases from 0.5, transitions $\eta = +1 \rightarrow -1$ begin at $\Sigma = 0^-$.

6. In Figure 4.18 we show A from the solution to the EMT equation, Eq. (4.31), in the form $\log(A)$ vs. Σ. In implementing Eq. (4.31) we used $B = A/3$. The elastic constant is hysteretic varying from $A \approx 0.016$ at negative stress to $A \approx 0.22$ at positive stress. The change in A with Σ follows closely the change in $\langle \eta \rangle$ with Σ since for the model under discussion the probability

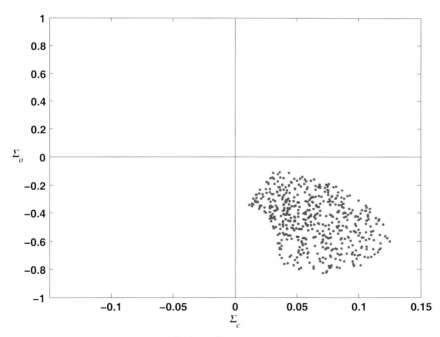

Fig. 4.16 (Σ_c, Σ_o) space. The pairs of values of (Σ_c, Σ_o) are in the lower right quadrant of a (Σ_c, Σ_o) Preisach space. Each value of (Σ_c, Σ_o) corresponds to a particular value of (D_0, p). The convention that compression is positive and tension is negative leads to the difference in signs here compared to Figure 4.14.

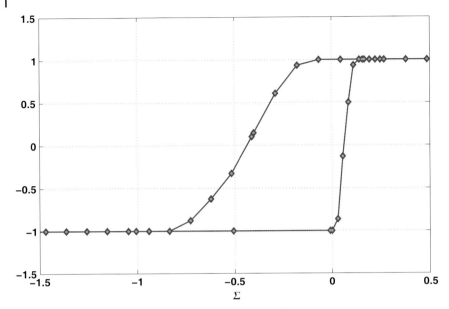

Fig. 4.17 $\langle \eta \rangle$ vs. stress. The average value of the state variable, η, as a function of applied stress. For $\Sigma \ll 0$, $\langle \eta \rangle \rightarrow -1$, and for $\Sigma \gg 0$, $\langle \eta \rangle \rightarrow +1$.

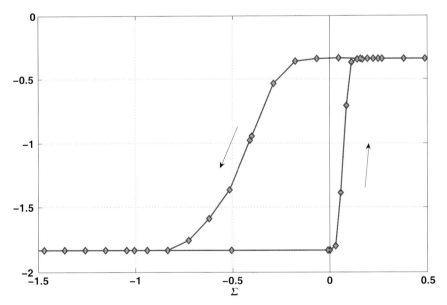

Fig. 4.18 Equation of state of van der Waals material. The compressive elastic constant as a function of Σ for van der Waals material from solution to Eq. (4.31).

density for (A, η) is approximately $\delta(A - A_{(a)})\delta_{\eta,-1} + \delta(A - A_{(b)})\delta_{\eta,+1}$, where $A_{(a)}$ and $A_{(b)}$ are the average of A over the weak and strong part of the elastic constant spectrum, respectively; see 2 above.

7. In Figure 4.19 we show $\langle D \rangle = \langle b_0 - b \rangle$ as a function of Σ, the EOS. The softness of the foot dominates the behavior for $\Sigma < 0$. The *strain stays in* on stress reversal if the van der Waals part of the encounter between tiles is involved.

8. In Figure 4.20 we show the nonlinear coefficient

$$\beta = \frac{\partial A}{\partial \Sigma} \tag{4.44}$$

as a function of Σ. The nonlinear coefficient, a measure of the change in dynamic modulus brought about by a change in stress, is largest where the dynamic modulus changes most rapidly with stress. For this model the changes in the dynamic modulus are brought about by a change of state. The most rapid change of state with stress is as Σ advances to above $\Sigma = 0$. This nonlinear coefficient is necessarily a quasistatic quantity. Should an elastic wave impinge on the system at ambient stress, where the quasistatic nonlinear coefficient is large, it will initially cause changes of state of the elastic elements. But successive stress oscillations cannot undo these changes, see Figure 4.16, so that the dynamic nonlinear modulus will be very different from the quasistatic nonlinear modulus. Because of the quadratic approxi-

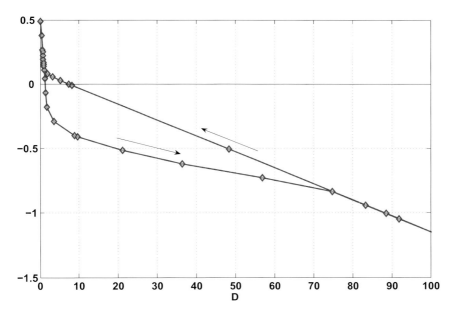

Fig. 4.19 EOS of van der Waals material. The displacement–stress relationship for van der Waals material, that is, $\langle D \rangle$ vs. Σ.

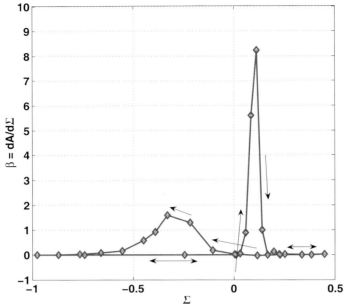

Fig. 4.20 Nonlinearity vs. Σ. The nonlinear parameter, $\beta = \partial A/\partial \Sigma$ as a function of Σ, Eq. (4.44).

mation to the potential in Eq. (4.41), the dynamic nonlinear modulus is zero (see Chapter 10).

4.4.3
Generalization and Caveats

The results described above are for two particular models. In fact, in one case they are for an approximation to the model. They have a number of features that are model independent and noteworthy. Turning a model elastic element into a set of elastic constants for a material, via the EMT equations above, requires a solution to an auxiliary problem. If the elastic elements are springs, the auxiliary problem is simple. For van der Waals surfaces the auxiliary problem is slightly more complex but nothing untoward. Regardless of the complexity of the auxiliary problem, when the elastic element is characterized by a state, there is an important caveat. The elastic constants are calculated at a fixed state. Thus they are appropriate to a dynamic stress disturbance, at fixed ambient stress, that does not cause a change in the state of the elastic elements. This is in contrast to a situation we will encounter below in which a change of state will be importantly involved in the behavor of dynamic disturbances.

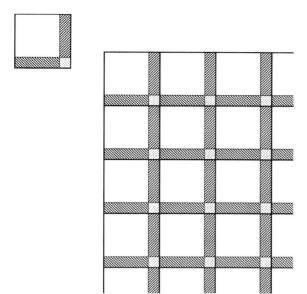

Fig. 4.21 Bricks-and-mortar tile. For a bricks-and-mortar system it must be possible to surround a brick with mortar. This can be done by choosing a tile with unit cell having four elastic pieces. The three shaded pieces can be assigned soft elastic constants relative to those of the brick, unshaded.

The two examples we looked at in some detail do not span the very broad range of possibilities we initially considered. However, the idea of how to proceed is clear and technical detail is all that is called for in doing more complex problems. As an illustration of one such problem we show in Figure 4.21 an example of the sort of tile that would be used in a bricks-and-mortar situation, that is, a tile that allows one to surround the brick with soft material. We will come upon this example in the next chapter where we discuss auxiliary fields.

References

1 Landau, L.D. and Lifshitz, E.M. (1999) *Theory of Elasticity*, Butterworth and Heinemann, Oxford.

2 Mindlin, D.R. and Deresiewicz, H. (1953) Elastic spheres in contact under varying oblique forces. *J. Appl. Mech.*, **20**, 327–344.

3 Perssons, B.N.J. (2000) *Sliding Friction*, Springer, New York.

4 Johnson, K.L. (1985) *Contact Mechanics*, Cambridge Univ. Press, Cambridge.

5 Isrealachvili, J. (1992) *Intermolecular and Surface Forces*, Academic Press, New York.

6 Johnson, D.L. and Norris, A.N. (1997) Rough elastic spheres in contact; memory effects and the transverse force. *J. Mech. Phys. Solids*, **45**, 1025–1036.

7 Elata, D. and Berryman, J.G. (1996) Contact force-displacement laws and the mechanical behavior of random packs of identical spheres. *Mech. Mater.*, **24**, 229–240.

8 Greenwood, J.A. and Williamson, J.B.P. (1966) Contact of Nominally Flat Surfaces. *Roy. Soc. London A*, 300–319.

9 Aleshin, V. and Van Den Abeele, K. (2006) *AIP Conference Proceedings*, **838**, 104.

10 Hirschfelder, J.O., Curtiss, C.F., and Bird, R.B. (1964) *Molecular Theory of Gases and Liquids*, John Wiley & Sons, Inc., New York.

11 Tutuncu, A.N., Sharma, M.M., and Podio, A.L.: Hysteresis observed in stress-strain diagrams and its relation to adhesion hysteresis and stick-slip sliding in sedimentary granular rocks, 35th US Symposium on Rock Mechanics, Lake Tahoe, Belkema.

12 O'Connell, R.J. and Budiansky, B. (1974) Seismic velocities in dry and saturated cracked solids. *J. Geophys. Res.*, **79**, 5412–5426.

13 Wang, G. and Li, S.F. (2004) A penny-shaped cohesive crack model for material damage. *Theor. Appl. Fract. Mech.*, **42**, 303–316.

14 Nihei, K.T., Hilbert, L.B., Cook, N.G.W., Nakagawa, S., and Myer, L.R. (2000) Frictional effects on the volumetric strain of sandstone. *Int. J. Rock Mech. Min. Sci.*, 121–132.

15 Dvorkin, J. and Nur, A. (1993) Dynamic poroelasticity: A unified model with the squirt and the Biot mechanisms. *Geophysics.*, **58**, 524–533.

16 Berryman, J.G. (1995) Mixture theory of rock properties, in: *Rock Physics and Phase Relations*, AGU Reference Shelf 3.

17 Torquato, S. (1991) Random heterogeneous media: Microstructure and improved bounds on effective properties. *Appl. Mech. Rev.*, **44**, 37–76.

18 Korteoja, M.J., Lukkarinen, A., Kaski, K., and Niskanen, K.J. (1995) Model for plastic deformation and fracture in planar disordered materials. *Phys. Rev. E*, 1055–1058.

19 Berryman, J.G. (2005) Bounds and estimates for elastic constants of random polycrystals of laminates. *Int. J. Solids and Struct.*, **42**, 3730–3743.

20 Preisach, F. (1935) On magnetic aftereffect. *Z. Phys.*, **94**, 277–302.

21 Mayergoyz, I.D. (1985) Hysteresis models from. the mathematical and control points of view. *J. Appl. Phys.*, **57**, 3803–3805.

5
Auxiliary Fields

In Chapter 4 we developed and worked through examples of an EMT that would let us assess the macroscopic elastic consequences of a set of mesoscopic elastic elements. As final preparation for going on to discuss mesoscopic elastic systems we introduce the set of auxiliary fields to which these systems may be coupled. These fields are internal fields, the temperature field, the saturation field, and a third field that we denote X. Here we establish some features of the quasistatic behavior of the auxiliary fields. (Their dynamics is *slow dynamics* and dealt with in Chapters 7, 10, and 11.) We employ two models that illustrate further generalization of the ideas developed in Chapter 4. In Section 5.1 we describe the temperature field and the set of internal stress/strains brought about by an anisotropic thermal expansion.

In Section 5.2 we describe the saturation field using a bricks-and-mortar model (alluded to at the close of Chapter 4). The saturation field, the fluid configuration in a pore space, is a hysteretic function of the chemical potential. Unambiguous description of the saturation field requires a chemical potential protocol. Thus a *bookkeeping space*, with chemical potential as the control variable, is necessary to describe the saturation field, Section 5.2.1. The saturaton field can be an inhomogeneous field that produces inhomogeneous internal stress/strain fields in a material, Section 5.2.2.

The X field is introduced in Section 5.3.

5.1
Temperature

The model system is a two-dimensional lattice of elastic elements (tiles), Figure 5.1. For illustrative purposes we allow displacement motion in only one direction, the y-direction [1]. The elastic energy of this system is expressed as a sum over the compressional and shear strain fields of the elastic elements, Eq. (4.14):

$$E = E_L + E_T + E_{NL}$$

$$E_L = \sum_\alpha \frac{A}{2} \left(\Delta \gamma_\alpha^c \right)^2 + \sum_\alpha \frac{B}{2} \left(\Delta \gamma_\alpha^s \right)^2,$$

Nonlinear Mesoscopic Elasticity: The Complex Behaviour of Granular Media including Rocks and Soil. Robert A. Guyer and Paul A. Johnson
Copyright © 2009 WILEY-VCH Verlag GmbH & Co. KGaA, Weinheim
ISBN: 978-3-527-40703-3

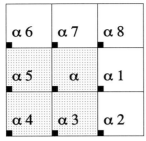

Fig. 5.1 Tile notation 1. An elastic element (tile) labeled α, with displacement y_α located in the lower left corner, participates in the compressional and shear energy of four elastic elements (shaded) and is coupled to the displacement of eight neighboring elastic elements. These displacements are compactly denoted as $y_{\alpha,n}$, where $n = 1, \ldots, 8$ as here.

$$E_T = -\sum_\alpha A\lambda_\alpha \left(\Delta y_\alpha^c\right)\delta T \,,$$

$$E_{NL} = \sum_\alpha A\beta \left(\Delta y_\alpha^c\right)^3 + \sum_\alpha A\gamma \left(\Delta y_\alpha^s\right)^2 \left(\Delta y_\alpha^c\right) \,, \tag{5.1}$$

where α runs over elastic elements and λ_α is the thermal expansion of elastic element α. There are three contributions to the energy, a linear elastic energy that is the same for each elastic element, E_L, a nonlinear elastic energy that is the same for each elastic element, E_{NL}, and a coupling between the elastic system and the change in temperature, $\delta T = T - T_0$, that is different for each elastic element because of λ_α, E_T [2]. In principle the linear and nonlinear elastic constants for the elastic elements, A, B, β, and γ, are found by some scheme, for example, the EMT scheme from Chapter 4, and are appropriate for equilibrium at temperature T_0. The system is uniform in all respects except for the thermal expansion, which is the subject of the present discussion. For a material like quartz the thermal expansion is strongly anisotropic [3, 4]. Thus the elastic elements, should they represent quartz grains with different orientation, would have very different thermal expansions. We take this to be the case.

Let us begin with the equation of motion for y_α:

$$m\ddot{y}_\alpha = -\frac{\partial E_L}{\partial y_\alpha} + A\delta T \left[\lambda_{\alpha,3} + \lambda_{\alpha,4} - \lambda_{\alpha,5} - \lambda_\alpha\right], \tag{5.2}$$

$$= -\boldsymbol{M} \cdot \boldsymbol{y} + A\delta T \boldsymbol{\Lambda} \,, \tag{5.3}$$

where we have neglected the nonlinear terms, the first term on the RHS is given by Eq. (4.19), and the notation $\lambda_{\alpha,j}$ denotes the thermal expansion of the elastic element, which is the jth neighbor of elastic element α as in Figure 5.1. In the static limit, $d/dt = 0$, we have

$$\boldsymbol{y} = A\delta T \boldsymbol{M}^{-1} \cdot \boldsymbol{\Lambda} \,. \tag{5.4}$$

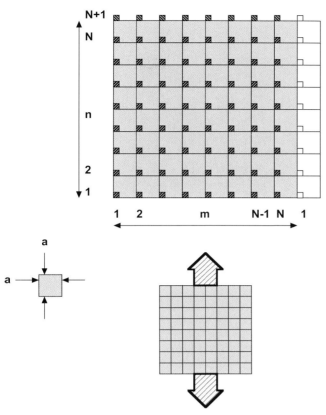

Fig. 5.2 Thermal forces. A system with $m = 1 \ldots M = 40$ and $n = 1 \ldots N = 20$, made up of square elastic elements that are $a \times a$, is subject to the forces from E_T in Eq. (5.1). As each of these thermal forces works on adjacent elastic elements, the net effect of all of them can be replaced by forces at the top and bottom of the system, those on rows $n = N + 1$ and $n = 1$, respectively.

From $\boldsymbol{\gamma}$ we can form $\Delta \gamma_\alpha^c$ and $\Delta \gamma_\alpha^s$ for each tile. An illustration of what is found is shown in Figures 5.3–5.5 for the example of a system of size $(M, N) = (40, 20)$, $A = 1$, $B = 1/4$, $\delta T = 0.1$, and $\lambda = 1 + 0.5\,\mathrm{sign}(r)$ with r uniformly distributed from -1 to $+1$, Figure 5.2. In Figure 5.3 we show the configuration of the system; elastic elements with $\lambda > 1$ are colored red and those with $\lambda < 1$ are colored white. The system is expanded from its equilibrium height 20 by approximately $N \langle \lambda \rangle \delta T \approx 2$, $\langle \lambda \rangle = 1.110$. Because of the anisotropy of the thermal expansion there is a spectrum of compressional strains, $\Delta \gamma^c$, Figure 5.4, and shear strains, $\Delta \gamma^s$, Figure 5.5. The compressional strains are broadly distributed with an average of 0.2022 (a factor of 2 more than $\langle \lambda \rangle \delta T$ because of the definition of $\Delta \gamma^c$). (Here $\langle \lambda \rangle$ is the average thermal expansion that might be found by extending the EMT ideas

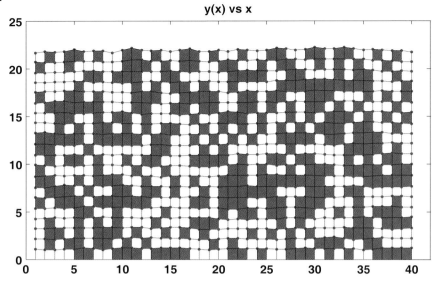

Fig. 5.3 Thermal configuration. An anisotropic thermal expansion, $\lambda_\alpha = 1 + 0.5\text{sign}(r_\alpha)$, is assigned to each elastic element ($\lambda > 1$ red and $\lambda < 1$ white). This produces an average expansion of the system, as well as compressional and shear forces on individual elastic elements. The spectrum of these internal forces is shown in Figures 5.4 and 5.5. (Please find a color version of this figure on the color plates)

from Chapter 4 to this case.) The shear strains are broadly distributed with an average of 0.00022, essentially zero. There is no external manifestation of these shear strains. The stresses associated with the compressional and shear strains are $A\Delta y^c$ and $B\Delta y^s$, respectively. Compressional stresses arise in part from the thermal expansion within each elastic element $\lambda_\alpha \Delta y^c_\alpha$. For shear stresses there is no direct temperature/shear strain coupling. Shear stresses arise from forces exerted on an elastic element by its neighboring elements.

At a bare minimum the temperature bathes the system in a uniform internal compressional stress field $A\langle\lambda\rangle\delta T$ about which there are fluctuations $A(\lambda_\alpha - \langle\lambda\rangle)\delta T$.

In an elastic system that is nonuniform, say one having a sprinkling of elastic elements with weak shear elastic constants, there could be a system of internal strains much larger than one could infer from external observation. If the sprinkling of elastic elements is of sufficient volume, their unusual strain would be seen in a neutron scattering experiment [5].

Should the large strain on an elastic element elicit an anomalous elastic response, the temperature would be seen to drive such a response.

Return to Eq. (5.1) and use $\Delta\gamma_\alpha = \overline{\Delta\gamma_\alpha} + \delta\gamma_\alpha$, where $\overline{\Delta\gamma_\alpha}$ is from the solution to Eq. (5.4). Then we have a total energy in the form

$$E = E_0 = \sum_\alpha \frac{A}{2}\left(1 + 6\beta\overline{\Delta\gamma_\alpha^c}\right)(\delta\gamma_\alpha^c)^2$$
$$+ \sum_\alpha \frac{B}{2}\left(1 + 2\gamma\overline{\Delta\gamma_\alpha^c}\right)(\delta\gamma_\alpha^s)^2 \qquad (5.5)$$
$$+ \sum_\alpha B\gamma\overline{\Delta\gamma_\alpha^s}\delta\gamma_\alpha^s\delta\gamma_\alpha^c,$$

where we have lumped a set of constant energies involving $\overline{\Delta\gamma_\alpha}$ into E_0, used Eq. (5.4), and dropped terms involving β and γ that are linear in $\delta\gamma_\alpha$ ($\delta\gamma_\alpha^c = \delta\gamma_{\alpha7} + \delta\gamma_{\alpha8} - \delta\gamma_{\alpha1} - \delta\gamma_\alpha$ and $\delta\gamma_\alpha^s = \delta\gamma_{\alpha8} + \delta\gamma_{\alpha1} - \delta\gamma_{\alpha7} - \delta\gamma_\alpha$), Figure 5.1. The thermal expansion leads to a modified set of elastic constants for the elastic elements. Replace $\overline{\Delta\gamma_\alpha^c}$ and $\overline{\Delta\gamma_\alpha^s}$ with the averages found above.

1. There would be no coupling of $\delta\gamma^c$ to $\delta\gamma^s$, see Figures 5.4 and 5.5.
2. As typically $\beta < 0$, there would be a reduction *softening* of the compressive elastic constant with a temperature increase.
3. The sign of γ is in general unknown so the effect of thermal expansion on the shear modulus is in general unknown.

If we were to replace $\overline{\Delta\gamma_\alpha^c}$ and $\overline{\Delta\gamma_\alpha^s}$ with the averages we found above, there would be no coupling of $\delta\gamma^c$ to $\delta\gamma^s$, see Figures 5.4 and 5.5.

While what we have here is a set of phenomena not much different from those implied by the discussion of traditional nonlinear elasticity, for example, see

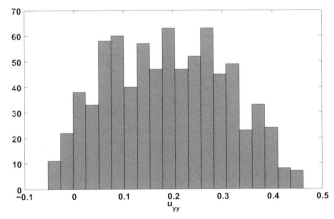

Fig. 5.4 Compressional forces, temperature. The probability density (unnormed) for the compressional strain of the elastic elements, $p(\Delta\gamma_\alpha^c)$, as a function of $\Delta\gamma_\alpha^c$ for the circumstance described below Eq. (5.4). The units used in the calculations and this figure carry dimensions. This strain is made dimensionless by dividing by $2a$. It is scaled to physical units by equating the average compressional strain to $\langle\lambda\rangle\delta T$. Application of EMT as in Chapter 4 to the thermal expansion yields $\langle\lambda\rangle = (\sum\lambda_\alpha)/NM$.

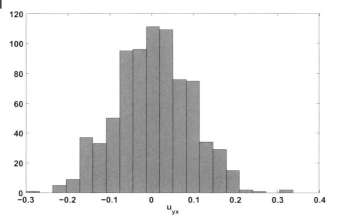

Fig. 5.5 Shear forces, temperature. The probability density (unnormed) for the shear strain of the elastic elements, $p(\Delta \gamma_a^s)$, as a function of $\Delta \gamma_a^s$ for the circumstance described below Eq. (5.4). The units used in the calculations and this figure carry dimensions. This strain is made dimensionless by dividing by $2a$. It is scaled to physical units by equating the compressional average strain to $\langle \lambda \rangle \delta T$. The shear response to the forces of thermal expansion yields $\langle \Delta \gamma^s \rangle = 0$ since δT does not directly drive $\Delta \gamma^s$.

Eqs. (2.28), the important qualitative point is that a change in temperature delivers a complex system of internal stress fields to the elastic elements in a system. In response to these internal stress fields the elastic elements, conceivably more exotic than simple nonlinear elements considered here, are possibly the source of unexpected behavior [6, 7].

5.2
Saturation

5.2.1
Saturation/Strain Coupling

Consider the case of an elastic system that is threaded by a pore space in which a fluid can have a variety of configurations. We model this system with a generalization of the model above. In place of a single elastic element per unit cell, Figure 5.1, we use four elastic elements in a bricks-and-mortar geometry, Figure 5.6. One idea behind this physical picture is that of an elastic system built of relatively rigid elastic units (bricks) that are separated from one another by an elastic system that is relatively soft (mortar). Examples are rock, soil, granular media, ... Contiguous with the mortar of this system we place a pore space that is to be filled with pore fluid according to a prescribed chemical potential protocol. We have to describe the elastic system, the pore space, the configurations of the pore fluid and the set of forces on the elastic system from the configurations of the pore fluid.

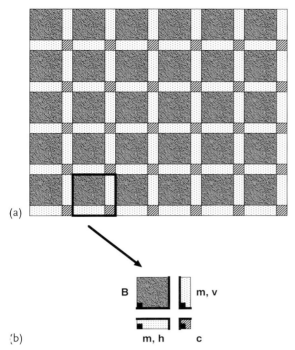

(a)

B   m, v

(b)   m, h   c

Fig. 5.6 Tile notation 2. (a) To describe the more complicated situation associated with an elastic system sharing a space with a pore fluid, a bricks-and-mortar model is used. (b) ⁻he unit cell has four elastic elements (tiles) denoted **B** (brick), **h** (horizontal mortar), **v** (vertical mortar), and **c** (contact). These elastic elements have size $a \times a$, $a \times b$, $b \times a$, and $b \times b$ for **B**, **h**, **v**, and **c**, respectively, $b = a/10$. The unit cells are glued to one another along their common boundaries, as indicated by the solid line surrounding the representative unit cell in (a). In addition, the four elastic elements in a unit cell are glued together along their common boundaries (solid line in (b)). There is a displacement variable associated with the lower left corner of each elastic element (the small squares in (b)), $\gamma_{\alpha,\nu}$, $\nu = B, h, \nu$, and c. The system is periodic in m and the lower edge of the bricks in the unit cells in row $n = 1$ are fixed at $\gamma = 0$. The elastic elements have elastic constants $A_B = 1$, $A_h = A_\nu = A_c = A_B/100$, $B_n = A_n/4$. This choice of elastic constants gives a bulk elastic constant of approximately $A_B/10$, that is, the bulk elastic system is soft compared to the bricks because of the mortar.

1. Elastic System. The elastic unit cell has four elastic elements (tiles), Figure 5.6, a brick with relatively large elastic constants, **B**, horizontal and vertical mortar, **h**, **v**, that are elastically soft, and a contact, **c**, that is elastically soft. The elastic energy of this system is written as

$$E_L = \sum_\nu \sum_\alpha \frac{A_\nu}{2} \left(\Delta \gamma_{\alpha,\nu}^c \right)^2 + \sum_\alpha \frac{B_\nu}{2} \left(\Delta \gamma_{\alpha,\nu}^s \right)^2 , \tag{5.6}$$

where α denotes the unit cell and ν goes over the four elastic elements in the unit cell. As above the cells are glued to one another along their common boundary, and the elastic elements in a cell are similarly glued. Thus a single displacement variable, taken to be the lower left corner of each elastic element, suffices for a description of the elastic energy.

2. Pore Space and Fluid Configurations. The pore space is taken to be contiguous with the mortar/contact elastic elements and to have their geometry, Figure 5.7. The pores are assigned the properties of a set of cylinders of varying radius [8]. Thus associated with each pore is a chemical potential pair, (μ_F, μ_E), where μ_F is the chemical potential at which a pore, having a thin fluid film on its walls, makes a transition to being full of fluid on chemical potential increase (capillary condensation) and μ_E is the chemical potential at which a pore, full of fluid, makes a transition to having a thin fluid film on its walls on chemical potential decrease. This latter process, pore emptying, is subject to the access condition [9], leading to invasion percolation [10] and inhomogeneous fluid configurations in the pore space. When a pore has a thin fluid film on its walls, it is in the empty, E, state. When capillary

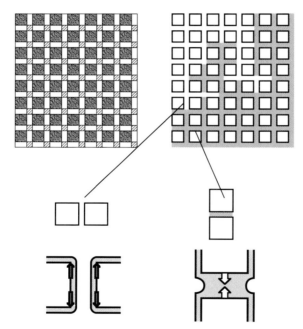

Fig. 5.7 Tile notation 3, the pore space. The pore space is contiguous with the mortar/contact system. Each pore is assigned a chemical potential pair (μ_E, μ_F) that determines the chemical potential at which the pore makes a change in fluid state. The pore space is occupied by fluid according to a chemical potential protocol. When a pore is empty (there is a thin fluid film on the pore wall) there are forces of tension exerted on the pore walls. When a pore is full, there are forces of compression trying to pull the pore walls into the pore.

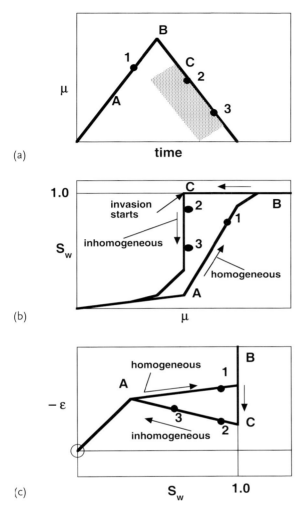

Fig. 5.8 Chemical potential protocol and ...
(a) The chemical potential (related to the vapor pressure) goes from large negative values for which there is only a thin film of liquid on the pore wall) to the value corresponding to bulk liquid (all the pores are full of fluid) back to large negative values. (b) The saturation of the pore space increases slowly as the thin film on the pore walls thickens and increases relatively rapidly when capillary condensation begins to occur (after A). Eventually all of the pores are full of liquid and the saturation ceases to change (at B). On lowering the chemical potential, because of the access condition for vapor invasion, the pore space remains full of liquid until the *invasion percolation* event occurs (at C). In the domain in which there are pores full of liquid, the saturation is a hysteretic function of the chemical potential and has endpoint memory [8]. (c) The average strain that attends these changes in liquid configurations in the pore space. The precise numbers are a function of details. But the qualitative idea is that on average the forces that liquid configurations bring to bear tend to be tensionlike, and the average size of a sample increases with increases in fluid content.

condensation has occurred for a pore it is in the full, *F*, state. Since chemical potential scales with vapor pressure ($\mu \propto \ln(P_v / P_{v,\text{sat}})$, where P_v and $P_{v,\text{sat}}$ are the fluid vapor pressure and saturated vapor pressure, respectively), we can state this evolution in terms of vapor pressure increase, $E \to F$, and vapor pressure decrease, $F \to E$. The evolution of a fluid configuration in the pore space is taken to be that associated with an adsorption isotherm; the chemical potential (vapor pressure) goes from large negative values (very low pressure) to the chemical potential of bulk liquid (saturated vapor pressure) and back to large negative values (very low pressure).

3. Coupling of the Elastic System to the Pore Fluid. As we are illustrating matters of principle we take a very simple model for the forces that a fluid configuration exerts on the elastic system in which it resides [8].

a. When a pore is empty, the primary effect is for the fluid to reduce the surface energy of the pore walls. This is equivalent to forces of tension

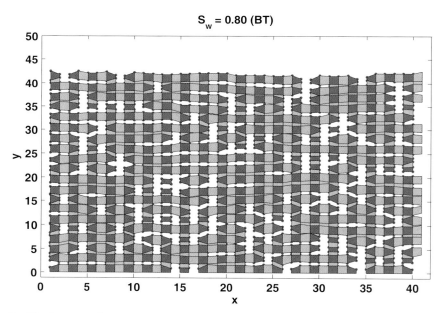

Fig. 5.9 System configuration, $S_w = 0.70$, filling. Displacement of the elastic elements in a 20 cell by 20 cell realization of the elastic system with $S_w = 0.70$. The bricks are red. The space allocated to the mortar and contacts is treated as the pore space and filled blue (mortar) and green (contact) if there is fluid in the associated pore and white otherwise. All elastic elements are shown as if they are of the same size, see the caption to Figure 5.6. This configuration, formed on chemical poten-tial increase, has homogeneous filling of the pore space. Elastic elements adjacent to filled vertical pores feel forces of tension tending to elongate the elastic element. Elastic elements adjacent to filled horizontal pores feel forces tending to pull the elastic element into the pore space. This appears as the pulling of bricks toward the pore. See Figure 5.12. (Please find a color version of this figure on the color plates)

that elongate the pore. In the simple elastic model used here, these forces work to elongate the surfaces shared by the v-mortar and the bricks. While the strength of these forces for a particular pore depend on the pore radius and the film thickness, we ignore this subtlety and choose a single number for the forces of tension for all pores.

b. When a pore is full, the primary effect is for the fluid to pull the pore walls into the pore space. In the simple elastic model here, these forces work to pull the bricks toward the h-mortar. As with the forces of tension we assign a single number for these forces independent of pore radius and chemical potential (pressure in the pore fluid).

c. Note that the one-dimensional motion permitted by the model we use means that the v-mortar, which exerts forces of tension when the pore associated with it is empty, exerts no force when the pore is full of fluid. Similarly, the h-mortar, which exerts forces of compression when the pore associated with it is full, exerts no force when the pore is empty.

d. In this model the state of a pore is an Ising-like variable that can be followed in a (μ_F, μ_E) Preisach space. The forces that are approximately

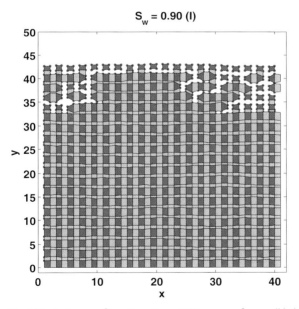

$S_w = 0.90$ (I)

Fig. 5.10 System configuration, $S_w = 0.90$, emptying. Displacement of the elastic elements in a 20 cell by 20 cell realization of the elastic system with $S_w = 0.90$ as the chemical potential is being lowered. This configuration occurs near the onset of invasion percolation. As the vapor invades the pore space from the surface, well below the interface between full and partially empty pores, the pore space is uniformly filled. Shear forces occur only near the outer edge of the system. See Figure 5.13. (Please find a color version of this figure on the color plates)

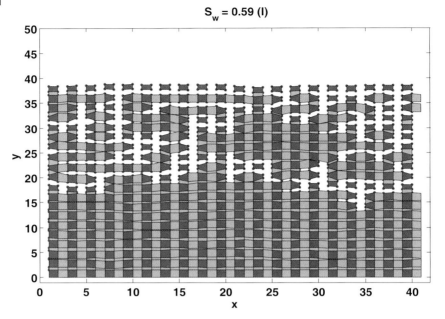

Fig. 5.11 System configuration, $S_w = 0.59$, emptying. Displacement of the elastic elements in a 20 cell by 20 cell realization of the elastic system with $S_w = 0.59$ as the chemical potential is being lowered. See Figure 5.14. (Please find a color version of this figure on the color plates)

Ising-like are strictly Ising-like here because of the extreme simplicity of the force model we use.

Because of the nature of this model and of the approximations we make to illustrate the major points, a number of important details are lost. We will come upon these later on and remark about them accordingly.

5.2.2
Saturation/Strain Response

We have taken the pore space of this elastic system through the chemical potential/saturation protocol shown in Figure 5.8 for a particular realization of (μ_F, μ_E) Preisach space. At several points along the chemical potential protocol we find the forces exerted by the fluid configuration on the elastic system and solve for the displacements $y_{a.v}$ and the set of strains on the elastic elements. In the lower two panels of Figure 5.8 we show schematically the saturation as a function of chemical potential and the total compressional strain across the elastic system as a function of the saturation. We report the configuration of the system and the probability density of the internal strains in detail. In Figures 5.9, 5.10, and 5.11 we show

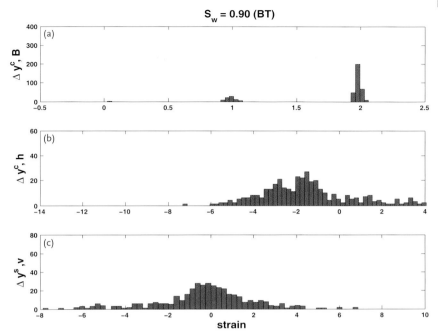

Fig. 5.12 Strain probability density, $S_w = 0.90$, filling. For the case of $S_w = 0.90$ on chemical potential increase the unnormed probability density of the (a) compressional strain in the bricks, (b) the compressional strain in the h-mortar, and (c) the shear strain in the v-mortar is plotted as a function of the strain. As most of the bricks are surrounded by liquid, the probability density for Δy^c is approximately a delta function. The shear strains that arise from an unbalanced fluid configuration near the v-mortar are broadly distributed about $\Delta y^s = 0$.

the configuration of the fluid and of the elastic elements. The first of these figures, corresponding to point 1 in Figure 5.8, is at a point encountered on filling the pore space to $S_w \approx 70\%$. For purposes of presentation the unstrained contacts and unstrained mortar are one unit in size, the same size as the bricks. Vertical brick surfaces adjacent to empty pores are pulled together. Horizontal brick surfaces adjacent to filled pores are pulled together. The complex system of strains that the fluid causes in the elastic elements is shown in Figure 5.12 for the bricks, the h-mortar and the v-mortar. On filling the pore space the fluid configuration is spatially uniform. This is in contrast to what is seen in Figure 5.10 for the case corresponding to point 2 on Figure 5.8. This configuration occurs on emptying the pore space [10]. As the emptying process corresponds to *invasion* of the pore space (by air) from the surface inward, the fluid configuration is very inhomogeneous. The system of strains corresponding to this case is shown in Figure 5.13. The probability density for various strains is not strikingly different between the two cases shown in Figures 5.12 and 5.13. But the location of these forces is very different. For $S_w \approx 0.90$ in the invasion region all of the shear forces are near the surface of the elastic system and all of the compressional forces are in the fluid-filled interior. As the invasion process continues, Figure 5.11, the domain of shear forces moves toward the interior of the elastic system. Compare the distribution of strains in Figures 5.13 and 5.14. At a chemical potential well below that of the onset of the invasion process, the fluid configurations are qualitatively similar to those found on initial chemical potential increase.

If we were to add a set of nonlinear elastic terms to the energy of the system, as we did for the case of internal forces due to temperature, the last line in Eq. (5.1), we would find a shift in the effective elastic constants analogous to those in Eq. (5.5) with the strains $\overline{\gamma_{\alpha,\nu}}$ due to the fluid configuration. When the fluid configuration is inhomogeneous, the nonlinear elasticity is similarly inhomogeneous.

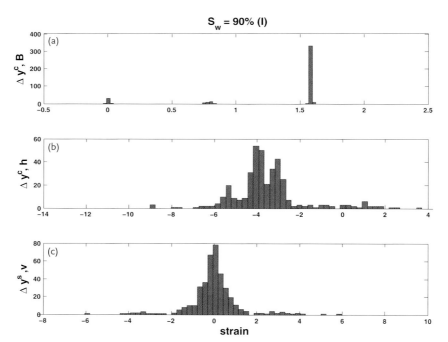

Fig. 5.13 Strain probability density, $S_w = 0.90$, emptying. For the case of $S_w = 0.90$ on chemical potential decrease the unnormed probability density of the (a) compressional strain in the bricks, (b) the compressional strain in the h-mortar, and (c) the shear strain in the v-mortar is plotted as a function of the strain. As most of the bricks are surrounded by liquid, the probability density is approximately a delta function. Compare to the similar delta function in Figure 5.12. The forces of tension, all near the top of the system, shift the location of the delta function away from the value seen in Figure 5.12. The shear strains that arise form an unbalanced fluid configuration near the v-mortar are broadly distributed about $\Delta y^s = 0$ and are due to shear forces near the surface of the elastic system.

What we want to take away from this discussion is

1. the fluid configurations in a pore space are complex, a function of the chemical potential protocol with which they are created, sometimes spatially homogeneous and sometimes spatially inhomogeneous;
2. these fluid configurations deliver a set of internal forces that can have features unlike any that can be delivered by an external force [11];

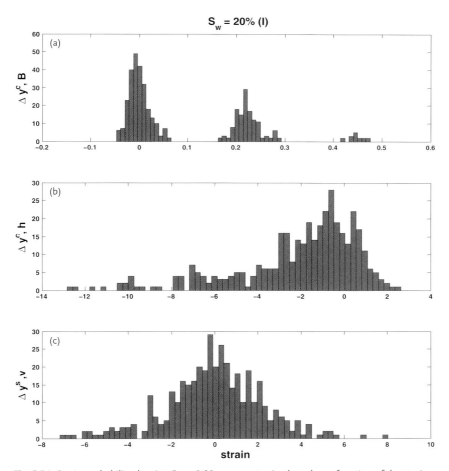

Fig. 5.14 Strain probability density, $S_w = 0.20$, emptying. For the case of $S_w = 0.20$ on chemical potential decrease the unnormed probability density of the (a) compressional strain in the bricks, (b) the compressional strain in the h-mortar, and (c) the shear strain in the v-mortar is plotted as a function of the strain. The compressional forces on the bricks are relatively small. Many bricks feel almost no force from the liquid configuration. A broadly distributed system of shear forces occurs throughout the elastic system.

3. should the elastic elements have an unusual elastic response, the forces delivered by fluid configurations will elicit this response in much the same way as an external force [12].

5.3
The Conditioning Field, X

In addition to the auxiliary fields associated with temperature and saturation, there is evidence for an additional field (or fields) of unknown nature that couples to the elastic system [13]. We will have more to say about this when we discuss slow dynamics in Chapters 7, 10, and 11.

References

1 Korteoja, M.J., Lukkarinen, A., Kaski, K., and Niskanen, K.J. (1995) Model for plastic deformation and fracture in planar disordered materials. *Phys. Rev. E*, **51**, 1055–1058.

2 Landau, L.D. and Lifshitz, E.M., *Theory of Elasticity*, Butterworth and Heinemann, Oxford, (1999).

3 Raz, U., Girsperger, S., and Thompson, A.B., http://e-collection.ethbib. ethz.ch/ecol-pool/bericht&nr=184, 1 June 2009.

4 Nye, J.F. (1957) *Physical Properties of Crystals*, Clarendon Press, Oxford.

5 Todd, R.I., Bourke, M.A.M., Borsa, C.E., and Brook, R.J. (1997) Neutron diffraction measurements of residual stresses in alumina/SiC nanocomposites. *Acta Mater.*, **45**, 1791–1800.

6 Ulrich, T.J. (2005) (thesis), University of Nevada, Reno. Ulrich, T.J. (2005) (thesis), University of Nevada, Reno.

7 Ide, J.M. (1937) The velocity of sound in rocks and glasses as a function of temperature. *J. Geol.*, **45**, 689–716.

8 Guyer, R.A. (2005) *The Science of Hysteresis III* (eds Bertotti, G., Mayergoyz, I.) Academic Press, San Diego.

9 Rouquerol, R., Rouquerol, J., and Sing, K. (1999) *Adsorption by Powders and Porous Solids*, Academic Press, London.

10 Wilkinson, D.J. and Willemsen, J. (1983) Invasion percolation: a new form of percolation theory. *J. Phys. A*, **16**, 3365–3376.

11 Amberg, C.H. and McIntosh, R. (1952) A study of absorption hysteresis by means of length changes of a rod of porous glass. *Can. J. Chem.*, **30**, 1012–1032.

12 Carmeliet, J. and van den Abeele, K. (2002) Application of the Preisach-Mayergoyz space model to analyze moisture effects on the nonlinear elastic response of rock. *Geophys. Res. Lett.*, **29** 48.1–48.4.

13 Vakhnenko, O.O., Vakhnenko, V.O., and Shankland, T.J. (2005) Soft-ratchet modeling of end-point memory in the nonlinear resonant response of sedimentary rock. *Phys. Rev B*, **71**, 174103(14).

6
Hysteretic Elastic Elements

In this chapter we describe the essentials of a phenomenology for assessing the influence of hysteretic elastic elements on various types of elastic behavior. We do this using an extreme picture in which the hysteretic elastic elements have very rudimentary motions and in which these motions have no dynamics. The elastic elements are called *finite displacement elastic elements*, FDEE, a name intended to emphasize their properties. These elastic elements are introduced in Section 6.1 where their behavior under quasistatic stress is established. One of the virtues of these elastic elements is that quasistatic data, which is influenced by them, can be inverted to learn something of their nature. This is discussed in Section 6.2. The behavior of FDEE in dynamic response is discussed in Section 6.3, the resonant bar in Section 6.3.1, and wave mixing in Section 6.3.2. To have an equation into which we can easily introduce an equation of state (EOS) for the FDEE, we transform the wave equation into a *lumped element* equation, Section 6.3.1.1. We develop an EOS for the FDEE for a resonant bar in Section 6.3.1.2. The analysis of the resonant bar is carried through in Section 6.3.1.3. Section 6.3.2 is devoted to wave mixing. In Section 6.4 we suggest that most models of hysteresis are equivalent to the FDEE model. (A detailed demonstration appears in the appendix, Section 6.6.) Consequently, a *form* of universality obtains. Summarizing remarks are in Section 6.5. In the appendix, Section 6.6, we examine several model of hysteretic elastic systems: shear contacts with friction (Section 6.6.1.1), an fcc lattice of Hertzian contacts (Section 6.6.1.2), the Masing rules (Section 6.6.2), and the endochronic formalism (Section 6.6.3).

6.1
Finite Displacement Elastic Elements; Quasistatic Response

6.1.1
Finite Displacement Elastic Elements: The Model

The original picture of hysteretic elastic elements is that of McCall and Guyer, MG [1]. This picture was anticipated by Holcomb [2], even if not all of the details had been worked through. Suppose we have an elastic material built up of elastic

Nonlinear Mesoscopic Elasticity: The Complex Behaviour of Granular Media including Rocks and Soil. Robert A. Guyer and Paul A. Johnson
Copyright © 2009 WILEY-VCH Verlag GmbH & Co. KGaA, Weinheim
ISBN: 978-3-527-40703-3

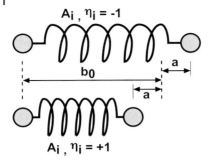

Fig. 6.1 Two-state compressive elastic element. A model for a two-state elastic element is a spring, spring constant A, that enforces two different displacements between its endpoints, $b_0 \pm a = b_0 + \eta a$, according to its state, $\eta = \pm 1$.

elements, shown schematically in Figure 6.1, having elastic constants A_i that attempt to enforce two relative displacements between their ends according to the state $\eta_i = \pm 1$, that is, $b_i = b_0 + \eta_i a$. There is a known rule for establishing the state η_i. The discussion of van der Waals surfaces in Chapter 4 provides an example of how such rules might arise. And, in accordance with the language of that discussion, index i runs over the set of instances of the elastic element. However, initially we want to proceed without regard for the the details of a particular model. We formulate an EMT description of this material. (The discussion of hysteresis, here in the context of elasticity, resides in a much more general context that can be visited in the three volumes by Berttoti and Mayergoyz [3].)

In EMT (Chapter 4) the elastic element would be associated with a tile with elastic energy

$$E_i = \frac{1}{2} A_i \left(\Delta \gamma^c - b_i \right)^2 , \tag{6.1}$$

and it would be placed in a medium with tiles of elastic energy

$$E_m = \frac{1}{2} A \left(\Delta \gamma^c - b_0 \right)^2 . \tag{6.2}$$

Analogously to Eq. (4.16), we would write

$$E = E_0 + \frac{1}{2} \delta A_i \left(\Delta \gamma_{00}^c - b_0 \right)^2 + A(\Delta \gamma_{00}^c - b_0) \Delta b_i , \tag{6.3}$$

where $\delta A_i = A_i - A$, $\Delta b_i = b_i - b_0$. We have kept only first-order terms in departure from E_m, used $E_0 = \Sigma E_m$, and dropped all but the compressional energy contribution. In the spirit of EMT we would solve for $\Delta \gamma_{00}^c$ and find, analogous to Eq. (4.27),

$$
\Delta \gamma_{00}^c = \mathcal{S} \left(D_{\mu\nu}^{c*} \frac{1}{M_{\mu\nu}} F_{\mu\nu} \right) - \delta A_i \Delta \gamma_{00}^c \mathcal{S} \left(D_{\mu\nu}^{c*} \frac{1}{M_{\mu\nu}} D_{\mu\nu}^c \right)
$$
$$
- A \Delta b_i \mathcal{S} \left(D_{\mu\nu}^{c*} \frac{1}{M_{\mu\nu}} D_{\mu\nu}^c \right) . \tag{6.4}
$$

This is one step before applying the central dogma of EMT. We take the simplest model, one in which the different instances of an elastic element involve different b_i only, $A_i = A \; \forall i$. We call these elastic elements finite displacement elastic elements (FDEEs). From the central dogma

$$\langle \Delta \gamma_{00}^c \rangle = \mathcal{S} \left(D_{\mu\nu}^{c*} \frac{1}{M_{\mu\nu}} F_{\mu\nu} \right) , \tag{6.5}$$

and $\langle \Delta b_i \rangle = 0$ or

$$b_0 = \langle b_i \rangle . \tag{6.6}$$

This is the EOS of the MG model. Briefly, the LHS is the size of the system and the RHS is an average that depends on the state of the FDEE, which in turn depends on the stress protocol. The simplicity of this equation is deceptive. We are going to discuss it at some length.

Aside.
Before we do that let us consider the issue from a slightly more general perspective. Suppose we have elastic elements that attempt to enforce both two-state compressive and two-state shear displacements, Figure 6.2. We would write

$$E_m = \frac{1}{2} A \left(\Delta \gamma^c - b_0 \right)^2 + \frac{1}{2} B \left(\Delta \gamma^s - c_0 \right)^2 , $$
$$E_i = \frac{1}{2} A \left(\Delta \gamma^c - \eta_i b_i \right)^2 + \frac{1}{2} B \left(\Delta \gamma^s - \tau_i c_i \right)^2 , \tag{6.7}$$

where the state of the compressive (shear) part of the elastic element is determined by the state variable $\eta_i(\tau_i)$. A prescribed stress protocol drives (η_i, τ_i). In setting $A_i = A$ and $B_i = B$ for all instances we concentrate on the contribution of the change in displacement to the elasticity. In Eq. (6.3) we would have

$$E = E_0 + A \Delta \gamma_{00}^c \Delta b_i + B \Delta \gamma_{00}^s \Delta c_i , \tag{6.8}$$

where $\Delta b_i = b_i - b_0$ and $\Delta c_i = c_i - c_0$. We calculate the average of γ_{00}^c and γ_{00}^s over instances, apply the central dogma of EMT to $\langle \gamma_{00}^c \rangle$ and $\langle \gamma_{00}^s \rangle$ separately, and find

$$\langle \Delta \gamma_{00}^c \rangle = \mathcal{S} \left(D_{\mu\nu}^{c*} \frac{1}{M_{\mu\nu}} F_{\mu\nu} \right) , $$
$$\langle \Delta \gamma_{00}^s \rangle = \mathcal{S} \left(D_{\mu\nu}^{s*} \frac{1}{M_{\mu\nu}} F_{\mu\nu} \right) , \tag{6.9}$$
$$\langle \Delta b \rangle A \Lambda_{cc} + \langle \Delta c \rangle B \Lambda_{cs} = 0 , $$
$$\langle \Delta b \rangle A \Lambda_{sc} + \langle \Delta c \rangle B \Lambda_{ss} = 0 , $$

where $\mathcal{S}$ stands for the sum, as in Eq. (4.27), and $\Lambda_{\alpha\beta}$ are integrals over the excitation spectrum, analogously to Eq. (4.29), that produce numbers that are nonzero. Consequently, we have $\langle \Delta b \rangle = 0$ and $\langle \Delta c \rangle = 0$, or

$$b_0 = \langle b_i \rangle , $$
$$c_0 = \langle c_i \rangle . \tag{6.10}$$

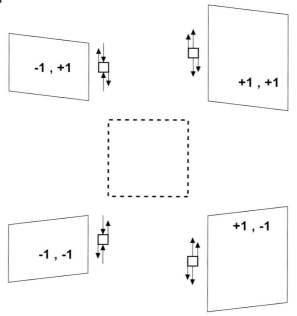

Fig. 6.2 Two-state compressive/shear elastic element. In analogy to the compressive-only elastic element illustrated in Figure 6.1, there can be elastic elements that are described by two-state variables, (η, τ). These elastic elements enforce two values of the compressive displacement, $b_0 \pm a = b_0 + \eta a$, and two values of the shear displacement, $c_0 \pm d = c_0 + \tau d$, $\tau = \pm 1$. The four states of a tile (EMT) are shown.

The compressional and shear equations of state are independent [4, 5]. This is in contrast to the EMT relationship between the compressive moduli and shear moduli, which are strongly coupled. See, for example, Eq. (4.2) (Hertzian contact model).

End Aside.

6.1.2
Finite Displacement Elastic Element: Implementing the Model

When the important response to a stress protocol has to do with the state-dependent aspects of the energy, we obtain the EOS given by Eq. (6.6). To turn this equation into something useful, we need (a) specification of the displacement associated with the instances of b, the b_i, and (b) a rule for determining the elastic state of each instance. We will begin with the simplest model, that of McCall and Guyer [1]. This model is intentionally nonspecific; it uses a description that might apply to many possible models of an elastic element. Its ingredients are:

1. two displacements that are the same for all instances of the elastic element, when $\eta_i = -1$, $b_i = L_0$ and when $\eta_i = +1$, $b_i = L_0 - \Delta L$;
2. a stress pair (Σ_c^i, Σ_o^i), for each instance, i, of the elastic element, that sets the state of the elastic element
   a. if $\eta_i = -1$ and Σ increases to above Σ_c, $\eta_i \rightarrow +1$,
   b. if $\eta_i = +1$ and Σ decreases to below Σ_o, $\eta_i \rightarrow -1$.
   There is a Preisach space associated with the (Σ_c^i, Σ_o^i) pairs, and in principle there is a density $\varrho(\Sigma_c, \Sigma_o)$ of these pairs in this space [6, 7]. Where the context makes clear what we mean we replace the cumbersome $\varrho(\Sigma_c, \Sigma_o)$ with $\varrho(X, Y)$. In general

$$\varrho(X, Y) = \sum_i \delta\left(X - \Sigma_c^i\right) \delta\left(Y - \Sigma_o^i\right) , \tag{6.11}$$

which may have an analytic representation.

To illustrate the qualitative behavior of this extreme model, we will look at several examples. We need three further ingredients: (1) specification of $\varrho(X, Y)$, (2) specification of the initial value of the η_i, and (3) a stress protocol. We take the case in which ϱ has a diagonal part, $\propto \delta(X - Y)$, and a constant off-diagonal part in a triangular region of the Preisach space,

$$\varrho(X, Y) = \frac{p}{2S}\delta(X - Y) + \frac{1 - p}{2S^2} , \tag{6.12}$$

for $-S \leq X \leq +S$, $-S \leq Y \leq S$. We consider two stress protocols, shown in Figure 6.3.

Case 1
The stress is taken from below $-S$ to above $+S$ and then back to below $-S$, upper left of Figure 6.3. Initially $\eta_i = -1 \ \forall i$. Thus $b_0 = L_0$ until Σ exceeds $-S$. As Σ increases to above $-S$ the change in b_0 can be found using

$$b_0^{\uparrow}(\Sigma) = \langle b_i \rangle = L_0 - \Delta L \int_{-S}^{\Sigma} dX \int_{-S}^{X} dY \, \varrho(X, Y) ,$$

$$b_0^{\uparrow}(\Sigma) = L_0 - \Delta L \left[p\left(\frac{\Sigma + S}{2S}\right) + (1 - p)\left(\frac{\Sigma + S}{2S}\right)^2 \right] , \tag{6.13}$$

$-S \leq \Sigma \leq +S$. As Σ advances to above $+S$ there is no further change in b_0 as there is no further change in the state of the instances, $b_0 = b_0^{\uparrow}(+S) = L_0 - \Delta L$. Now as Σ is reduced and passes to below $+S$, b_0 is given by

$$b_0^{\downarrow}(\Sigma) = b_0^{\uparrow}(+S) - \Delta L \int_{\Sigma}^{+S} dX \int_{X}^{+S} dY \, \varrho(X, Y) ,$$

$$b_0^{\downarrow}(\Sigma) = L_0 - \Delta L + \Delta L \left[p\left(\frac{S - \Sigma}{2S}\right) + (1 - p)\left(\frac{S - \Sigma}{2S}\right)^2 \right] , \tag{6.14}$$

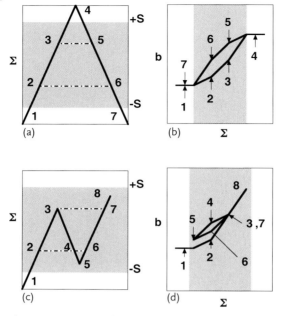

Fig. 6.3 Stress protocol and strain response. A simple quasistatic stress protocol is shown in (a), and a stress protocol that tests endpoint memory is shown in (b). The x-axis is "time". (c) The displacement response to stress protocol (a). (d) The displacement response to stress protocol (b). In stress protocol (b) there are two revisits to the stress value denoted 2, (4,6) and one revisit to the stress value 3, (7). While there are three different displacements at stress 2, 4, 6, there is only one displacement at stress 3, 7. See Figures 6.4 and 6.5.

$-S \leq \Sigma \leq +S$. Note $b_0^{\downarrow}(-S) = b_0^{\uparrow}(-S) = L_0$. The moves described here can be followed in Preisach space. A schematic of such a space is shown in Figure 6.4. The instances in state $\eta_i = -1$ are shaded light gray; the instances in the state $\eta_i = +1$ are shaded darkly. The qualitative behavior of shaded areas in Preisach space translate directly into qualitative behavior in the b_0 vs. Σ plot (upper right panel of Figure 6.3).

The average strain of an elastic element is sensibly defined as $\varepsilon(\Sigma) = (b_0(\Sigma) - L_0)/L_0$. Further, by the EMT construction this is the strain of the material. This strain is hysteretic (upper right panel of Figure 6.3). To exhibit the hysteresis in a simple way, form the sum and difference of $b_0^{\uparrow}$ and $b_0^{\downarrow}$. We have, $b = b_+ + b_-$,

$$
\begin{aligned}
b_+ &= \frac{b_0^{\uparrow} + b_0^{\downarrow}}{2} = L_0 - \Delta L \frac{S + \Sigma}{2S} , \\
b_- &= \frac{b_0^{\downarrow} - b_0^{\uparrow}}{2} = -\Delta L(1 - p)\frac{S^2 - \Sigma^2}{4S^2} , \\
\varepsilon(\Sigma)_{\pm} &= \frac{-\Delta L}{L_0}\left(\frac{S + \Sigma}{2S} - s(\dot{\Sigma})(1 - p)\frac{S^2 - \Sigma^2}{4S^2}\right) ,
\end{aligned}
\qquad (6.15)
$$

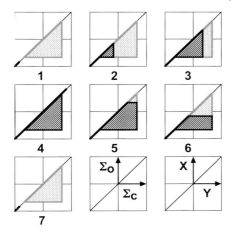

Fig. 6.4 Stress protocol in Preisach space. The stress protocol in (a) of Figure 6.3 can be followed in Preisach space. The
elastic elements at (X, Y), shaded gray, have $\eta = -1$, and those
at (X, Y), shaded black, have $\eta = +1$. Points (2,6) are at the
same stress but correspond to different displacements because
of the different state of the elastic elements at these points. So
also for the points (3, 5). See Eqs. (6.13) and (6.14).

where $s(\dot{S})$ is the sign of dS/dt, as is the subscript on ε. This strain has two components; the first, independent of $s(\dot{S})$, is a typical result from elasticity theory (in this
particular case linear elasticity) and the second is a hysteretic term that depends on
the sign of the change in stress. The compliance, defined by $k = -\partial\varepsilon/\partial\Sigma$, is given
by

$$k = \frac{\Delta b L}{L_0} \frac{1}{2S} \left(1 + s(\dot{\Sigma})(1 - p)\frac{S}{\Sigma} \right). \tag{6.16}$$

The strain, Eq. (6.15), and the compliance, Eq. (6.16), are shown as a function of Σ
in Figures 6.6 and 6.7 for the case $S = 1$ and $p = 0.25$.

Case 2
The stress is taken from below $-S$ to $\Sigma_3 < +S$, reversed and taken to $\Sigma_5 > -S$,
reversed again and taken to Σ_8, $\Sigma_3 < \Sigma_8 < S$, lower left of Figure 6.3. Initially $\eta_i =
-1$ $\forall i$. As above $b_0 = L_0$ until Σ exceeds $-S$. The strain evolution that accompanies
this stress protocol is shown in the lower right panel of Figure 6.3. This stress
protocol is too elaborate to make analytic treatment useful. There are two qualitative
points that can be gleaned from looking at the relationship between the state of
instances in Preisach space and b_0 vs. Σ.

    1. The stresses Σ_2, Σ_4, and Σ_6 are the same, but the displacement b_0 is differ-
        ent in all three cases. This is because the state of the instances in Preisach
        space is different for each of these cases, as illustrated in the lower left of
        Figure 6.5.

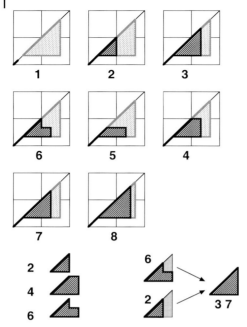

Fig. 6.5 Endpoint memory stress protocol. The stress protocol in (b) of Figure 6.3 can be followed in Preisach space. The three visits to the stress at 2 involve different fillings of Preisach space and consequently different displacements (strains). See the three "fillings" at the lower left. The filling of Preisach space at 3 and 7 is the same. The approach to this state 3 = 7 is from 2 initially and 6 secondly. The number of elastic elements changing state for the stress move 2 → 3 is different from the number changing state for the stress move 6 → 7. See the "fillings" at the lower right. Thus the slopes on the two approaches to 3 = 7 are different. The slope on leaving 3 = 7 is independent of how 3 = 7 is approached and is the same as the slope for the protocol 2 → 3 → 8; 3 = 7 is an endpoint of the segment of the stress protocol in (b) of Figure 6.3 until 3 = 7 is passed.

2. For Σ_3 and Σ_7 the state of the instances in Preisach space is the same, but the slope 2 – 3 is different from the slope 6 – 7. This is because the approach to 2 from 3 involves a change in the state of the instances that differs from the change involved in the approach to 7 from 6, lower right of Figure 6.5. The displacement on continuation of the stress from 7 to 8 is the same as on continuation of the stress from 3 to 8. This is an illustration of endpoint memory [8]. One might say that the state of the material at 3 was remembered in that the stress moves in going from 3 to 7 leave the system unchanged; at 7 the state of the instances is the same as it was at 3.

The two cases examined here illustrate important qualitative properties of a material in which the important displacements are those associated with the change in state of elastic elements. This is not all materials, nor is it one material all of the time. One could surround the domain of application of this extreme model with

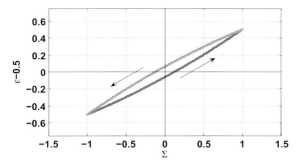

Fig. 6.6 Stress vs. strain. The strain as a function of stress for stress protocol (a) from Figure 6.3. The strain, calculated from the displacement as in Eq. (6.15), is shifted by 0.5 to exhibit symmetry about $(\Sigma, \varepsilon) = (0, 0)$. This symmetry is specific to the particular model used for $\varrho(X, Y)$. The sense in which the hysteresis loop is traversed is shown; it is described by *the strain stays in*. See also the hysteresis loops in Figure 6.3.

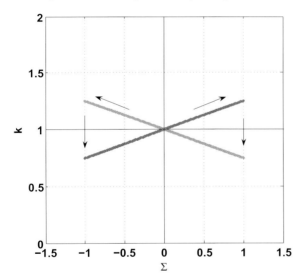

Fig. 6.7 Compliance vs. stress. The compliance, calculated from Eq. (6.16), as a function of the stress. The sense in which this curve is traversed is shown. Its qualitative properties can be read from Figure 6.6.

lots of caveats. But the fact is it works where it works and it doesn't work where it doesn't work. Where it works, that is, one finds macroscopic hysteresis with end-point memory, it is reasonable to assume that the physics underlying these observations is a set of elastic elements with hysteretic displacements, possibly mesoscopic, possibly microscopic. The nature of these elastic elements is not a part of the model. There are many potential candidates for them, the van der Waals surfaces from Chapter 4, the Hertz–Mindlin contacts of Johnson and Norris [9], the frictional contacts of Nihei and coworkers [10], surfaces with asperities [11], etc. It is unlikely that one of these candidates provides the answers to all questions.

Before we go on to explore this model further, we want to describe another of its virtues, that is, that it is invertible. Because it is, one has in principle the means to interrogate a hysteretic elastic system about the elastic elements that reside in it.

6.2
Finite Displacement Elastic Elements: Inversion

The principle associated with the idea of inverting to learn the nature of the elastic elements comes from Eq. (6.15) or, equivalently, Eqs. (6.13) and (6.14). From Eq. (6.13)

$$\Delta \varepsilon_+ = -\frac{\Delta L}{L_0} \left[\int_{-S}^{\Sigma} dY \varrho(\Sigma, Y) \right] \Delta \Sigma, \tag{6.17}$$

and from Eq. (6.14)

$$\Delta \varepsilon_- = -\frac{\Delta L}{L_0} \left[\int_{\Sigma}^{S} dX \varrho(X, \Sigma) \right] \Delta \Sigma. \tag{6.18}$$

As illustrated in Figure 6.8, the changes in strain $\Delta \varepsilon$ are related to integrals over the density $\varrho(\Sigma_c, \Sigma_o)$ in columns or rows in Preisach space [12]. Thus a suitable experimental exploration, with the stress/strain protocol tailored to probe regions of Preisach space, could reveal the strengths of the stresses that cause the state changes of the elastic elements. This philosophy is not limited to the simple suggestion here. As illustrated in Figure 6.9, more imaginative stress/strain protocols can focus very specifically on elastic elements in particular regions of Preisach space.

Let us not oversell the point. What a suitable experiment can do is reveal the relative number of elastic elements that have hysteretic behavior characterized by a particular (Σ_c, Σ_o). To the degree that this information can inform choices among competing models of elastic elements it can be very useful. Applications of these ideas are discussed in Chapter 10.

6.3
Finite Displacement Elastic Elements: Dynamic Response

Let us look at finite displacement elastic elements (FDEE) in dynamics: (a) in a resonant bar experiment and (b) in a wave mixing experiment.

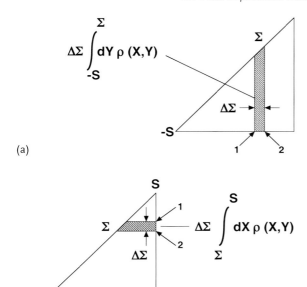

Fig. 6.8 Inversion in Preisach space. (a) An infinitesimal upward stress step $\Delta\Sigma$ at Σ, $(1 \to 2)$, yields a change in strain due to the elastic elements in the strip $(\Sigma + S)\Delta\Sigma$ and given by the integral in Eq. (6.17). (b) An infinitesimal downward stress step $\Delta\Sigma$ at Σ, $(1 \to 2)$, yields a change in strain due to the elastic elements in the strip $(S - \Sigma)\Delta\Sigma$ and given by the integral in Eq. (6.18).

6.3.1
Finite Displacement Elastic Element: Resonant Bar

We consider the case of a resonant bar experiment as shown schematically in Figure 6.10. As discussed above the idea is to study the low-lying resonances of a resonant bar as a function of the amplitude. The response of interest is the Fourier component of the displacement of the bar end, $u(L)$, at the drive frequency [13], as a function of the amplitude of the drive. For the displacement field in a resonant bar we have the equation of motion [Eq. (3.54)]

$$\varrho\ddot{u} + \varrho\frac{1}{\tau}\dot{u} = \frac{\partial\Sigma}{\partial x} + f(t)\frac{L}{2}\left[\delta\left(x + \frac{1}{2}L\right) - \delta\left(x - \frac{1}{2}L\right)\right] . \qquad (6.19)$$

Before we turn to the use of the EOS of the FDEE we replace the resonant bar equation of motion by a lumped element equation that simplifies doing this.

6.3.1.1 Lumped Element Model
We consider the fundamental mode, in which the RHS of the bar moves mirrorlike with respect to the LHS of the bar, and integrate Eq. (6.19) over x, $-L/2 \le x \le 0$, to

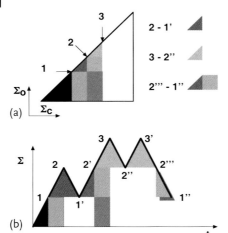

Fig. 6.9 Inversion in Preisach space; elaborate stress protocols. The regions of Preisach space, shown with color coding in (a), are swept over by the stress protocol in (b) with the same color coding. The elastic elements in the red region (of Preisach space) are responsible for the strain as the stress evolves $2 \rightarrow 1'$ ($1' \rightarrow 2'$). The elastic elements in the blue region are responsible for the strain as the stress evolves $3 \rightarrow 2''$ ($2'' \rightarrow 3'$).

The elastic elements in the blue, yellow, and red regions are responsible for the strain as the stress evolves $3' \rightarrow 2''' \rightarrow 1''$. By manipulating the stress protocol the strain due to elastic elements in a particular region of Preisach space can be found. In more physical terms, the strains due to elastic elements that respond to particular stresses can be determined. (Please find a color version of this figure on the color plates)

find

$$\varrho \ddot{U}_L(t) + \varrho \frac{1}{\tau} \dot{U}_L(t) = \Sigma(0, t) + f(t), \tag{6.20}$$

where

$$U_L(t) = \frac{2}{L} \int_{-L/2}^{0} dx \, u(x, t). \tag{6.21}$$

The equation for the RHS of the bar is the same except for the sign of the terms on the RHS. Thus $U_R(t) = -U_L(t)$. To find $U_L(t)$ we need the stress at the bar center, $\Sigma(0, t)$. We imagine we know $\Sigma(0, t)$ as a function of the strain at the bar center. We close this equation by relating $U_L(t)$ to the strain at the bar center. To do this we use the amplitude relations among displacement and strain appropriate to a simple linear system at the fundamental resonance

$$u = A \, \sin(kx), \tag{6.22}$$

$$u'(0) = kA, \tag{6.23}$$

$$U_R = \frac{2}{\pi} A, \tag{6.24}$$

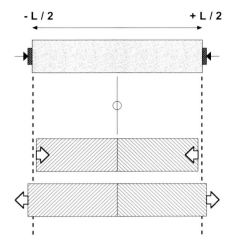

Fig. 6.10 Resonant bar → lumped element. The equation of motion of a resonant bar driven symmetrically, so that the center of mass stays fixed at $x = 0$, can be replaced approximately by the equation of motion for a lumped element. The lumped element is the average position of the RHS of the bar. The RHS vibrates against its opposite, the average position of the LHS. The forces on the lumped element are the net force, for example, applied force at $x = L/2$, and the stress at the bar center, Eq. (6.26). The internal forces away from the bar center do not appear.

$$ e = \frac{2}{L}(U_R - U_L) = \frac{8}{\pi}\frac{A}{L} = \frac{8}{\pi^2}u'(0) \approx u'(0) \,, \tag{6.25} $$

where $k = \pi/L$. To within a numerical factor of order 1 the average strain defined by e and the strain at the bar center are equal. If we define ε as the strain at the bar center and relate U_L to ε using Eqs. (6.25), we have

$$ \ddot{\varepsilon} + \frac{1}{\tau}\dot{\varepsilon} = -\phi(\varepsilon) - F(t) \,, \tag{6.26} $$

where $\phi = \pi^2\Sigma/4L\varrho$ and $F = \pi^2 f/4L\varrho$. This equation, a lumped element equation, is for the motion of the strain field at the bar center in terms of the stress at the bar center. To this point there is no input that is specific to FDEE.

6.3.1.2 Stress-Strain for Finite Displacement Elastic Element

To find Σ for use in Eq. (6.26) we need a specific model for $\varrho(X, Y)$ in Preisach space. Since the stress/strain field involved in a resonant bar experiment are typically several orders of magnitude lower than those used in quasistatic measurements, say less than one atmosphere (0.1 MPa), we take a model appropriate to a small local region of Preisach space

$$ \varrho(X, Y) = A\delta(X - Y) + \alpha \,, \tag{6.27} $$

where A and α are parameters to be set below. Let us assume that $F(t)$ is a sinusoidal drive, amplitude F_0, that sweeps the stress at the bar center back and forth between

$\pm C$, $C \propto F_0$. We can use the same scheme as illustrated above, Eqs. (6.13) and (6.14), to find $\varepsilon_\uparrow = (b_\uparrow - L_0)/L_0$ and $\varepsilon_\downarrow = (b_\downarrow - L_0)/L_0$ with the result

$$\varepsilon = \varepsilon(0) + \varepsilon_0 \left[(A + \alpha C)\Sigma - s(\dot{\Sigma}) \frac{\alpha}{2}(C^2 - \Sigma^2) \right] , \tag{6.28}$$

where $\varepsilon(0) = AC + \alpha C^2$ and $\varepsilon_0 = \Delta b/L$. We shift ε by $\varepsilon(0)$ and scale it by ε_0. Then, solving for Σ we have

$$\Sigma = K_0 \left(\varepsilon - \alpha C \Sigma + s(\dot{\Sigma}) \frac{\alpha}{2} (C^2 - \Sigma^2) \right) , \tag{6.29}$$

where $K_0 = 1/A$. An iterative solution for Σ in terms of ε leads to

$$\Sigma = K_0 \left(\varepsilon - \gamma \varepsilon_m \varepsilon + s(\dot{\varepsilon}) \frac{\gamma}{2}(\varepsilon_m^2 - \varepsilon^2) + \dots \right) , \tag{6.30}$$

where $\gamma = \alpha K_0^2$ and $\varepsilon_m = C/K_0$ is the maximum strain at the bar center. Using this stress/strain relation in Eq. (6.26) results in

$$\ddot{\varepsilon} + \frac{1}{\tau}\dot{\varepsilon} = -\omega_0^2 \left(\varepsilon - \gamma \varepsilon_m \varepsilon + s(\dot{\varepsilon}) \frac{\gamma}{2}(\varepsilon_m^2 - \varepsilon^2) + \dots \right) - F(t) , \tag{6.31}$$

where we have combined numbers and physical constants to form ω_0^2, which must necessarily be the resonant frequency of the fundamental mode of the bar at low strain. We have in this equation a lumped element equation of motion for a bar with EOS given by Eq. (6.28). As a review of the procedure for getting to this equation will show, it is an equation for the strain at the bar center that contains two parameters that characterize the linear, ω_0^2, and nonlinear, γ, elastic properties of the bar in the approximation that the important strains are due to FDEE, the FDEE of the MG model.

1. The strength of the nonlinear terms is measured by $\gamma = \alpha K_0^2$. The background density, α in Eq. (6.27), has units $(stress)^{-2}$. An approximate measure of the size of γ can be made from quasistatic data, like that described above, which provides an estimate of α. Of course, K_0 is of order 10^{10}–10^{12} dyne/cm^2. This is discussed in detail in Chapter 10.
2. The amplitude of the strain field ε_m enters the equation of motion importantly. This is because the number of FDEE contributing to the nonlinear response depends on the area of Preisach space swept over by the stress at the bar center. This number could be characterized by the amplitude of the stress, C, or by the strain this stress induces.
3. We expect the linear elastic response to be several orders of magnitude greater than the nonlinear elastic response. Consequently terms of higher than the leading order are handled approximately but consistently.

6.3.1.3 Resonant Bar Response

We wish to find the frequency response of the resonant bar described by Eq. (6.31). More precisely, we want the Fourier component of ε at ω, the frequency of the drive, as a function of ω, at fixed drive amplitude, for various drive amplitudes. To do this we will introduce a computational procedure, useful beyond this specific domain of application, that will let us get to what we want quickly.

1. Assume $F(t)$ is of the form

$$F(t) = F_0 \sin(\omega t + \phi), \tag{6.32}$$

where by introducing the unknown phase ϕ we are free to choose the strain to have zero phase,

$$\varepsilon(t) = R \sin(\omega t), \quad \varepsilon_m = R. \tag{6.33}$$

2. Substitute $F(t)$ and $\varepsilon(t)$ into Eq. (6.31) and project the resulting equation onto $(\mathcal{S}, \mathcal{C}) = (\sin(\omega t), \cos(\omega t))$. Since $\ddot{\varepsilon} = -\omega^2 R \mathcal{S}$ and $\dot{\varepsilon} = \omega R \mathcal{C}$, we have

$$
\begin{aligned}
-\omega^2 R &= -\omega_0^2 (1 - 2\Lambda) R - F_0 \cos \phi - \omega_0^2 R \Lambda \langle \mathcal{S} | s(\mathcal{C}) \mathcal{C}^2 \rangle \\
\frac{\omega}{\tau} R &= -F_0 \sin \phi - \omega_0^2 R \Lambda \langle \mathcal{C} | s(\mathcal{C}) \mathcal{C}^2 \rangle,
\end{aligned}
\tag{6.34}
$$

where $\Lambda = \gamma R / 2$, $T = 2\pi / \omega$, and

$$\langle a | b \rangle = \frac{1}{T} \int_0^T dt \; a(t) \, b(t). \tag{6.35}$$

The projections $\langle \mathcal{S} | s(\mathcal{C}) \mathcal{C}^2 \rangle$ and $\langle \mathcal{C} | s(\mathcal{C}) \mathcal{C}^2 \rangle$ have the numerical values 0 and $4/(3\pi)$, respectively. Thus we have

$$
\begin{aligned}
-\omega^2 R + \omega_0^2 (1 - 2\Lambda) R &= -F_0 \cos \phi, \\
\frac{\omega}{\tau} R + \omega_0^2 \frac{4}{3\pi} \Lambda R &= -F_0 \sin \phi,
\end{aligned}
\tag{6.36}
$$

from which R and ϕ are to be found. We square these equations, add, and rearrange to find R as a function of $\Omega = \omega/\omega_0$ in the form

$$R = \frac{1}{\sqrt{\left(\Omega^2 - (1 - \gamma R)^2 \right)^2 + \left(\frac{\Omega}{Q_0} + \frac{4}{3\pi} \gamma R \right)^2}} F_0, \tag{6.37}$$

where $Q_0 = \omega_0 \tau$ and factors of ω_0 have been absorbed into F_0.

3. When $R = 0$ on the RHS of this equation, we find the expected linear result, $\Omega = 1$ and $Q = Q_0$. At finite R we find a resonant frequency shift to lower frequencies, proportional to γR,

$$\delta \Omega = -\frac{\gamma R}{2}, \tag{6.38}$$

and an increased attenuation proportional to γR:

$$\delta \frac{1}{Q} = \frac{1}{Q} - \frac{1}{Q_0} = \frac{4}{3\pi} \gamma R, \quad \Omega \approx 1. \tag{6.39}$$

These results are shown in Figures 6.11 and 6.12.

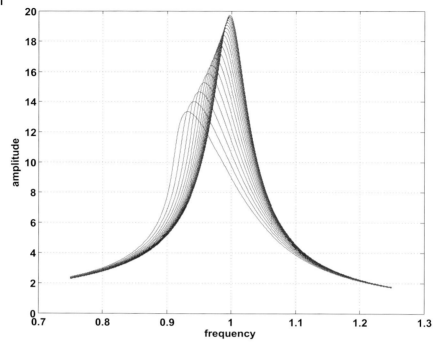

Fig. 6.11 Resonant response of a bar with hysteretic elastic elements. The amplitude, in the form A/F_0, as a function of the driving frequency, $f = \omega/(2\pi)$, for a range of values of F_0, $0.001 \le F_0 \le 0.05$. As the amplitude of the drive increases, the resonant frequency shifts to a lower frequency and the apparent atten- uation increases. The attenuation estimate is from the behavior of the maximum of the res- onance curves, $A_{max} \approx QF_0$, $1/Q \approx F_0/A_{max}$. Both the frequency shift and the added atten- uation are a function of the strain in the bar, which is measured by the amplitude at the resonance maximum, A_{max}.

4. Let us argue for these results so that their status, independent of details, can be established.

   a. $\delta\Omega$. The FDEE are such that when the stress sweeps over Σ there is a dis- placement (strain) beyond the linear displacement, $\varepsilon^{(1)} \approx \Sigma/K_0$, propor- tional to the number of elastic elements involved, that is, $\varepsilon^{(2)} \approx n\Delta L/L$, $n \approx \alpha\Sigma^2$. Thus

$$\Delta\varepsilon \approx \varepsilon^{(1)} + \varepsilon^{(2)} \approx \frac{\Sigma}{K_0} + \alpha\Sigma^2 \frac{\Delta L}{L}\,. \tag{6.40}$$

   This is more strain per unit of stress, that is, a softening of the modulus. To achieve something like the result in Eq. (6.37), use the definition of γ, replace one factor of Σ/K_0 in the second term with $\Delta b\varepsilon$, and rearrange

$$\Delta\varepsilon \approx \frac{1}{1 - \gamma(\Delta\varepsilon)\frac{\Delta L}{L}} \cdot \frac{\Sigma}{K_0} = \frac{\Sigma}{K}\,, \tag{6.41}$$

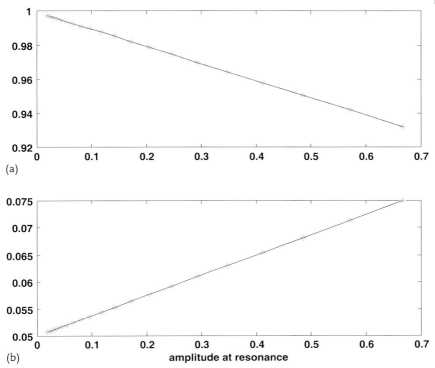

Fig. 6.12 Frequency shift and $1/Q$. (a) The frequency at the maximum of the resonance curves, from Figure 6.11, is plotted as a function of the amplitude at resonance. (b) The inverse of Q, defined by $1/Q = F_0/A_{max}$, is plotted as a function of the amplitude at resonance. The amplitude at resonance, A_{max}, is a measure of the strain field in the bar.

$K = K_0(1 - \gamma(\Delta\varepsilon)\Delta L/L)$. The important involvement of $\gamma(\Delta\varepsilon)$ is the same as above, Eq. (6.38). The different involvement of $\Delta L/L$ has to do with the scaling of ε by ε_0 above. What is important is the number of FDEE affected by the sweep of Σ and the size of the displacement associated with each.

b. $\Delta(1/Q)$. The quantity $1/Q$ is related to the ratio of the work done per cycle to the energy stored per cycle, $\Delta W/W_0$. For W_0 we have $K_0(\Delta\varepsilon)^2 \approx \Sigma^2/K_0$. For the additional work done per cycle we again look at the $n \approx \alpha\Sigma^2$ FDEE swept over by the stress. Strain steps $\Delta L/L$ occur at Σ_c and Σ_o, $\Delta w \approx (\Sigma_c - \Sigma_o)\Delta L/L \approx \Sigma\Delta L/L$. Thus $\Delta W \approx n\Delta w$ and

$$\Delta\frac{1}{Q} = \frac{\Delta W}{W_0} \approx \gamma(\Delta\varepsilon)\frac{\Delta L}{L}. \tag{6.42}$$

Compare to Eq. (6.39).

5. A general set of scaling relations for ΔK and for $\Delta(1/Q)$ can be constructed following the argument above. We have

$$\frac{\Delta K}{K_0} = n(\Sigma) \left\langle \frac{\Delta b}{b_0} \right\rangle \left(\frac{\Sigma}{K_0} \right)^{-1} ,$$

$$\Delta \frac{1}{Q} = \frac{n(\Sigma)\langle \Sigma_c - \Sigma_o \rangle \langle \Delta b / b_0 \rangle}{\Sigma^2 / K_0} . \tag{6.43}$$

That is, the frequency shift and shift in Q^{-1} depend on the number of elastic elements swept over by a stress field of amplitude Σ, $n(\Sigma)$, the average displacement of each elastic element, $\langle \Delta b \rangle$, and the average of the difference between the two values of the stress at which the displacement steps occur, $\langle \Sigma_c - \Sigma_o \rangle$. Assume we have the scaling relations

$$n(\Sigma) \sim \alpha K_0^2\, x^\nu = \gamma\, x^\nu ,$$

$$\left\langle \frac{\Delta b}{b_0} \right\rangle \sim x^\mu , \tag{6.44}$$

$$\langle \Sigma_c - \Sigma_o \rangle \sim K_0\, x^\eta ,$$

where $x = \Sigma / K_0$. Then

$$\frac{\Delta K}{K_0} \sim \gamma\, x^{\mu+\nu-1} ,$$

$$\Delta \frac{1}{Q} \sim \gamma\, x^{\mu+\nu+\eta-2} , \tag{6.45}$$

$$\frac{K_0}{\Delta K} \Delta \frac{1}{Q} \sim x^{\eta-1} .$$

The MG model has $\nu = 2$, $\mu = 0$, and $\eta = 1$, or $\Delta K / K_0 \sim \Delta(1/Q) \sim \Sigma \sim \Delta\varepsilon$. Other results are possible.

6.3.2
Finite Displacement Elastic Element: Wave Mixing

Finally, let us turn to an example of the workings of hysteretic elastic elements in a wave mixing scenario. We consider a case similar to one explored in Chapter 3 involving traditional nonlinear elasticity: two compressional waves moving colinearly. We begin with the equation of motion, Eq. (3.30),

$$\ddot{u} + \frac{1}{\tau_0}\dot{u} = c_L^2 \frac{\partial^2 u}{\partial x^2} + \Lambda[\Sigma] + \lambda_1 f_1(t)\delta(x) + \lambda_2 f_2(t)\delta(x) , \tag{6.46}$$

where sources f_1 and f_2 at $x = 0$ launch two compressional waves, angular frequencies ω_1 and ω_2, having different (or not) amplitude and phase, and $\Lambda[\Sigma]$ is a hysteretic nonlinearity. We construct a specific model for $\Lambda[\Sigma]$ based on the physical picture of how the FDEE work,

$$\Lambda[\Sigma] = c_L^2 \frac{\Delta b[\Sigma(x')]}{2a} \left(\delta(x - x' - a) - \delta(x - x' + a) \right) , \tag{6.47}$$

where Δb, a displacement, is a function of the stress field at x', $\Delta b = \Delta b_0 n[\Sigma(x')]$. To construct $n[\Sigma(x')]$, we use $\varrho(X, Y) = \alpha$ and

$$n(t) = n(0) + \iint_{[\Sigma]} \varrho(X, Y)\eta(X, Y)\, dX\, dY\,, \tag{6.48}$$

where here and above the square bracket, $[\Sigma]$, is a reminder that the quantity involved is a function of the history of Σ. We have made this choice of Λ so that in the quasistatic limit ($f_1 = f_2 = 0$, $d/dt = 0$) if $\partial u/\partial x = 0$ for $x < x' - a$ and $\partial u/\partial x = 0$ for $x > x' + a$, then $\partial u/\partial x = \Delta b/(2a)$ for $-a \le x - x' \le +a$. Thus, there is a displacement between $x' + a$ and $x' - a$, $u(x' + a) - u(x' - a) = \Delta b_0 n[\Sigma(x')]$, a function of the history of the stress at x'. We solve Eq. (6.46) using what is by now a standard procedure: perform a Fourier analysis of u, f_1, f_2, and Δb, write $u_\omega = \lambda_1 u_\omega^{(1)} + \lambda_2 u_\omega^{(2)} + \lambda_1 \lambda_2 u_\omega^{(3)}$, and find

$$D_\omega^{-1} u_\omega^{(1)} = f_\omega^{(1)}\,,$$
$$D_\omega^{-1} u_\omega^{(2)} = f_\omega^{(2)}\,, \tag{6.49}$$
$$D_\omega^{-1} u_\omega^{(3)} = c_L^2 \frac{\Delta b_\omega^{(3)}}{2a}(\delta(x - x' - a) - \delta(x - x' + a))\,,$$

where to construct $\Delta b_\omega^{(3)}$ we form $\Delta b(t)$ using $\Sigma(x', t) = K_0(\varepsilon^{(1)}(x', t) + \varepsilon^{(2)}(x', t))$ from $u_\omega^{(1)}$ and $u_\omega^{(2)}$. We focus on $u_\omega^{(3)}$, the amplitude broadcast from $x' \pm a$, the location of the hysteretic nonlinearity. Inverting D_ω and using the delta functions on the RHS of the $u_\omega^{(3)}$ equation we have

$$u_\omega^{(3)}(x) = c_L^2 \frac{G_\omega(x|x' + a) - G_\omega(x|x' - a)}{2a} \Delta b_\omega^{(3)} = \frac{1}{2} e^{ik_\omega(x - x')} \Delta b_\omega^{(3)}\,, \tag{6.50}$$

where the representation on the right comes from using the appropriate Green function, Eq. (3.112), and assuming $a \to 0$. Except for the phase factor $\exp ik_\omega(x - x')$ the amplitude $u_\omega^{(3)}$ is the amplitude $\Delta b_\omega^{(3)}$.

The amplitude $\Delta b_\omega^{(3)}$ results from the stress at x'. This stress in turn is due to the superposition of the strain fields $\varepsilon^{(1)}$ and $\varepsilon^{(2)}$ at x'. Without loss of generality we can take these strain fields to produce a stress at x' of the form

$$\Sigma(t) = \sin(\omega_1 t) + r \sin(\omega_2 t + \theta)\,, \tag{6.51}$$

where ω_1 and ω_2 are the frequencies of sources f_1 and f_2, respectively, and, since our concern is the frequency structure of Δb, we have a phase and r to measure the relative amplitude of the two frequency components. The amplitude $\Delta b_3(t)$, constructed from $\Sigma(t)$ and the "transformation" represented by Eq. (6.48), is Fourier analyzed to produce $\Delta b_\omega^{(3)}$. In general this calculation cannot be carried out manually. We illustrate the outcome in a sequence of figures in which we show, for various choices of ω_1, ω_2, r, and θ, the stress-displacement relationship, $\Sigma(t)$ vs. $\Delta b(t)$, and the power spectrum of $\Delta b_\omega^{(3)}$. Let us look at these figures. There are four choices of the frequency pair $(f_1, f_2) = (\omega_1, \omega_2)/(2\pi)$, $(f_1, f_2) = (1, 1)$, $(1, 14/16)$, $(1, 7/16)$, and $(1, 2\pi/16)$. For most of these cases we look at the $\Delta b(t)$ and $\Sigma(t)$ vs. t, the "phase portrait" $\Delta b(t)$ vs. $\Sigma(t)$ and the power spectrum of $\Delta b(t)$.

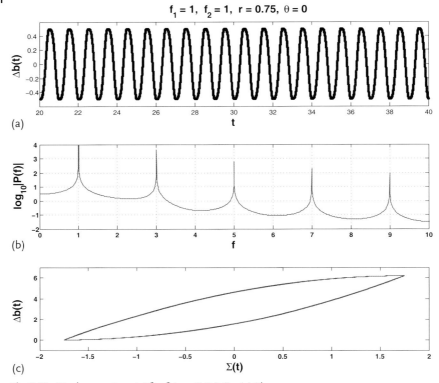

Fig. 6.13 Displacement vs. t, $(f_1, f_2) = (1.0, 1.0)$. (a) The displacement $\Delta b(t)$ from Eq. (6.49) as a function of time. The two frequencies are commensurate. (b) The power spectrum of $\Delta b(t)$ as a function of frequency, $f = \omega/(2\pi)$. (c) Displacement vs. stress. The displacement $\Delta b(t)$ from Eq. (6.49) as a function of the stress $\Sigma(t)$ from Eq. (6.51).

1. $(f_1, f_2) = (1, 1)$. The two strain fields at x' are of the same frequency, amplitude, and phase. The strain fields add simply, $r = 1$, $\theta = 0$ in Eq. (6.51), and a simple $\Delta b(t) - \Sigma(t)$ relationship results, Figure 6.13. The persistence of the strain on stress reversal, for example, at $\Sigma \approx \pm 2$, is seen directly in Figure 6.13, $\Delta b(t)$ vs. $\Sigma(t)$ at $t = \ldots 0.5$, 1.0, 1.5 $\ldots$ It is this discontinuous rate of strain evolution that is responsible for the dominance of odd harmonics in the power spectrum of $\Delta b(t)$, Figure 6.13b.
2. $(f_1, f_2) = (1, 7/8)$. In this case the two strain fields have similar frequencies so at x' a phase evolution occurs with the strain amplitude at x' alternating between adding and subtracting. Since f_1 and f_2 are commensurate, $8f_2 = 7f_1$, over time a repeating pattern occurs, Figure 6.14. The $\Delta b(t)$ vs. $\Sigma(t)$ relationship is much more complicated than that for $(f_1, f_2) = (1, 1)$, Figure 6.14c. The power spectrum of $\Delta b(t)$ continues to be dominated by

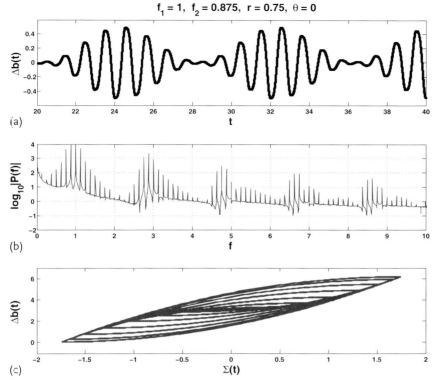

$$f_1 = 1, \ f_2 = 0.875, \ r = 0.75, \ \theta = 0$$

Fig. 6.14 Displacement vs. stress, $(f_1, f_2) = (1.0, 7/8)$. (a) The displacement $\Delta b(t)$ from Eq. (6.49) as a function of time. The two frequencies are commensurate. (b) The power spectrum of $\Delta b(t)$ as a function of frequency, $f = \omega/(2\pi)$. (c) Displacement vs. stress. The displacement $\Delta b(t)$ from Eq. (6.49) as a function of the stress $\Sigma(t)$ from Eq. (6.51).

the basic "odd" harmonic structure seen for $(f_1, f_2) = (1, 1)$. There is detail in the power spectrum that represents the amplitude modulation that occurs as the two strains go in and out of phase.

3. $(f_1, f_2) = (1, 7/16)$. Again f_1 and f_2 are commensurate, $16 f_2 = 7 f_1$, and over time a repeating pattern occurs, Figure 6.15 (there is just one repetition of the pattern in time in this figure). The complex, but repeatable, phase portrait is shown in Figure 6.15c. The power spectrum of $\Delta b(t)$, Figure 6.15b, is relatively complex, with the superficial vestiges of "odd" harmonics much less apparent. The basic physical event, persistence of $\Delta b(t)$ on stress reversal, continues, but these events occur in a complex way as time evolves. Thus the "ragged" power spectrum.

4. $(f_1, f_2) = (1, \pi/8)$. In this case f_1 and f_2 are incommensurate. As a consequence the time evolution of $\Delta b(t)$ is followed out to a longer time. Nonethe-

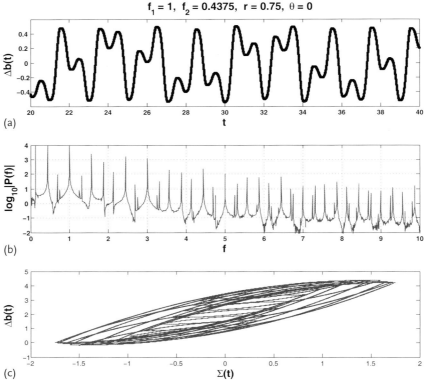

Fig. 6.15 $(f_1, f_2) = (1.0, 7/16)$. (a) The displacement $\Delta b(t)$ from Eq. (6.49) as a function of time. The two frequencies are commensurate. (b) The power spectrum of $\Delta b(t)$ as a function of frequency, $f = \omega/(2\pi)$. (c) Displacement vs. stress. The displacement $\Delta b(t)$ from Eq. (6.49) as a function of the stress $\Sigma(t)$ from Eq. (6.51).

less, direct observation of $\Delta b(t)$ vs. t is not very informative; a small segment is shown in Figure 6.16a. The phase portrait, Figure 6.16c, reveals a complex structure that is suggestive of chaotic motion. However, the power spectrum of $\Delta b(t)$ is relatively simple. Over a long period of time the fundamental frequencies in the motion come to dominate the relatively slowly varying features that make the phase portrait so complex.

The discussion here reveals that a rudimentary hysteretic elastic element, for example, a localized set of FDEE with $\varrho(X, Y)$ constant, irradiated by a pair of strain fields broadcasts, a strain field with a complex time signature. Thus, in principle a careful study of such an elastic element with sets of (f_1, f_2) pairs provides the means to learn something of the nature of the elastic element. For a hysteretic elastic element that is spatially local, such as the one we have treated, this is an imag-

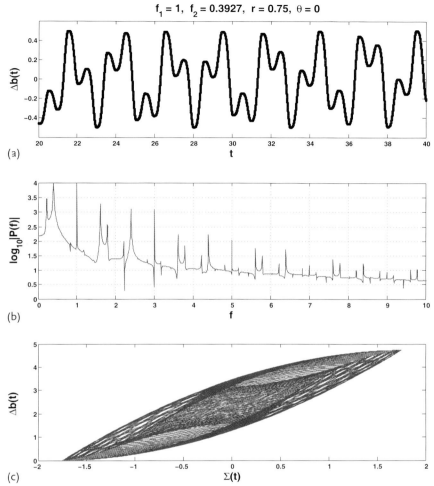

Fig. 6.16 $(f_1, f_2) = (1, \pi/8)$. (a) The displacement $\Delta b(t)$ from Eq. (6.49) as a function of time. The two frequencies are incommensurate. (b) The power spectrum of $\Delta b(t)$ as a function of frequency, $f = \omega/(2\pi)$. (c) Displacement vs. stress. The displacement $\Delta b(t)$ from Eq. (6.49) as a function of the stress $\Sigma(t)$ from Eq. (6.51).

inable scenario. So think about a crack, a single set of asperities, etc. For a spatially distributed set of hysteretic elastic elements we face the problem of the phase of strain field arriving at each space point and the problem of the coherent superposition of the signals broadcast from each space point. This is, of course, the bread and butter of calculations like those of Jones and Kobett. For the relatively complicated interaction like Λ in Eq. (6.47) such a program is not feasible with any generality.

We are content to call attention to the difficulty. The few calculations that have been carried out for specific models/circumstances will be mentioned below [14, 15].

There are a few principles. In general the sum of the strain amplitudes incident on a scatterer determine the amplitude of the scattered signal. The simple amplitude rules of classical nonlinear elasticity do not apply. Similarly, the dynamics of the scatterer cause the broadcast of frequencies not present in the incident strain fields, so the rules of frequency addition from classical nonlinear elasticity do not apply.

6.4
Models with Hysteresis

The essential feature of the model described above was the presence of hysteresis in the displacement of elastic elements and in the resulting equation of state. There are other studied models for the behavior of contacts/elastic systems that explicitly introduce hysteresis, for example, Hertzian contact models [10], the Masing rules [16], and the endochronic model [17]. We describe these in the appendix, Section 6.6, and we show their connection to the FDEE model. We do not examine what these models have to say in every imaginable context but rather consider the behavior of each in the context of a resonant bar experiment. In the context of a resonant bar all of these models are identical to the FDEE model. Thus we take the FDEE model as representative of a large class of models.

6.5
Summary

In this chapter we have looked at the behavior of hysteretic elastic elements in a number of experimental scenarios. We have done this using elastic elements that have a single property, two different displacements that they move between according to the stress that they must support, the FDEE. The rule for motion between displacements (states) is hysteretic. A Preisach space was introduced to aid in the bookkkeeping necessary to follow the history of the states (displacements) of the elastic elements. We considered the behavior of collections of such elastic elements in three scenarios: (1) quasistatic response to a stress protocol, (2) dynamic response to a sinusoidal stress drive in a resonant bar, and (3) wave mixing. The fundamental lesson from all of this is that there are important features in what is seen that depend critically on stress protocol. Scenarios that are relatively simple in continuum elasticity are often complex when hysteretic elastic elements are involved.

We have examined a number of other models for the behavior of an elastic system with hysteresis. We showed that all of these are essentially equivalent to the FDEE model. The single exception is the endochronic model, which has the same hysteretic behavior as the other models, but does not have the amplitude-dependent change in modulus that is found in the other models. Thus the FDEE model of McCall and Guyer captures the essential features of all of the hysteretic models examined. This is a form of *universality*.

In every instance in which we have considered the response of FDEE, we have taken it as given that they have no intrinsic dynamics. The FDEE respond instantaneously, that is, change state instantaneously, whether in quasistatic or dynamc scenario, to the demands of the stress that they feel. This may be far from the truth. In the next chapter we look at modeling of the dynamics of elastic elements, and in Chapters 10 and 11 we examine some experimental evidence of their dynamics.

6.6
Models with Hysteresis, Detail

6.6.1
Hertzian Contacts

In Chapter 4 we discussed single Hertzian–Mindlin contacts under conditions of simple forcing in some detail and we alluded to the possibility of complicated behavior under more elaborate forcing conditions. These complications arise when one adds to the mechanics of material distortion, the emphasis of our earlier discussion, the consequences of contact slipping in the presence of friction. A detailed discussion of this subject is beyond the scope of our inquiry. However, there are several results, for specific contact models, that are relevant here.

6.6.1.1 The Mindlin Model; Shear Contact with Friction [18]
For a Hertzian contact, with friction coefficient μ and under normal load N, the equation for the shear strain that results from an oscillatory shear force T is

$$\varepsilon_\uparrow^*(x, Y) = 1 + (1 - Y)^{2/3} - 2 \left(1 - \frac{Y}{2} - \frac{x}{2}\right)^{2/3} , \quad \dot{T} > 0 , \tag{6.52}$$

$$\varepsilon_\downarrow^*(x, Y) = -\varepsilon^*(-x, Y)_\uparrow , \quad \dot{T} < 0 , \tag{6.53}$$

where $x = T/(\mu N)$ and $-Y \le x \le +Y$. These equations have their sterile appearance because we have incorporated the elastic constant, the Poisson ratio, and numbers into the scaled form ε^*. We put this equation of state in standard form, Eq. (6.15) or (6.29),

$$\varepsilon^* = \varepsilon_B^* + s(\dot{x})\varepsilon_H , \tag{6.54}$$

$$2\varepsilon_B^* = \varepsilon_\uparrow^* + \varepsilon_\downarrow^* , \tag{6.55}$$

$$2\varepsilon_H^* = \varepsilon_\uparrow^* - \varepsilon_\downarrow^* , \tag{6.56}$$

where we might read the subscript B (H) as *backbone* and *hysteretic*. Assume $Y \ll 1$ and develop a series expansion for ε_B^* and ε_H^* in x and Y. Find

$$\varepsilon_B^* = \frac{2}{3}x + \frac{1}{9}Yx + \frac{1}{81}x^3 + \dots , \tag{6.57}$$

$$\varepsilon_H^* = -\frac{1}{18}(Y^2 - x^2) + \dots \tag{6.58}$$

Of particular interest are the two terms that depend on the amplitude of the shear force, Y. To emphasize these we write

$$\varepsilon^* = \dots + \frac{1}{18}\left[2Yx - s(\dot{x})(Y^2 - x^2)\right] + \dots \tag{6.59}$$

and compare to Eq. (6.28). As the analytic forms are identical, we expect the same qualitative/quantitative behavior from a system described by the Mindlin model as we found above for a system described by the MG model.

6.6.1.2 An fcc Lattice of Hertz–Mindlin Contacts [10]
Nihei and coworkers have developed a model for a rock that involves an fcc lattice of Hertzian contacts. For the case of uniaxial strain, $x = \varepsilon_{zz}$, taken through a periodic loop, they find

$$\sigma_\uparrow^* = (1 - \mu)x^{3/2} + 2\mu[(1 + K_2)X - K_2 x]^{3/2}, \quad \dot{x} > 0 , \tag{6.60}$$

$$\sigma_\downarrow^* = (1 + \mu)x^{3/2} + 2\mu[(1 + K_2)X - K_2 Z]^{3/2} \\ - 2\mu[(1 + K_3)Z - K_3 x]^{3/2}, \quad \dot{x} > 0 , \tag{6.61}$$

where the upper (lower) limit of the strain is X (Z), $Z \le x \le X$, and the constants K_n are $K_3 = -1 - K_2$, $K_2 = -(1 + \mu K_1)/(2\mu K_1)$, and $K_1 = (1 - \nu/2)/(1 - \nu)$. As with Eq. (6.53), various numbers and elastic constants are incorporated in the definition of the * quantities. We make one further scaling by dividing by $X^{3/2}$ and redefining σ^*. We have

$$\sigma_\uparrow^* = (1 - \mu)x^{3/2} + 2\mu[(1 + K_2) - K_2 x]^{3/2}, \quad \dot{x} > 0 , \tag{6.62}$$

$$\sigma_\downarrow^* = (1 + \mu)x^{3/2} + 2\mu[(1 + K_2) - K_2 Z]^{3/2} \\ - 2\mu[(1 + K_3)Z - K_3 x]^{3/2}, \quad \dot{x} > 0 , \tag{6.63}$$

with $Z = (Z/X) \le x = (x/X) \le 1$. Proceeding as above we put these equations in standard form:

$$\sigma^* = \sigma_B^* + s(\dot{x})\sigma_H , \tag{6.64}$$

$$2\sigma_B^* = \sigma_\uparrow^* + \sigma_\downarrow^* , \tag{6.65}$$

$$2\sigma_H^* = \sigma_\uparrow^* - \sigma_\downarrow^* . \tag{6.66}$$

We consider small strain change from the maximum strain $x = 1$ and develop σ^* in Taylor series, $x = 1 - \delta$, $Z = 1 - \Delta$, $0 \le \delta \le \Delta \ll 1$, with the result

$$\sigma_B = 1 + C_{10}\delta + A_{11}\Delta\delta + C_{20}\delta^2 + C_{30}\delta^3 + \ldots \tag{6.67}$$

$$\sigma_H = A_{11}\delta(\Delta - \delta) + \ldots \tag{6.68}$$

The coefficients C_{nm} lead with the numerical factor appropriate to the expansion of $(1 - x)^{3/2}$ and have corrections proportional to μ:

$$C_{10} = -\frac{3}{2} + 3\mu\left(K_2 + \frac{1}{2}\right),$$

$$C_{20} = \frac{3}{8} - \frac{3}{4}\mu\left(K_2 + \frac{1}{2}\right), \tag{6.69}$$

$$C_{30} = \frac{1}{16} - \frac{1}{16}\mu\left(1 + 3K_2 + 3K_2^2 + 2K_2^3\right).$$

The coefficient giving the linear frequency shift in σ_B as well as the leading term in σ_H is

$$A_{11} = \mu\frac{3}{4}K_2(1 + K_2). \tag{6.70}$$

For the choice $\nu = 0.16$, $\mu = 0.2$, $K_2 = -2.7826$, and $A_{11} = 0.74405$. If we measure δ from $\Delta/2 = \bar{u}$, $\delta = (\Delta/2) + u$, we have

$$\sigma^* = \cdots + A_{11}[2\bar{u}u - s(\dot{u})(\bar{u}^2 - u^2)] + \ldots \tag{6.71}$$

This is to be compared to Eqs. (6.59), (6.29), and (6.28).

An fcc lattice of Hertzian contacts is a specific mechanical model for an interganular bond that has exactly the same properties as the MG model. The MG model, deliberately constructed without microscopic/mesoscopic underpinnings, leads to behaviors discussed above that we can immediately take over as the behaviors of an fcc lattice of Hertzian contacts because of Eqs. (6.59) and (6.71).

There is a final pedagogic point. The result in Eq. (6.58), for shear forces only, has no term in x^2. A term that scales as x^2 in the force/strain relation would come from a term in the energy that scales as x^3 and would permit a $t + t \rightarrow t$ wave mixing process. There is no such process in continuum elasticity, and the mechanical model is consistent with this. On the other hand, for the force/strain relation for an fcc lattice of Hertzian contacts there is a term in δ^2, Eq. (6.67), a $l + l \rightarrow l$ process is allowed and consistent with continuum elasticity. See the discussion in Chapter 3.

6.6.2
The Masing Rules [16]

The Masing rules are a model used for elastic systems with complex quasistatic stress/strain relations. This model is qualitatively similar to the MG model but different in detail as it uses the memory of only the most recent strain reversal:

1. If the strain reversed from decreasing to increasing at strain/stress point $(\varepsilon_1, \sigma_1)$, then the stress/strain relationship is

$$\sigma_\uparrow = \sigma_1 + G \frac{\varepsilon - \varepsilon_1}{1 + \frac{L}{2}(\varepsilon - \varepsilon_1)} , \tag{6.72}$$

while $\dot{\varepsilon}$ remains greater than 0.

2. If the strain reversed from increasing to decreasing at strain/stress point $(\varepsilon_2, \sigma_2)$, then the stress/strain relationship is

$$\sigma_\downarrow = \sigma_2 + G \frac{\varepsilon - \varepsilon_2}{1 - \frac{L}{2}(\varepsilon - \varepsilon_2)} , \tag{6.73}$$

while $\dot{\varepsilon}$ remains less than 0.

Here G is the linear elastic constant, $\varepsilon = \varepsilon_1 + \delta\varepsilon$ in Eq. (6.72), with $L\delta\varepsilon \ll 1$, and L is related to the stress at large strain; $\varepsilon \to \infty$ in Eq. (6.72) gives $\sigma \to 2G/L$. A strain protocol with a sequence of strain reversals results in stress given by alternate use of Eqs. (6.72) and (6.73).

For the case of a resonant bar in steady state we have $(\varepsilon_1, \sigma_1) = -(\varepsilon_2, \sigma_2) = (\varepsilon_0, \sigma_0)$. When σ_B and σ_H are calculated as above, Eq. (6.64), the result is

$$\sigma_B = G\varepsilon \frac{1}{\left(1 + \frac{L}{2}(\varepsilon_0 + \varepsilon)\right)\left(1 + \frac{L}{2}(\varepsilon_0 - \varepsilon)\right)} , \tag{6.74}$$

$$\sigma_H = G \frac{L}{2} \frac{\left(1 + \frac{L}{2}\varepsilon_0\right)}{(1 + L\varepsilon_0)} \frac{\varepsilon_0^2 - \varepsilon^2}{\left(1 + \frac{L}{2}(\varepsilon_0 + \varepsilon)\right)\left(1 + \frac{L}{2}(\varepsilon_0 - \varepsilon)\right)} . \tag{6.75}$$

The small numbers are $L\varepsilon_0$ and $L\varepsilon$, $-L\varepsilon_0 \leq L\varepsilon \leq L\varepsilon_0$. Carrying out the Taylor series expansion in these variables leads to

$$\sigma_B = G\varepsilon \left[1 - \gamma + \left(1 - \frac{P}{4}\right)\gamma^2 + \ldots\right] \tag{6.76}$$

and

$$\sigma_H = \frac{1}{2}G\varepsilon_0 P \left[\gamma - \frac{3}{2}\gamma^2 + \ldots\right] , \tag{6.77}$$

where $\gamma = L\varepsilon_0$ and $P = 1 - \varepsilon^2/\varepsilon_0^2$. We have

$$\sigma = G \left(\varepsilon - L\varepsilon_0\varepsilon + s(\dot{\varepsilon})2L(\varepsilon_0^2 - \varepsilon^2) + O(\gamma^2)\right) , \tag{6.78}$$

which is to be compared with Eq. (6.30), to which it is identical. The Masing rules have the same physics as the MG model featured above. The measure of the strength of the nonlinear coupling L can be found from quasistatic measurements that probe large strain, see below Eq. (6.73).

6.6.3
The Endochronic Formalism [17]

Endochronic formalisms for the behavior of elastic systems are used widely in circumstances where plastic motion is prominent. Minster and coworkers have adapted this formalism for the description of geophysical materials, and we will sketch a simplified version of their treatment of a Berea sandstone. We begin with a set of three coupled equations for the stress, strain, and plastic strain, γ,

$$d\varepsilon = \frac{d\sigma}{K} + d\gamma , \tag{6.79}$$

$$\sigma = \sum_{i=1}^{M} q_i(\gamma : \dots) , \tag{6.80}$$

$$q_i = k_i \int_0^z e^{-\alpha_i(z-z')} s(\dot{\gamma}(z')) dz' , \tag{6.81}$$

where $s(\dot{\gamma})$ is the sign of $\dot{\gamma}$ and

$$dz = |d\gamma| = s(\dot{\gamma}) d\gamma . \tag{6.82}$$

The first of these equations describes the strain that results from the stress in the usual way, σ/K, plus an additional additive strain, the plastic strain γ. The plastic strain is the analog of the discontinuous displacements of the MG model, that is, it is a displacement that appears without apparent additional stress. The second equation states that the stress is built up as a sum over stress elements q_i that are found from the third equation. It is the third equation that requires attention. The derivative of q_i with respect to z can be written in the form

$$\frac{dq}{d\gamma} = -\alpha[s(\dot{\gamma})q - \Lambda] , \tag{6.83}$$

where $\Lambda = k/\alpha$ and we have dropped the index i until required. This equation has the same formal structure as the equation of motion for a strain element in the model of slow dynamics introduced by Lomnitz, Eq. (7.2). *Chronic* is in the description of this formalism because of the analogy between z in Eqs. (6.81) or γ in Eq. (6.83) and time.

While in general the endochronic formalism is hard to work with, in the special case of a resonant bar, one can get quite far analytically. When Eqs. (6.79)–(6.81) are used to describe the material in a resonant bar, ε, γ, and σ undergo cyclic motion $(-\varepsilon_0, -\Delta, -\sigma_0) \rightarrow (\varepsilon_0, \Delta, \sigma_0) \rightarrow (-\varepsilon_0, -\Delta, -\sigma_0)$. Thus we integrate the two forms of Eq. (6.83) ($dq = -\alpha q d\gamma + k d\gamma$ from $-\Delta$ to γ and $dq = \alpha q d\gamma + k d\gamma$ from Δ to γ) to find

$$q_\uparrow(\gamma) = \Lambda - \left(\Lambda + q(\Delta)\right) e^{-\alpha\Delta} e^{-\alpha\gamma} , \quad \dot{\gamma} > 0 , \tag{6.84}$$

$$q_\downarrow(\gamma) = -\Lambda + \left(\Lambda + q(\Delta)\right) e^{-\alpha\Delta} e^{+\alpha\gamma} , \quad \dot{\gamma} < 0 . \tag{6.85}$$

Setting $q_\uparrow(\gamma = \Delta) = q(\Delta)$ in Eq. (6.85) leads to $q(\Delta) = \Lambda \tanh(\alpha\Delta)$,

$$q_\uparrow(\gamma) = \Lambda \left[1 - \mathrm{sech}(\alpha\Delta)e^{-\alpha\gamma} \right] , \tag{6.86}$$

$$q_\downarrow(\gamma) = \Lambda \left[-1 + \mathrm{sech}(\alpha\Delta)e^{-\alpha\gamma} \right] , \tag{6.87}$$

and

$$\sigma_\uparrow(\gamma) = \sum_{i=1}^{M} \Lambda_i \left[1 - \mathrm{sech}(\alpha_i\Delta)e^{-\alpha_i\gamma} \right] , \tag{6.88}$$

$$\sigma_\downarrow(\gamma) = \sum_{i=1}^{M} \Lambda_i \left[-1 + \mathrm{sech}(\alpha_i\Delta)e^{+\alpha_i\gamma} \right] . \tag{6.89}$$

These are equations for the stress on strain increase and strain decrease as a function of the strength of the plastic strain, γ, which is unknown. To proceed we find an analogous set of equations for the strains $\varepsilon_\uparrow$ and $\varepsilon_\downarrow$.

Carry out integrations of Eq. (6.79) to find

$$\varepsilon_\uparrow(\gamma) = -\varepsilon_0 + \gamma + \Delta + \frac{1}{K}(\sigma_\uparrow + \sigma_0) , \tag{6.90}$$

$$\varepsilon_\downarrow(\gamma) = \varepsilon_0 + \gamma - \Delta + \frac{1}{K}(\sigma_\downarrow - \sigma_0) . \tag{6.91}$$

The requirement $\varepsilon_\uparrow(\gamma = \Delta) = \varepsilon_0$ leads to $\sigma_0 = K(\varepsilon_0 - \Delta)$. Note $q_\uparrow(\gamma) = -q_\downarrow(-\gamma)$, $\sigma_\uparrow(\gamma) = -\sigma_\downarrow(-\gamma)$, and $\varepsilon_\uparrow(\gamma) = -\varepsilon_\downarrow(-\gamma)$.

To find ε as a function of σ, solve Eq. (6.88) for γ as a function of σ to be used in Eq. (6.90) [and Eq. (6.89) for use with Eq. (6.91)]. It is convenient to solve Eq. (6.88) from the Taylor series expansion of the RHS (to second order in γ as this is the leading NL term),

$$\begin{aligned} \sigma_\uparrow &= \sum_{i=1}^{M} k_i \left[\gamma + \alpha_i \frac{1}{2}(\Delta^2 - \gamma^2) + O(\gamma^3) \right] , \\ &= \overline{K} \left[\gamma + \frac{a_1}{2}(\Delta^2 - \gamma^2) + O(\gamma^3) \right] , \end{aligned} \tag{6.92}$$

where $\overline{K} = \sum k_i$ and $a_n = \sum k_i \alpha_i^n / \overline{K}$. Rearrange

$$\gamma = \frac{\sigma}{\overline{K}} - \frac{a_1}{2}(\Delta^2 - \gamma^2) + O(\gamma^3) \tag{6.93}$$

and iterate to find

$$\gamma = \frac{\sigma}{\overline{K}} - \frac{1}{2}a_1 \left(\frac{\sigma_0^2}{\overline{K}^2} - \frac{\sigma^2}{\overline{K}^2} \right) + O(\gamma^3) , \tag{6.94}$$

where we have used $\Delta = \sigma_0/\overline{K}$. [The result $\Delta = \sigma_0/\overline{K}$ is the leading approximation to $\sigma(\Delta) = \sigma_0 = \sum \Lambda \tanh(\alpha\Delta)$ from above Eq. (6.86).] Using this equation for γ in Eq. (6.90) leads to

$$\varepsilon_\uparrow = -\varepsilon_0 + \frac{1}{K_0}(\sigma + \sigma_0) - \frac{1}{2}a_1 \left(\frac{\sigma_0^2}{\overline{K}^2} - \frac{\sigma^2}{\overline{K}^2} \right) , \tag{6.95}$$

where

$$\frac{1}{K_0} = \frac{1}{K} + \frac{1}{\overline{K}} \,. \tag{6.96}$$

Similar treatment of Eqs. (6.89) and (6.91) leads to

$$\varepsilon_\downarrow = \varepsilon_0 + \frac{1}{K_0}(\sigma - \sigma_0) + \frac{1}{2}a_1 \left(\frac{\sigma_0^2}{\overline{K}^2} - \frac{\sigma^2}{\overline{K}^2} \right) \,. \tag{6.97}$$

With the "up" and "down" components of the stress/strain relation in hand we proceed as above and form the "backbone" and hysteretic parts:

$$\varepsilon = \varepsilon_B + s(\dot{\varepsilon})\varepsilon_H \,,$$
$$2\varepsilon_B = \varepsilon_\uparrow + \varepsilon_\downarrow \,, \tag{6.98}$$
$$2\varepsilon_H = \varepsilon_\uparrow - \varepsilon_\downarrow \,,$$

with the results

$$\varepsilon_B = \frac{\sigma}{K_0} + O(\gamma^3) \tag{6.99}$$

and

$$\varepsilon_H = -\frac{1}{2}\hat{a}_1 \left(\frac{\sigma_0^2}{K_0^2} - \frac{\sigma^2}{K_0^2} \right) + O(\gamma^3) \,, \quad \hat{a}_1 = a_1 K_0^2/\overline{K}^2 \,. \tag{6.100}$$

Thus we have

$$\varepsilon = \frac{1}{K_0} \left[\sigma - s(\dot{\varepsilon})\frac{1}{2}\frac{\hat{a}_1}{K_0}\left(\sigma_0^2 - \sigma^2 \right) + O(\gamma^3) \right] \,, \tag{6.101}$$

to be compared to Eqs. (6.59), (6.29), and (6.28). In contrast to the models above, the MG model, models based on the behavior of Hertzian contacts, and the Masing rule models, while the endochronic model yields attenuation proportional to the strain field, it has no first-order change in the modulus proportional to the strain field.

Before we leave the endochronic model it is useful to interpret the results we have found. From Eqs. (6.99) and (6.96) it is apparent that the elasticity of the system is that of a parallel arrangement of hysteretic strain elements in series with a conventional strain element. The individual hysteretic elastic elements have stress that depends on the plastic strain:

$$q_i = k_i \left[\gamma + s(\dot{\gamma})\frac{\alpha_i}{2}\left(\Delta^2 - \gamma^2 \right) + O(\gamma^3) \right] \,. \tag{6.102}$$

This interpretation might suggest a number of models, able to be built up from considering the arrangement of various elastic elements and having the same physics as the endochronic model, that are based on a less abstract formalism.

References

1 McCall, K.R. and Guyer, R.A. (1994) Equation of state and wave propagation in hysteretic nonlinear elastic materials. *J. Geophys. Res.*, **99**, 23887–23897.

2 Holcomb, D.J. (1981) Memory, relaxation, and microfracturing in dilatant rock. *J. Geophys. Res.*, **86**, 6235–6248.

3 Bertotti, G., Mayergoyz, I. (eds) (2005) *The Science of Hysteresis I–III*, Academic Press, San Diego.

4 Boudjema, M., Santos, I., McCall, K.R., Guyer, R.A., and Boitnott, G.N. (2003) Linear and nonlinear modulus surfaces in stress space, from stress-strain measurements on Berea sandstone. *Nonlinear Proc. Geophys.*, **10**, 589–597.

5 Helbig, K. and Rsaolofosaon, P.N.J. (2000) *Anisotropy* 2000, Society of Expl. Geophys. Tulsa.

6 Preisach F. (1935) On magnetic aftereffect. *Z. Phys.*, **94**, 277–302.

7 Mayergoyz, I.D. (1985) Hysteresis models from the mathematical and control points of view. *J. Appl. Phys.*, **57**, 3803–3805.

8 Guyer, R.A. and McCall, K.R. (1996) Capillary condensation, invasion percolation, hysteresis, and discrete memory. *Phys. Rev. B*, **54**, 18–21.

9 Johnson, D.L. and Norris, A.N. (1997) Rough elastic spheres in contact: Memory effects and the transverse force. *J. Mech. Phys. Solids*, **45**, 1025–1036.

10 Nihei, K.T., Hilbert, L.B., Cook, N.G.W., Nakagawa, S., and Myers, L.R. (2000), Frictional effects on the volumetric strain of sandstone. *Int. J. Rock Mech. Min. Sci.*, **37**, 121–132.

11 Aleshin, V. and van den Abeele, K. (2005) Micro-potential model for stress-strain hysteresis of micro-cracked materials. *J.Mech. Phys. Solids*, **53**, 795–824.

12 Guyer, R.A., McCall, K.R., and Boitnott, G.N. (1995) Hysteresis, discrete memory, and nonlinear wave propagation in rock: a new paradigm. *Phys. Rev. Lett.*, **74**, 3491–3494.

13 http://en.wikipedia.org/wiki/ Homodyne_detection, 15 May 2009.

14 Van den Abeele, E.-A., Johnson, P.A., Guyer, R.A., and McCall, K.R. (1997) On the quasi-analytic treatment of hysteretic nonlinear response in elastic wave propagation. *J. Acoust. Soc. Am.*, **101**, 1885–1898.

15 Gusev, V. (2002) Theory of non-collinear interactions of acoustic waves in an isotropic material with hysteretic quadratic nonlinearity. *J. Acoust. Soc. Am.*, **111**, 80–94.

16 Puzrin, A., Frydman, S., and Talesnick, M. (1995) Normalized nondegrading behavior of soft clay under cyclic simple shear loading. *J. of Geotech. Eng.*, **121**, 836–843.

17 Xu, H., Day, S.M., and Minster, B. (2000) Hysteresis and two-dimensional nonlinear wave propagation in Berea Sandstone. *J. Geophys. Res.*, **105**, 6163–6175.

18 Laurent, J. and Jia, X. Université Paris-Est Marne-la-Vallee (manuscript in preparation).

7
The Dynamics of Elastic Systems; Fast and Slow

In this chapter we take up the problem of the dynamics of elastic elements. In this context, *fast dynamics,slow dynamics,* and *conditioning* appear. As we proceed these words and possibly related jargon will be given careful meaning. We begin with a discussion of fast and slow dynamics for the elastic elements in linear elastic systems, Section 7.1; quasistatic response is treated in Section 7.1.1 and dynamic response in Section 7.1.2. In Section 7.2 we allow nonlinear elastic elements, the FDEE of Chapter 6, to have finite time dynamics and extend the ideas from Section 7.1 to these. *Fast* and *slow* dynamics are defined in these discussions in terms of the response time of elastic elements. A second use of *slow dynamics* is in the description of the time evolution of auxiliary fields that influence the elastic state of a system, temperature, saturation, and the *conditioning* field, a field caused by a large amplitude AC pump. This is discussed in Section 7.3. In Section 7.3.1 we describe the dynamics of the conditioning field. In Section 7.3.2 we describe the dynamics of response to the temperature field (Figure 7.1). (Further, extensive discussion of *fast dynamics*, in the context of resonant bar measurements, appears in Chapter 11.) Our summarizing remarks are in Section 7.4.

7.1
Fast/Slow Linear Dynamics

For the discussion of fast and slow *linear* dynamics of elastic elements we use the treatment of Lomnitz [1]. While this does not give the most general description of phenomena, it has the great advantage of being based on a physical idea that survives generalization well. Based on the EMT result from Chapter 5, we assume that the strain in a material at stress Σ can be built up as the sum of the strains in a set of instances of an elastic element, that is,

$$\varepsilon(\Sigma) = \langle \varepsilon_i(\Sigma) \rangle = \frac{1}{N} \sum_{i=1}^{N} \varepsilon_i(\Sigma), \qquad (7.1)$$

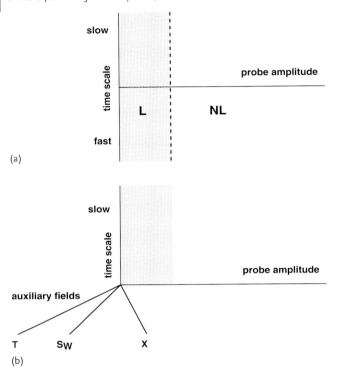

Fig. 7.1 Fast/Slow Dynamics. Fast and slow dynamics occur in two distinct contexts: (a) when the dynamics of elastic elements is fast/slow (this may be linear or nonlinear) and (b) when the elastic elements respond to a slow auxiliary field as in the case of "conditioning" (where the slow auxilliary field X is created by a large-amplitude fast drive), temperature, and saturation.

where the index i labels instances of the elastic element. Following Lomnitz we assume that each instance of an elastic element obeys an equation of motion

$$\frac{d\varepsilon_i}{dt} = -\frac{1}{\tau_i}\left(\varepsilon_i - \frac{\Sigma}{A}\right) , \tag{7.2}$$

where all instances of an elastic element have the same elastic constant, A, but different relaxation times, τ_i.

7.1.1
Quasistatic Response

When the material is subject to a step in stress, $\Sigma(t) = \Sigma_0 \theta(t)$, $\varepsilon_i(t < 0) = 0$, we have

$$\varepsilon_i(t) = \frac{\Sigma_0}{A}\left(1 - e^{-t/\tau_i}\right) , \tag{7.3}$$

and

$$\varepsilon(t, \Sigma_0) = \frac{\Sigma_0}{A} \frac{1}{N} \sum_{i=1}^{N} \left(1 - e^{-t/\tau_i}\right) = \frac{\Sigma_0}{A} \frac{1}{N} \sum_{i=1}^{N} B_i(t) = \frac{\Sigma_0}{A} \langle B(t) \rangle, \qquad (7.4)$$

$$\langle B(t) \rangle = \frac{1}{N} \sum_{i=1}^{N} \left(1 - e^{-t/\tau_i}\right), \quad t \geq 0. \qquad (7.5)$$

1. Assume that the τ_i are distributed between $\tau_<$ and $\tau_>$ according to

$$p(\tau) = \frac{C}{\tau^{\mu}}, \qquad (7.6)$$

where $\mu > 1$, $\int d\tau \, p(\tau) = 1$, and $C = (1 - \mu)/(\tau_>^{1-\mu} - \tau_<^{1-\mu})$.

2. For $\tau_< < t < \tau_>$ make the simple approximation to B_i in Eq. (7.5):

$$\begin{aligned} B_i &= 1, \quad t > \tau_i, \\ B_i &= 0, \quad t < \tau_i. \end{aligned} \qquad (7.7)$$

3. Using Eq. (7.7) in Eq. (7.5) gives

$$\begin{aligned} \frac{\varepsilon}{\varepsilon_{\infty}} = \langle B(t) \rangle &= \int_{\tau_<}^{t} p(\tau) d\tau, \\ &= \frac{1 - r^{\mu-1}}{1 - r_0^{\mu-1}}, \quad \mu > 1, \\ &= \frac{\ln(1/r)}{\ln(1/r_0)}, \quad \mu = 1, \end{aligned} \qquad (7.8)$$

where $\varepsilon_{\infty} = \Sigma_0/A$, $r = \tau_</t$, $\tau_< \leq t \leq \tau_> \rightarrow \tau_</\tau_> = r_0 \leq r \leq 1$. In Figure 7.2 we show an example of $\langle B(t) \rangle$ as a function of t for $\tau_< = 0.0001$, $\tau_> = 1.0$, $2 \leq \mu \leq 1.1$, and $\mu = 1$, which leads to $\log(t)$ behavior.

Several remarks are in order.

1. On a time scale faster than any of the relaxation times, $t \ll \tau_<$, none of the elastic elements can respond to the applied stress, $B_i \approx 0$, $\forall i$, and $\varepsilon \sim 0$.

2. On a time scale slower than any of the relaxation times, $t \gg \tau_>$, all of the elastic elements can respond to the applied stress $B_i \approx 1$, $\forall i$, and $\varepsilon \sim \varepsilon_0$.

3. The important time scales for seeing evidence of the dynamics of the elastic elements are times within the domain of the relaxation time spectrum. Then what one sees depends importantly on the nature of the relaxation time spectrum.

   a. For $\mu = 2$ there are relatively more short relaxation times than long, and the evolution of ε to ε_{∞} occurs early in the time interval that covers the relaxation time spectrum. See Figure 7.2.

   b. For $\mu \rightarrow 1$ there are equal numbers of relaxation times in all decades of the relaxation time spectrum, and the evolution of ε to ε_{∞} occurs uniformly over all times that cover the relaxation time spectrum. See Figure 7.2.

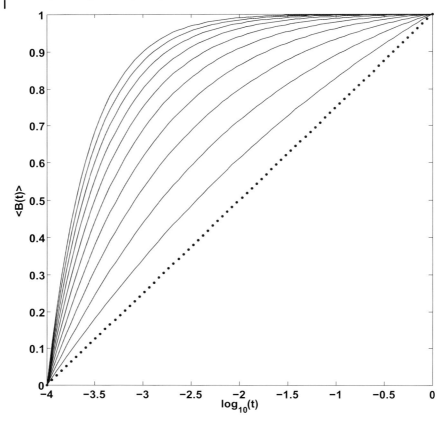

Fig. 7.2 Strain response to Transient Stress. The value of the average strain, Eq. (7.8), as a function of time for $p \propto \tau^{-\mu}$, $0.0001 \leq \tau \leq 1$ and $2 \leq \mu \leq 1.1$. For $\mu = 2$, $\langle B(t) \rangle$ rises quickly to 1, the upper curve. For $\mu \rightarrow 1$, $\langle B(t) \rangle \sim \log(t)$, the lowest, dashed, curve.

    c. The ideas that are descriptive here can be made formal by using the cumulative probability, $P(\tau) \propto \int^{\tau} p(z)dz$. In the case of $p \propto 1/\tau$, $P(\tau) \propto \ln(\tau)$ and is uniform in all decades.

In the present context we see that when there is a set of instances of elastic elements, with response characterized by a broad spectrum of time scales, some of the elastic elements are *fast* and respond to an applied stress, and some of the elastic elements are *slow* and are unable to notice an applied stress. A given elastic element is judged *fast/slow* depending on the relationship of the time scale on which it responds to a time scale of observation. In the example here on time scale $t \ll \tau_<$, all of the elastic elements show evidence of what is termed *slow dynamics*. On time scale $t \gg \tau_>$ all of the elastic elements show evidence of what is termed *fast dynamics.* The important point is that the notions of *fast/slow dynamics* are defined relative to a careful statement of the circumstance of observation. Roughly, when the time scale of observation influences the outcome of observation, one is seeing evidence of *slow dynamics*. For example, in an early measurement Lomnitz found the strain relaxation of a sandstone to be approximately $\log(t)$ for $1 \text{ s} \leq t \leq 10^6 \text{ s}$ [2].

7.1.2
AC Response

Suppose the elastic elements, described by Eq. (7.2), are subject to a sinusoidal drive, $\Sigma(t) = \Sigma_0 \sin \omega t$. Then in steady state, that is, after all transients have died away, the strain ε_i is

$$\varepsilon_i = a_i \sin \omega t + b_i \cos \omega t \,,$$

$$a_i = \frac{1}{1 + (\omega \tau_i)^2} \varepsilon_\infty \,,$$

$$b_i = \frac{-\omega \tau_i}{1 + (\omega \tau_i)^2} \varepsilon_\infty \,,$$

(7.9)

where $\varepsilon_\infty = \Sigma_0/A$. From the equation for the *in phase* response we define an effective frequency-dependent compliance using the average over instances

$$\frac{1}{A(\omega)} = \int d\tau \, p(\tau) \frac{1}{A_i} \,,$$

$$\frac{1}{A_i} = \frac{a_i}{A} = \frac{1}{1 + (\omega \tau_i)^2} \frac{1}{A} \,.$$

(7.10)

From the energy lost per period (per unit volume) we have

$$\Delta W = \int d\tau \, p(\tau) \Delta W_i \,,$$

$$\Delta W_i = \int_0^{T_0} \Sigma(t) d\varepsilon_i(t) = -\pi \omega \Sigma_0 b_i$$

$$= \frac{\omega \tau_i}{1 + (\omega \tau_i)^2} \pi \Sigma_0 \varepsilon_\infty \,,$$

(7.11)

where $T_0 = 2\pi/\omega$. Carrying out the average over instances for the case $p(\tau) \propto 1/\tau$ and using the definition of $1/Q$ above, Eq. (6.42), we have

$$\frac{A}{A(\omega)} = 1 + \frac{\ln\left(\frac{1 + (\omega \tau_>)^2}{1 + (\omega \tau_<)^2}\right)}{\ln\left(\frac{\tau_>}{\tau_<}\right)} \,,$$

$$\frac{1}{Q} = \frac{\arctan(\omega \tau_>) - \arctan(\omega \tau_<)}{\ln\left(\frac{\tau_>}{\tau_<}\right)} \,.$$

(7.12)

In Figures 7.3 and 7.4 we show $A/A(\omega)$ vs. ω and $1/Q$ vs. ω, respectively.

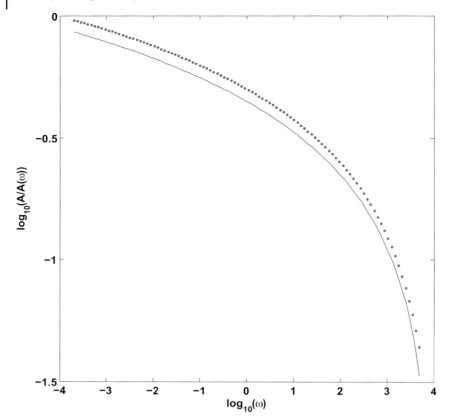

Fig. 7.3 Strain response to AC stress, elastic constant. The value of the effective elastic constant, $A(\omega)$, as a function of ω for the case $\mu = 1$ and $0.0001 \leq \tau \leq 1$, Eq. (7.12). As $\omega \to 0$, $A(\omega)$ decreases because more and more elastic elements are able to follow the force; there is more displacement per unit force.

From Figure 7.3 we see that $A(\omega)$ decreases, becomes softer, as $\omega \to 0$. This is understood by the same type of argument as that offered above. In the current context the frequency ω sets the time scale. As $\omega \to 0$ the time that separates slow from fast becomes larger and larger, and more and more elastic elements are able to respond in phase with the AC drive. There is more strain (displacement) per cycle; the system is elastically softer.

From Figure 7.4 we see that $1/Q$ is essentially independent of ω except near the edges of $p(\tau)$. The reason for this is that $b(\tau)$, Eq. (7.9), has a maximum at $\tau = \omega^{-1}$, where its value is $1/2$. More than 50% of the contribution to $1/Q$ comes from τ close to $1/\omega$, that is, in the range $1/(3\omega) \leq \tau \leq 3/\omega$. As ω changes, the values of τ that contribute importantly to $1/Q$ change, but not $1/Q$; it is independent of ω.

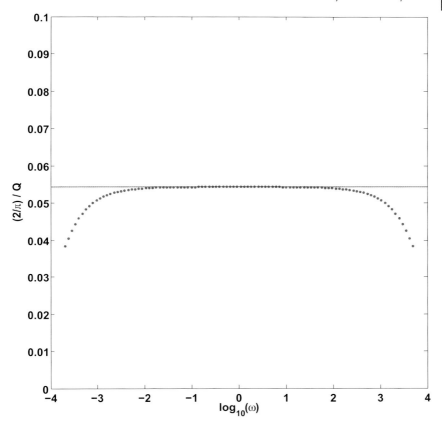

Fig. 7.4 Strain response to AC stress, damping. The value of the damping constant, $1/Q$, as a function of ω for the case $\mu = 1$ and $0.0001 \leq \tau \leq 1$, Eq. (7.12). Because the primary contribution to the attenuation comes from strains that respond on time scale ω^{-1}, the attenuation is independent of ω for $\tau_< \leq \omega^{-1} \leq \tau_>$.

Let us restate these results in different language. From Eq. (7.9) we can write ε_i in terms of an amplitude and a phase

$$\varepsilon_i = \varepsilon_\infty R_i \sin{(\omega t - \phi_i)}\,, \tag{7.13}$$

where $R_i = 1/\sqrt{1 + (\omega\tau_i)^2}$ and $\phi_i = \omega\tau_i R_i$. At $\omega\tau_i = 1$, $R_i = 1/\sqrt{2}$ and $\phi_i = \pi/4$. At fixed ω in the interior of the range of $1/\tau$ there are three groups of elastic elements, Figure 7.5:

1. $\tau_< < \tau < 1/(3\omega)$ ($\phi_i \ll \pi/4$): the motion of these elastic elements is essentially **in phase** with the drive. These elastic elements exhibit *fast dynamics*.
2. $3/\omega < \tau < \tau_>$ ($\phi_i \gg \pi/4$): the motion of these elastic elements is essentially **out of phase** with the drive. These elastic elements exhibit *slow dynamics*.

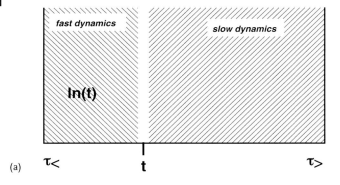

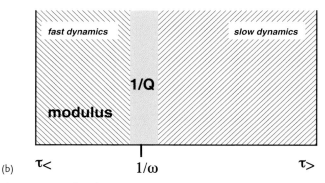

Fig. 7.5 Linear slow dynamics. (a) At time t following a step in stress the elastic elements with time scale less than t are able to respond (those with time scale greater than t are essentially inert). So the strain builds up proportional to the number of such elastic elements, as $\log(t)$ for $\mu = 1$. (b) When the system is subject to an AC drive at frequency ω, the argument in (a) applies to elastic elements with time scale less than ω^{-1}. The attenuation involves primarily elastic elements that respond on time sale ω^{-1} and is independent of ω. Elastic elements with $\tau < t$ ($\tau < \omega^{-1}$) are exhibiting "fast dynamics", and those with $\tau > t$ ($\tau > \omega^{-1}$) are exhibiting "slow dynamics".

3. $1/(3\omega) \le \tau \le 3/\omega$ ($\phi_i \approx \pi/4$): the motion of these elastic elements, neither *slow* nor *fast*, contributes most of the energy loss.

Often one finds that the distribution of relaxation rates is more complex than Eq. (7.6), for example, of the form

$$q(\tau) = (1 - f)\delta(\tau) + f\,p(\tau),\tag{7.14}$$

where $p(\tau)$ is given by something like Eq. (7.6). That is, there is a finite fraction, $1 - f$, of essentially infinitely fast relaxation rates. So that in an experiment there is an instantaneous response, fast dynmaics, independent for all practical purposes of time scale. The elastic elements so responding contribute nothing to the attenuation, $1/Q$, and nothing to the frequency shift and appear as an instantaneous net strain in a transient measurement of amplitude $(1 - f)\varepsilon_\infty$. Thus a factor f in each of Eqs. (7.8), (7.10), and (7.11) accounts for what happens.

The treatment of the dynamics of linear elastic elements, due to Lomnitz, has a number of generalizations [3]. The results of these generalizations can be cast as a quantitative relationship between Q, the frequency shift, and the amplitude of $\log(t)$. So also can Eqs. (7.8) and (7.12) [4, 5].

7.2
Fast Nonlinear Dynamics

In the discussion of the dynamical response of finite displacement only elastic elements (FDEE) in Chapter 6 we called attention to the explicit assumption of instantaneous dynamics. That is, the elastic elements changed state instantaneously and the end-to-end displacement across an elastic element followed suit. The elastic elements we are concerned with are typically associated with mesoscopic physical structures. Indeed on some time scale their behavior is effectively instantaneous. There are other time scales on which one would expect to capture the motion of the elastic elements. In this section we describe a particular example of dynamical generalization of the model of FDEE.

Consider the one-dimensional chain of elastic elements described by the equation

$$m\ddot{u}_k + \frac{m}{\tau}\dot{u}_k = \gamma\left[(u_{k+1} - u_k - b_k) - (u_k - u_{k-1} - b_{k-1})\right] + F_k, \qquad (7.15)$$

where $b_k = b_0 + \eta_k\Delta b$ and $\eta_k = \pm 1$. This is a linear chain of elastic elements, like those in Eq. (6.1), that enforce a separation between their ends according to the state of the elastic element, specified by η, Figure 7.6. The displacement variables are associated with a set of sites separated by b_0 and the state variables are associated with the bonds to the right of the sites. In the picture from Chapter 6 the dynamics of the η_k is the instantaneous dynamics given by the rules above Eq. (6.11) that is followed in a (Σ_c, Σ_o) Preisach space. The physics in the rules is that an elastic element undergoes a finite displacement at particular values of the stress. This stress is the stress carried by the elastic element, $\Sigma_k \propto \gamma(u_k - u_{k-1} - b_0)$ in leading approximation. One way to proceed is to give η_k a dynamics that is driven by an energy landscape is Σ_k. A model called the dynamic McCall–Guyer model (DMG model) does this [6]. We will look at this model in some detail.

In the DMG model, Eq. (7.15) for u_k is complemented by an equation of motion for η_k:

$$\dot{\eta}_k = \frac{1}{\tau_\eta}\left(-\alpha_k + \beta_k\eta_k - \eta_k^3\right), \qquad (7.16)$$

where

$$\beta_k = \frac{T_c^k - T_o^k}{2}, \qquad (7.17)$$

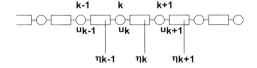

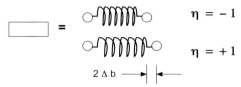

Fig. 7.6 Dynamic hysteretic elastic elements. Hysteretic elastic elements may enforce two displacements between their ends. The state variable that determines which displacement is to be enforced can have a dynamics, Eq. (7.16), that involves motion in an energy landscape that depends on the force carried by the elastic element (Figure 7.7).

$$\alpha_k = \frac{T_c^k + T_o^k}{2}, \tag{7.18}$$

$$T_c^k = \tanh\left(\kappa(f_c^k - f_k)\right), \tag{7.19}$$

$$T_o^k = \tanh\left(\kappa(f_o^k - f_k)\right), \tag{7.20}$$

and $f_k = \gamma(u_{k+1} - u_k - b_0)$. The state variable η_k moves in an energy landscape shaped by f_k, the force carried by the elastic element. Of course, in this case we have an energy landscape characterized by two force values, (f_c, f_o), instead of two stress values. The parameter κ controls the sharpness of the response to f_k. In the absence of other understanding, the state variable is given Brownian dynamics, Eq. (7.16).

The hysteretic response of the elastic element is determined by the behavior of the state of the elastic element, η. The state of the elastic element is taken to depend on the internal force, $f_k = \gamma(u_{k+1} - u_k - b_0)$, that it is required to support. In mechanical equilibrium we have $f_k = F$, where F is the applied force. Thus $u_{k+1} - u_k = b_0 + F/\gamma$. The force and state conventions are that $F \gg 0$ is a force of tension, to the right, and that $\eta = +1$ under large applied tension. Similarly, under a large applied compression, $F \ll 0$, $\eta = -1$. Thus the state, an Ising-like variable, has a sign that is the sign of the applied force F.

The equation of motion for η is complicated by the involvement of the internal force, f_k, in the determination of α and β. The essential content of this equation of motion can be seen by replacing the internal force in the element by a force prescribed from outside, $f_k = F$. Then the RHS of the equation for η can be regarded as coming from a potential. We write Eq. (7.16) as

$$\tau_\eta \dot{\eta} = -\frac{\partial V(\eta, F)}{\partial \eta}, \tag{7.21}$$

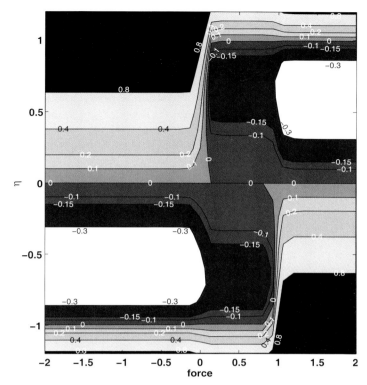

Fig. 7.7 Energy landscape. The energy landscape for $(f_c, f_o) =$ (1.0, 0.0), see Eq. (7.20) and the discussion below, Eq. (7.22). (Please find a color version of this figure on the color plates)

where

$$V(\eta, F) = \alpha(F)\eta - \beta(F)\frac{\eta^2}{2} + \frac{\eta^4}{4} \qquad (7.22)$$

and $\alpha(F)$ and $\beta(F)$ are found from Eqs. (7.16)–(7.20) with f_k replaced by F. The potential $V(\eta, F)$ is the sum potential of a particle in an external field of strength $\alpha(F)$ and a potential symmetric in $\eta \rightarrow -\eta$, a ϕ^4 potential. Because $f_o < f_c$, $\beta(F)$ is always greater than or, at most, equal to zero ($f_o = f_c$). For η near 0 the β term in the potential always pushes η away from 0. For large $|F|$, $T_c \approx T_o \approx -\text{sign}(F)$, $\alpha(F) \rightarrow -\text{sign}(F)$, $\beta(F) \rightarrow 0$, and $V(\sigma, F) \rightarrow -\text{sign}(F)\eta + \eta^4/4$. At large negative F (the elastic element is under compression) η is pushed toward -1, and at large positive F (the elastic element is under tension) η is pushed toward $+1$.

In Figure 7.7 we show a contour plot of the potential $V(\sigma, F)$ as a function of F and η for the case $(f_c, f_o) = (0.0, 1.0)$ and $\kappa = 16.0$. This potential, an energy landscape, has a well at $\eta = +1$ for $\eta > 0$ that terminates abruptly near $F \approx f_o = 0.0$ and a second well for $\eta < 0$ that terminates abruptly near $F \approx f_c = 1.00$.

In Figure 7.8 we show η as a function of time (lower panel) from the solution to Eq. (7.16) for the case that F is a function of time that "chirps" up in frequency (upper panel). The parameters for the force on η are those for the energy landscape in Figure 7.7; we use $\tau_\eta = 1/16$ and the chirp is from period $T_1 = 1$ to $T_6 = 1/32$. In Figure 7.9 we plot η as a function of F. The trajectory in this plot could be overlaid on the energy landscape in Figure 7.7. What one sees is a sequence of hysteresis loops that begin with loops that have a range of ± 1, $(\omega\tau_\eta)/2\pi = 1/16$, and as the force moves more rapidly evolve toward loops that are at almost constant η, $(\omega\tau_\eta)/2\pi = 1.2$. This particular elastic element exhibits fast dynamics if driven at frequencies $(\omega\tau_\eta)/2\pi < 1/4$; it exhibits slow dynamics if driven at frequencies $(\omega\tau_\eta)/2\pi > 1/4$. Thus for the model in Eqs. (7.15)–(7.20) the system will exhibit fast nonlinear dynamics involving those elastic elements for which τ_η is such that $\omega\tau_\eta \ll 1$. Numerical simulations of Eqs. (7.15)–(7.20) yield the result shown in Figure 7.10. Analysis of these resonance curves for the frequency at resonance and for the dependence of the amplitude at resonance on amplitude, Eq. (6.39),

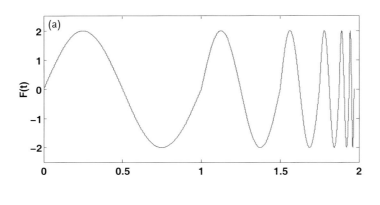

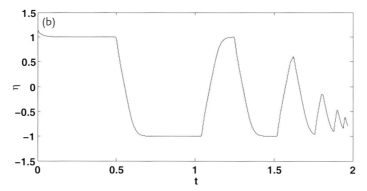

Fig. 7.8 Force protocol and η response. (a) The force protocol, a chirp as a function of time. (b) The behavior of η as a function of time, Eq. (7.16). See Figure 7.9.

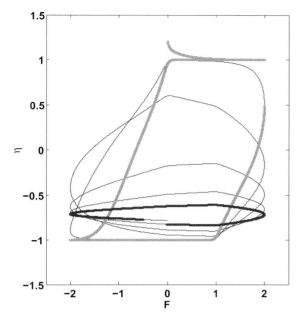

Fig. 7.9 Motion of the state variable. The motion of the state variable, η, in an energy landscape depends on the time scale of the force on η and on the time scale τ_η. Here the force is chirped, Figure 7.8, from $T_0 = 1$ (broad light gray) to $T_0 = 1/32$ (dark gray) for $\tau_\eta = 1/16$, where T_0 is the period of the force oscillation. The response of η goes from "fast" to "slow" as T_0 decreases.

find agreement with the results of the model in Chapter 6. These numerical results are achieved in the limit $(\omega\tau_\eta)/2\pi \gg 1$ for all elastic elements, that is, in the fast dynamics limit. This fast dynamics is the fast dynamics of the hysteretic elastic elements participating in the elastic response. Fast is defined by the frequency of the drive, for example, in a resonant bar by a frequency determined by the bar length and the linear elastic constant. Such a frequency has no fundamental meaning. Thus it remains to carry out investigations of hysteretic elastic elements on a time scale that exposes something of their intrinsic dynamics (see the discussion of quasistatic hysteresis in Chapter 10).

Just as there are hysteretic elastic elements that are fast compared to a typical resonant bar frequency, there may well be hysteretic elastic elements that respond slowly to the demands of an external force. In the picture here such a force induces change in the energy landscape in which the elastic element resides. There are few experiments to allow much more than speculation in this regard. However, as must be clear from the preceding discussion, a set of such elastic elements could cause slow time evolution in the response to a change in applied stress. This response

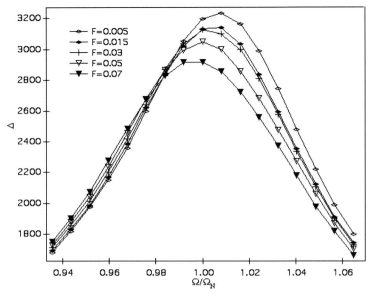

Fig. 7.10 Resonance of a Hysteretic Chain [6]. The amplitude of the response of a hysteretic chain, Eqs. (7.15)–(7.20), near resonance as a function of frequency for five values of the drive force. The amplitude is in the form of amplitude per unit force and the frequency is scaled by the resonance frequency at low amplitude. The chain has $N = 500$ elastic elements, each with intrinsic damping $Q_0 = 2000$ and $\tau_\eta = 1.0$. The chain has a natural frequency $\Omega_N = \pi/N \approx 0.0063 \ll 1/\tau_\eta$, so that all elastic elements are in the fast dynamics limit. The Preisach space for the elastic elements was filled uniformly with $0 \leq |f_c| \leq 1.5$ and $0 \leq |f_o| \leq 1.5$. The amplitude dependence of the resonance frequency and of the amplitude at resonance, $1/Q$, are in accord with the predictions of Eqs. (6.38) and (6.39).

would be distinguishable from that due to slow linear dynamics because it should have hysteretic features.

7.3
Auxiliary Fields and Slow Dynamics

In the discussion above the actors have been the elastic elements (linear elastic elements in Section 7.1, hysteretic elastic elements in Section 7.2). We turn now to a description of slow dynamics of the elastic state of a system that is not in the dynamics of the elastic elements but the dynamics of auxiliary fields to which the elastic elements couple. There are at least three experimental realizations of this, a slow dynamics in response to changes in temperature, a slow dynamics in response to changes in saturation, and a slow dynamics in response to a large-amplitude AC strain field.

Suppose that the displacement field of an elastic system in a resonant bar obeys the equation

$$\varrho \ddot{u} = K(X) \frac{\partial^2 u}{\partial x^2} + F(t), \tag{7.23}$$

where $F(t)$ is the external drive and $K(X)$ is an elastic constant that depends on an auxiliary field in the form

$$K = K_0 (1 + \lambda_X X). \tag{7.24}$$

Here λ_X is a constant and X the auxiliary field. For an example of how this form of $K(X)$ might arise in traditional nonlinear elasticity see the discussion surrounding Eq. (3.23). We first treat the case of an auxiliary field driven by a large-amplitude AC strain field, a pump [7, 8].

7.3.1
X = The Conditioning Field

To deal with an auxiliary field that responds to a large-amplitude AC strain field we rewrite Eq. (7.23) to include two applied fields, a large-amplitude pump field $F_P(t)$ at frequency ω_P (whose strain field drives the auxiliary field) and a low-amplitude probe field f_p at frequency ω_p (to test the elastic state of the system):

$$\varrho \ddot{u} = K(X) \frac{\partial^2 u}{\partial x^2} + F_P(t) + f_p(t). \tag{7.25}$$

We separate the displacement field into two parts:

$$\varrho \ddot{U} = K_0 \frac{\partial^2 U}{\partial x^2} + F_P(t), \tag{7.26}$$

$$\varrho \ddot{u} = K_0 (1 + \lambda_X X) \frac{\partial^2 u}{\partial x^2} + f_p(t), \tag{7.27}$$

where we have dropped the term in $1 + \lambda_X X$ in the equation for U to emphasize our interest in the effect of X on the probe field. The influence of X on the field causing X, the pump U, is of minor consequence. We look at the behavior of the probe field. The probe field, u, sees the presence of the pump field, U, through the way in which the pump field drives the field X. We employ a leaky ratchet model [9] for the coupling of X to U:

$$\frac{dX}{dt} = -r_1 (X - \beta U) \theta (\beta U - X) - r_2 X \theta (X - \beta U), \tag{7.28}$$

where β is a constant and the rates r_1 and r_2 obey $\omega_P \gg r_1 \gg r_2$. The way in which the terms on the RHS work is illustrated in Figure 7.11. When X is below βU, it is driven toward βU at the rate $r_1 \ll \omega_P$, that is, **slowly**, so that in a single cycle of the pump there is little change in X. When X is above βU, it relaxes toward 0 at the

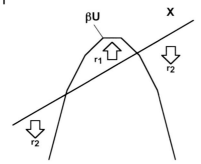

Fig. 7.11 Equation of motion of X, AC drive. The auxiliary field X seeks βU if $X < \beta U$ and it seeks 0 if $X > \beta U$. Because the decay to 0 for $X > \beta U$ is extremely slow, X ratchets up to that point at which the impulse received while $\beta U > X$ balances the decay while $\beta U < X$.

rate $r_2 \ll r_1 \ll \omega_P$, that is, **very slowly**, so that little of the increase in X that occurs during the part of the drive for which $\beta U > X$ is lost. Then what happens is that X ratchets up toward βU until a balance is reached between the gain in X during the small part of a pump cycle for which $\beta U > X$ and the loss in X over the rest of the pump cycle (most of it).

In Figure 7.12 we show the result of the solution to Eq. (7.28) for $\beta = 1$, $U = U_0 \sin(\omega_P t)$, $U_0 = 0.10$, and $(r_2, r_1, \omega_P) = (0.01, 0.2, 2\pi)$. Because of the choice of $\omega_P = 2\pi$, the units of time are the period of the pump. Early in time X climbs to an asymptotic value of approximately 0.078, where it stays until $t = 200$, at which time the pump is turned off. Thereafter X decays toward zero at the rate r_2. In the two insets the behavior of X in two time segments is shown, (a) early in time while X is ratcheting up and (b) later in time when X is in steady state. While X is in steady state, held there by the pump, the probe senses the presence of the auxiliary field through the frequency shift in a resonance probed by ω_p, Eq. (7.27). Decay of the resonance frequency from its shifted value is seen at times beyond $t = 200$ as X decays to zero.

In the treatment here we have been content to use an X field characterized by two time scales. In highly inhomogeneous systems one expects a spectrum of both time scales, that is, r_1^{-1} and r_2^{-1}, and nonexponential long time behavior.

From the form of Eq. (7.27) the changes in X feed back into the equation for U. However, in known examples of "conditioning", the sobriquet associated with the phenomena under discussion, the frequency shift due to X is modest. If one insists on using ω_P at a low-frequency resonance in order to have a large pump field U, and if one insists on maintaining U at constant amplitude, it will be necessary to account for the changes in K brought about by X. The initial "conditioning" experiments used ω_P on a resonance and ω_p swept over the same resonance to test the conditioned state. This is not necessary, and quite possibly it complicates the matter.

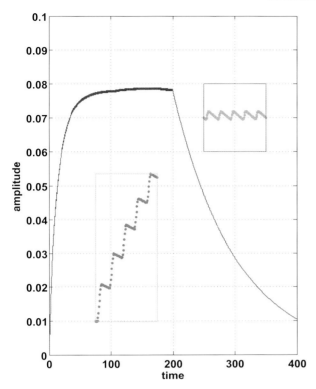

Fig. 7.12 Time evolution of X, AC drive. X as a function of t for a protocol in which $U = 0.10 \sin(2\pi t)$ for $0 \leq t \leq 200$ and zero thereafter, $r_1 = 0.2$, $r_2 = 0.01$. Two insets, early in time and in steady state, show the behavior of X vs. t in detail.

The basic structure of the problem of the auxiliary field X as formulated in Eq. (7.27) features the use of two displacement fields, one to manipulate X, the pump field U, and a second to test for the consequence of X, the probe field u. To the degree that these two displacement fields are independent, that is, to order the coupling due to $K(X)$, a number of NMR-style scenarios suggest how to learn about X. One uses the pump field to set the elastic state and the probe field to investigate the time evolution of the elastic state. In Chapter 11 the sequence of experiments that led to our present understanding of X are described.

Finally, in contrast to the discussion in Sections 7.1 and 7.2, the discussion here is not about the slow dynamics of elastic elements but about the slow dynamics of an auxiliary field that the elastic elements see. As formulated, it would seem that a fast dynamical displacement field, the pump field, causes the auxiliary field, the field with slow dynamics.

7.3.2
X = Temperature

In the continuum picture of an elastic system, the coupling of temperature to a dynamic displacement field involves two steps, as in Eq. (2.37). The temperature through the agency of the thermal expansion creates a quasistatic thermal strain field, $\varepsilon_T = \alpha \delta T$, to which a dynamic strain field couples via the cubic nonlinearity, $\beta \varepsilon_T \varepsilon$, as in Figure 3.1. Indeed the qualitative behavior seen in experiment is roughly as suggested by Eq. (2.37), but there are striking slow dynamics effects that commands one's attention [10]. One extreme model that captures some of these effects will be described here. It is appropriate to first make the point that temperature, unlike stress, cannot be imposed on a physical system instantaneously. For a stress field *instantaneously* means a time scale of order L/c, where L is the sample size and c is the velocity of sound, for example, $L/c \approx 10^{-5}$ s for $L = 1$ cm and $c = 10^5$ cm/s. For a temperature field *instantaneously* means a time scale of order L^2/D, where D is the nominal thermal diffusion constant, for example, $L^2/D \approx 100$ s for $L = 1$ cm and $D = 0.01$ cm^2/s for sandstone [11]. When we refer to slow dynamics involving temperature for a material, we mean phenomena on time scales at least long compared to L^2/D.

Suppose that the displacement field of an elastic system in a resonant bar obeys the equation

$$\varrho \ddot{u} = K(X)\frac{\partial^2 u}{\partial x^2} + F(t),$$

$$(7.29)$$

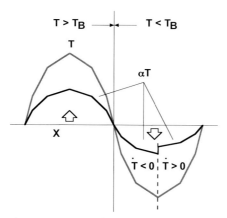

Fig. 7.13 Equation of motion of *X*, *T* drive. The auxiliary field *X* seeks $\alpha_>(T - T_B)$ for $T > T_B$, Eq. (7.32). It seeks $\alpha_<(T - T_B)$ for $T < T_B$ but in a way that depends on the sign of the rate of change of *T*. Approximately, the behavior is as if *X slips back* on reversal in the sign of $\dot{T}$.

where $F(t)$ is the external drive and $K(X)$ is an elastic constant that depends on an auxiliary field in the form

$$K = K_0(1 + \lambda_X X).\tag{7.30}$$

Here λ_X is a constant and X, the auxiliary field, is driven by temperature. While these equations are similar to Eqs. (7.23) and (7.24), this does not mean that we believe the auxiliary field here is the same auxiliary field as that discussed above. We have no evidence of a connection between the two [8].

For the equation of motion of the auxiliary field X we take

$$
\begin{aligned}
\frac{dX}{dt} = &-r_>[X - \alpha_>(T - T_B)]\theta(T - T_B) \\
&- r_<[X - \alpha_<^-(T - T_B)\theta(-\dot{T}) - \alpha_<^+(T - T_B)\theta(\dot{T})]\theta(T - T_B),
\end{aligned}\tag{7.31}
$$

$$\dot{T}_B = -r_B(T_B - T).\tag{7.32}$$

The terms on the RHS of this equation require explanation, see Figure 7.13.

1. The auxiliary field X is driven by the temperature in a way that depends on the sign of the temperature relative to a background temperature, T_B, that is very slowly varying. In the illustration here we will fix T_B at the ambient temperature T_0 from which an experiment starts.
2. For $T > T_B$ the field X approaches the value, $\alpha_> T$, on time scale set by $r_>$.
3. For $T < T_B$ the behavior is:
   a. for $\dot{T} < 0$, that is, T decreasing to below T_B, X approaches the value, $\alpha_<^- T$, on a time scale set by $r_<$. This value differs from that for $T > T_B$, $\alpha_<^\pm < \alpha_>$ and $r_< < r_>$;
   b. for $\dot{T} > 0$, that is, T increasing back toward T_B, X collapses relatively quickly toward T_B.
4. There are three time scales, $r_B \ll r_< < r_>$, and two thermal expansions, $\alpha_<^\pm < \alpha_>$.

There are two striking features of this model: (1) the asymmetry, here with respect to the sign of $T - T_B$, having an analog in Eq. (7.28) and (2) the rate dependence for $T < T_B$. One necessarily prescribes a T protocol in describing an experiment. As an illustration of the kind of behavior that is seen, we show the solution to Eqs. (7.31)–(7.32) in Figure 7.14: (a) the temperature protocol T vs. t and (b) $\Delta K / K_0 \propto -X$ vs. t. The parameters that produced this particular result are in the figure caption. The asymmetry between $T > T_B$ and $T < T_B$ is apparent in this result as is the dependence on rate for $T < T_B$. Look for behavior qualitatively similar to this in experiments (Chapter 10).

7.4
Summary

Mesoscopic elastic elements are by their nature large compared to microscopic elastic elements, which are atomic in scale. Rate processes involving microscop-

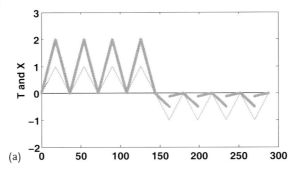

(a)

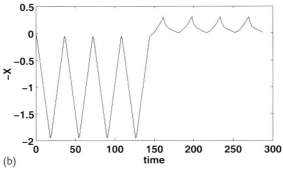

(b)

Fig. 7.14 Time evolution of X, T drive. X as a function of t for a protocol (a) in which T (light gray) is above T_B for four cycles and then below T_B for four cycles. (b) The behavior of $\lambda_X X$ ($\lambda_X = -1$) as a function of t (expect the frequency shift of an oscillator described by Eq. (7.29) to follow $\lambda_X X$). This result is for the case $(a_>, a_{<+}, a_{<-}) = (2.0, 0.5, 0.125)$ and $r_>, r_{<-}, r_{<+}) = (1.5, 0.125, 0.5)$.

ic elastic elements proceed at rates set by microscopic energy scales, for example, $\exp(-\varepsilon_0 / k_B T)$, and microscopic frequencies, for example, $\omega_D = k_B \Theta_D / \hbar$. Rate processes involving mesoscopic elastic elements can be much slower since mesoscopic energy scales are much larger than microscopic energy scales. Also, mesoscopic elastic elements are subject to fields that are of a different character from the fields on microscopic elastic elements. Consequently there are, for both the response of mesoscopic elastic elements and for the fields that affect mesoscopic elastic elements, a broad range of time scales. The discussion above calls attention to a number of models for these phenomena with particular emphasis on the slow dynamics behavior that might be expected.

References

1 Lomnitz, C. (1956) Creep measurements in igneous rock. *J. Geol.*, **64**, 473–479.

2 Lomnitz, C. (1957) Linear dissipation in solids. *J. Appl. Phys.*, **28**, 201–205.

3 Kjartansson, E. (1979) Constant Q-wave propagation and attenuation. *J. Geophys. Res.*, **84**, 4737–4748.

4 Pandit, B.I. and Savage, J.C. (1973) An Experimental test of lomnitz's theory of internal friction in rocks. *J. Geophys. Res.*, **78**, 6097–6099.

5 Lienert, B.R. and Manghnani, M.H. (1990) The relationship between Q^-_E and dispersion in extensional modulus, *E. Geophys. Res. Lett.*, **17**, 677–680.

6 Capogrosso-Sansone, B. and Guyer, R.A. (2002) Dynamic model of hysteretic elastic systems. *Phys. Rev. B*, **66**, 224101.1–220101.12.

7 TenCate, J.A. and Shankland, T.J. (1996) Slow dynamics in the nonlinear elastic response of Berea sandstone. *Geophys. Res. Lett.*, **23**, 3019–3022.

8 TenCate, J.A., Smith, D.E., and Guyer, R.A. (2000), Universal slow dynamics in granular solids. *Phys. Rev. Lett.*, **85**, 1020–1023.

9 Vakhnenko, O.O., Vakhnenko, V.O., and Shankland, T.J. (2005) Soft-ratchet modeling of end-point memory in the nonlinear resonant response of sedimentary rocks. *Phys. Rev. B*, **71**, 174103–174117.

10 Ulrich, T.J. (2005) (thesis), University of Nevada, Reno. Ulrich, T.J. (2005) (thesis), University of Nevada, Reno.

11 Mavko, G., Mukerji, T., and Dvorkin, J. (2003) *The Rock Physics Handbook: Tools for Seismic Analysis of Porous Media*, Cambridge Univ. Press, New York.

8
Q and Issues of Data Modeling/Analysis

The emphasis of the discussion to this point has been on the elastic properties of materials. By that we mean the character and description of the forces within a material that bring about the motion of the displacement field. Little has been said about attenuation and physical processes that degrade the amplitude of the displacement field. Our concern with attenuation focuses on what it can tell us about the material under investigation. This will lead us to discuss several methods of data analysis that are particularly appropriate to nonlinear systems. We begin in Section 8.1 with a short review of attenuation in linear elastic systems. We distinguish between wave vector dispersion, Section 8.1.1, in which the energy continues to reside in the elastic system (albeit in wave vector states different from those launched) and energy extraction, Section 8.1.2, in which the energy leaves the elastic system. The flow of energy out of the elastic system requires coupling the elastic system to another system having the capacity to take up energy and carry it away. Linear coupling to the temperature field, to fluid configurations, to variegated *dashpots* all do this, Section 8.1.3. We discuss nonlinear attenuation for Section 8.2.1, properly called energy dispersion, energy continues to reside in the elastic system, albeit at different frequencies from those launched. And hysteretic attenuation, in which the elastic system is coupled nonlinearly to a dynamical system having the capacity to take up energy and carry it away, is discussed in Section 8.2.2. The next two sections are about attenuation (and its surrogate, Q^{-1}) more generally: Section 8.3, what can be learned from measurement of Q^{-1}, Sections 8.4.1 and 8.4.2, how to measure Q^{-1} in linear and nonlinear systems, respectively.

We turn to a careful discussion of the resonant bar in nonlinear elasticity in Section 8.5. Two schemes of data processing that do not require a model of the elastic system are described in detail: constant field analysis in Section 8.5.2.1, and resonance template matched filtering (RTMF) in Section 8.5.2.2.

Nonlinear Mesoscopic Elasticity: The Complex Behaviour of Granular Media including Rocks and Soil. Robert A. Guyer and Paul A. Johnson
Copyright © 2009 WILEY-VCH Verlag GmbH & Co. KGaA, Weinheim
ISBN: 978-3-527-40703-3

8.1
Attenuation in Linear Elastic Systems

8.1.1
Wave Vector Dispersion

A tone burst of amplitude A_0 is launched from $\boldsymbol{x}$ at time t_0 in a space that has weakly inhomogeneous elasticity. The tone burst, of sufficient duration that it can be characterized by the single frequency ω_0, is detected at $\boldsymbol{x}'$, where its amplitude is A. If the amplitude at $\boldsymbol{x}'$ is less than the anticipated result, for example, $A_0/|\boldsymbol{x} - \boldsymbol{x}'|$, then the wave traveling at ω_0 is said to be attenuated. There are two qualitatively different sources for this attenuation [1]. The first of these, which might be termed *wave vector dispersion*, arises from the scattering of the elastic wave from *static* inhomogeneity in the medium through which it propagates. For example, the equation of motion for the case at hand can be written

$$\varrho \frac{\partial^2 u}{\partial t^2} = \nabla \cdot \left(K(\boldsymbol{x}) \nabla u \right) = \overline{K} \nabla^2 u + \nabla \cdot \left(\delta K(\boldsymbol{x}) \nabla u \right) , \tag{8.1}$$

where $\delta K(\boldsymbol{x}) = K(\boldsymbol{x}) - \overline{K}$ and $\overline{K}$ is the average of $K(\boldsymbol{x})$ over the region of interest. The first term on the RHS describes a uniform system with elastic constant $\overline{K}$ and sound velocity $\overline{c}$. The second term, possibly handled using perturbation theory, describes the scattering of the displacement field from the static $\delta K(\boldsymbol{x})$ field. This scattering of u from $\delta K(\boldsymbol{x})$ produces no change in frequency or loss of elastic energy. Rather the displacement field, at ω_0, is scattered to many wave vectors. Some of the scattered displacement field may be multiply scattered and eventually arrive late to $\boldsymbol{x}'$ as the coda. Some sensible measure of the amplitude of arrival of the elastic wave at $\boldsymbol{x}'$ at $t \approx |\boldsymbol{x} - \boldsymbol{x}'|/\overline{c}$ will find a reduction in amplitude to below $A_0/|\boldsymbol{x} - \boldsymbol{x}'|$, that is, evidence of attenuation.

8.1.2
Extracting Elastic Energy

The physical situation described by Eq. (8.1) is in marked contrast to that described by Eqs. (3.84) in Chapter 3. Equations (3.84) describe the coupling of the displacement field to a dynamic field, the temperature field [2]. This dynamic field is capable of taking energy delivered to it by the displacement field and carrying it away. Let us look in some detail at Eq. (3.84) to see this physical process at work. Equations (3.85)–(3.86), to linear order, are

$$\left(\ddot{u} + \frac{1}{\tau_0} \dot{u} \right) = c_L^2 \frac{\partial}{\partial x} \varepsilon - c_L^2 \Gamma_0 \frac{\partial \delta T}{\partial x} , \tag{8.2}$$

$$\delta \dot{T} = D_T \frac{\partial^2 \delta T}{\partial x^2} - r \Gamma_0 \dot{\varepsilon} . \tag{8.3}$$

The coupling between u and δT, arising from the $K_0 \alpha \varepsilon \delta T$ term in the free energy, Eq. (3.80), is symmetric in the involvement of the coupling constant, $\Gamma_0 \propto \alpha$. If one looks for a solution to these equations in the form $\delta T \propto u \propto \exp i(kx - \omega t)$, one finds

$$\omega^2 + i\frac{\omega}{\tau_0} = c_L^2 k^2 \left(1 + \Gamma_0 \frac{r}{1 + iM} \Gamma_0\right), \tag{8.4}$$

where $r = K_0/(C_0 T_0)$ and $M = k^2 D_T/\omega$. The nature of the solution is controlled by M, a measure of (1) the time to carry thermal energy over a wavelength of the displacement field, $1/(k^2 D_T)$, compared to (2) the time the displacement field is in place, approximately a period of $2\pi/\omega$. For $M \to 0$, the RHS of Eq. (8.4) is real and there is no attenuation due to the coupling of u to δT. However, there is a stiffening of the system, and the isothermal elastic constant goes over to the adiabatic elastic constant, $c_L^2 \to c_L^2(1 + \Gamma_0^2 r)$[4]. For finite $M > 0$, some energy can be carried away by the temperature field during a period of the wave, and there is attenuation beyond that associated with $1/\tau_0$,

$$\frac{1}{\tau_0} \to \frac{1}{\tau_0} + \Gamma_0^2 r k^2 D_T. \tag{8.5}$$

The physics is different from that above. Here energy is transferred out of the elastic system and fully or partially returned (depending on the dynamics of the system to which it is transferred). When it is only partially returned, there is extraction of energy from the elastic system, attenuation.

8.1.3
Other

The two attenuation mechanisms we have discussed above are linear mechanisms; the energy extracted per unit of stored energy is independent of the amplitude of the displacement field [5]. This observation follows from the form of the equations we have used to describe wave vector dispersion, Eq. (8.1), and extracting elastic energy, Eq. (8.3). Earlier we passed by another linear attenuation mechanism, Section 7.1, where a system of elastic elements moved with damped motion toward equilibrium [6]. This situation is similar to that above, $u \leftrightarrow \delta T$, except that no explicit description of the system carrying the energy out of the elastic system is provided. Behind equations like Eq. (7.2) is the notion of energy transferred out of the elastic system to the system of phonons or some surrogate that is in contact with the thermal reservoir in which the system resides.

In rock physics a well-established mechanism of attenuation is the *squirt-flow* mechanism [7]. A strain field in the framework that defines a pore space can distort the pore space and force the fluid within it to undergo rearrangement. The fluid flows that accomplish this are coupled to the framework through the fluid viscosity. Energy is transferred out of the elastic system to fluid flow and through the working of the viscosity to the thermal reservoir in which the system resides.

It is apparent from this discussion that the study of attenuation mechanisms involves a detailed understanding of energy transfer and transport phenomena, for example, thermal transport, fluid transport, etc. For that reason it is our plan to give an extensive discussion of neither linear or nonlinear attenuation. Rather we sketch the simple ideas that underlie an understanding of some aspects of this subject.

8.2
Nonlinear Attenuation

In Chapter 3 we found that traditional nonlinear elasticity admitted a process in which a second harmonic was generated from the coalescence of two fundamentals, Eq. (3.46). Energy flows from the fundamental to the second harmonic, damping the fundamental. In Chapters 6 and 7 we encountered several models in which hysteresis was prominent. We showed there that a nonlinear attenuation resulted.

8.2.1
Nonlinear Dampling: Traditional Theory

Let us examine the situation associated with Eq. (3.46). A strain field source at $x = 0$ (amplitude A_1, frequency ω_1, and wave vector $k_1 = \omega_1/c$) broadcasts into a material to the right of $x = 0$, which has cubic anharmonicity, the β term in Eq. (3.31). As a consequence, a wave at frequency $\omega_2 = 2\omega_1$ and amplitude A_2 builds up to the right of $x = 0$ with amplitude proportional to x, $A_2 \propto \beta(k_1 A_1)^2 x$. Let us assume that only the amplitudes A_1 and A_2 are present in the elastic system. Energy conservation requires that the energy density at x at t

$$e(x) = \frac{1}{2} K \varepsilon_1^2(x) + \frac{1}{2} K \varepsilon_2^2(x) , \tag{8.6}$$

(where K is the elastic constant, $\varepsilon_1 = k_1 A_1$ and $\varepsilon_2 = k_2 A_2$), be the same as the energy density at $x + dx$ at $t + dx/c$ as both waves move with the speed, c. Writing $A_1(x + dx) = A_1 + \delta A_1$ and $A_2(x + dx) = A_2 + \delta A_2$ we have

$$k_1^2 A_1 \delta A_1 = -k_2^2 A_2 \delta A_2 . \tag{8.7}$$

But the physics behind $A_2 \propto \beta(k_1 A_1)^2 x$ is that the amplitude A_2 builds up in dx by $\delta A_2 \propto \beta(k_1 A_1)^2 dx$. Thus $(k_2 = 2k_1)$

$$\frac{\delta A_1}{A_1} \sim -\beta^2 k_1^4 A_1^2 x dx . \tag{8.8}$$

Because A_1 is delivering energy to A_2, we write $A_1(x) = A_1(0) \exp\{- \int_0^x \alpha_1(z) dz\}$ or

$$\alpha_1(x) = \beta^2 k_1^4 A_1^2 x , \tag{8.9}$$

an attenuation coefficient for A_1 that scales as A_1^2 and with x. Results of this kind are typical in nonlinear acoustics. Nonlinear acoustics, in which the nature of the nonlinearity is not in doubt, is a highly developed and very successfully applied discipline [8].

In this simple example elastic energy is transferred among the modes of the elastic system while remaining within it. In analogy with the language suggested above, Eq. (8.1), we would call the processes caused by traditional nonlinear elasticity *frequency dispersion*.

8.2.2
Nonlinear Damping: Hysteretic Elasticity

In Chapter 6 we found that the model of McCall and Guyer [9] for FDEE gave rise to an equation of state (EOS) of the form, Eq. (6.30),

$$\frac{\Sigma}{K_0} = \varepsilon + \gamma \varepsilon_m \varepsilon + s(\dot{\varepsilon}) \frac{\gamma}{2} \left(\varepsilon_m^2 - \varepsilon^2 \right) + \dots \tag{8.10}$$

The second term in this EOS, $\propto s(\dot{\varepsilon})$, describes hysteresis and is responsible for the damping found in the discussion of the modes of a resonant bar, Eq. (6.39). This damping is due to the network made by the elastic elements. In the hysteretic force/displacment relations that characterize the elastic elements in the FDEE model, an infinitesimal stress change, $\sigma_c \to \sigma_c + \delta\sigma$, triggers finite displacement, $+a \to -a$, and an infinitesimal stress change, $\sigma_o \to \sigma_o + \delta\sigma$, triggers the reverse nite displacement, $-a \to +a$, $\sigma_o < \sigma_c$, Figure 6.1. In the dynamic version of this model, Chapter 7, the rate at which the finite displacement takes place is controlled by a viscous damping [10]. These models contain no evidence of the mechanism of energy transfer out of the elastic system. However, whether the simple or more elaborate model is involved, there is a set of excitations generated by friction that take up the energy that leaves the elastic system and carry it to the temperature reservoir in which the system resides.

From the discussion in Section 6.6 we know that the functional form of the nonlinear damping of most hysteretic models of nonlinear elasticity is essentially the same. A stress/strain relation like that in Eq. (8.10) describes most models, albeit with different numbers. In the case of the MG model [9] and the Masing rules model [11], there is a quantitative connection between quasistatic data and the amplitude of the nonlinear damping, see Section 6.6.2 and the discussion in Chapter 10. In the Hertzian contact models of hysteretic elastic elements, a coefficient of friction controls the numbers [12]. In the endochronic model [13], parameters that characterize the plastic strain, having no apparent physical interpretation, determine the numbers.

8.3
Why Measure Q?

Why do you want to know Q? (1) A measurement of Q provides a direct measurement of the thermal diffusvity, for example, Eq. (8.5). (2) The relationship between seismic signals assembled from many Fourier components and propagated from x to x' depends on the relative attenuation of the Fourier components. (3) Verification of a particular model for the behavior of an elastic system, including coupling to pore fluid, rests on confirmation of a prediction for the pressure dependence of the attenuation. (4) Progressive damage of a worked part makes an additive contribution to Q that scales with damage beyond some threshold. And so on. The reasons may be many. It may be necessary to learn the numerical value of Q as precisely as possible, (1). It may be of interest to learn how Q scales with frequency (2), pressure (3), etc. It may be of interest to monitor the behavior of Q over time, (4).

8.4
How to Measure Q

8.4.1
Measurement of Q in a Linear System

In principle Q is defined by [5]

$$\frac{1}{Q} = \frac{1}{4\pi} \frac{\Delta E}{\overline{E}},$$

(8.11)

where ΔE is the energy lost per period and $\overline{E}$ is the average stored energy in a period. As one cannot buy an *energy lost per period* meter or an *average stored energy* gauge at the hardware store, this formal definition, useful in evaluating theoretical models, has to be given practical meaning.

1. For a propagating wave at frequency ω in a linear material (for which the attenuation produces a small change in amplitude in a period), the energy difference in adjacent periods is

$$\Delta E = \overline{E}(A) - \overline{E}(A - \delta A) = \overline{E}\left[1 - \left(\frac{A - \delta A}{A}\right)^2\right] = 2\overline{E}\frac{\delta A}{A}$$

(8.12)

and

$$\frac{1}{Q} = \frac{1}{2\pi} \frac{\delta A}{A}.$$

(8.13)

Thus Q can be measured by measuring the time evolution of the amplitude. Adjacent periods occur in adjacent material segments of length $\lambda = cT$, $T = 2\pi/\omega$. For $A = A(0)\exp{-\alpha x}$ the amplitudes in adjacent material segments differ by $\delta A = \alpha\lambda A$, $\alpha\lambda \ll 1$, so that

$$\frac{1}{Q} = \frac{\alpha}{k},$$

(8.14)

where $k = 2\pi/\lambda$. This can be rearranged to read

$$\alpha = \frac{k}{Q}, \quad \alpha^{-1} = l_\alpha = \frac{1}{2\pi}Q\lambda.$$

(8.15)

It is this last rewriting that leads to the rough rule of thumb *a wave is attenuated in Q wavelengths*. Measurement of the decay of amplitude with distance from its source is a primary method of measuring Q. In a linear system this measurement is independent of where it is carried out.

2. For a resonant bar of linear material driven at frequency ω the material comes to a nonequilibrium steady state. It is in this steady state that the material is studied. In the nonequilibrium steady state the rate at which energy is delivered to the material by the external forces in one period, ΔE_{in}, exactly balances the rate of energy loss from the material in the period, ΔE_{out}. Using Eq. (8.11) we have

$$\overline{E} = \frac{1}{4\pi} Q(\Delta E_{in}) . \tag{8.16}$$

Thus in the nonequilibrium steady state the material has elastic energy of order $Q(\Delta E_{in})$; it takes of order Q periods to build up the nonequilibrium steady state and of order Q periods for the nonequilibrium steady state to decay away. These are two more rules of thumb.

To look at the nonequilibrium steady state in somewhat more detail, we consider the equation of motion of a linear resonant bar, Eq. (3.54),

$$\varrho \left(\ddot{u} + \frac{1}{\tau_0} \dot{u} \right) = K \frac{\partial^2 u}{\partial x^2} + \frac{F}{a^2} \{ \delta(x - L/2) - \delta(x + L/2) \} , \tag{8.17}$$

where a^2 is the cross-section of the bar end. Employing the lumped element procedure from Section 6.3.1 we have

$$\ddot{U} + \frac{1}{Q_0} \dot{U} = -U + \frac{F}{M/2} , \tag{8.18}$$

where $t \to \omega_0 t$, $Q_0 = \omega_0 \tau_0 = Q/2$ (from $k^2 = \omega^2 + i\omega/\tau_0$, $\omega_0 \tau_0 \gg 1$), ω_0 is the frequency of the fundamental resonance of the bar and M is the mass of the bar. For $F = F_0 \sin(\omega t + \phi)$ we have $U = A \sin(\omega t)$ with

$$A(\omega) = \frac{1}{\sqrt{(\omega^2 - \omega_0^2)^2 + \frac{\omega^2 \omega_0^2}{Q_0^2}}} \frac{2F_0}{M} . \tag{8.19}$$

The quantity of interest, Q_0, is related to two simple features of $A(\omega)$. For $Q_0 \gg 1$ the maximum amplitude of $A(\omega)$ is at resonance, $\omega = \omega_0 - 1/(2Q_0^2) \approx \omega_0$, and

$$\frac{1}{Q_0} = \frac{2}{M\omega_0^2} \frac{F_0}{A(\omega_0)} ; \tag{8.20}$$

Q_0 is related to the amplitude at resonance as in $F = ma$, that is, $Q_0 F_0 = (M/2)\omega_0^2 A(\omega_0)$. For $\omega_{\pm} = \omega_0(1 \pm \sqrt{3}/2Q_0)$, $A(\omega_{\pm}) = A(\omega_0)/2$. Thus the frequency width from ampitude $A(\omega_0)/2$ to the left of the resonance to amplitude $A(\omega_0)/2$ to the right of the resonance, $\Delta\omega$, is related to Q_0 by

$$\frac{1}{Q_0} = \frac{1}{\sqrt{3}}\frac{\Delta\omega}{\omega_0} \, . \tag{8.21}$$

Finally, one can learn the Q of a mode of a resonant bar by letting the mode *ring down*. This event is described by Eq. (8.17) with $F = 0$. Using the homogeneous solution of the corresponding lumped element equation we have $\omega = \omega_0 + i/(2\tau_0)$ for $Q_0 \gg 1$ and $A(t) = A(0)\cos(\omega_0 t)\exp(-t/(2\tau_0))$. Thus the ratio of the amplitude in two adjacent periods is

$$\frac{A(t+T)}{A(t)} = e^{-\frac{T}{2\tau_0}} = e^{-\frac{\pi}{Q_0}} \tag{8.22}$$

or

$$\frac{1}{Q_0} = -\frac{1}{\pi}\ln\left(\frac{A(t+T)}{A(t)}\right) \, , \tag{8.23}$$

cf. Eq. (8.13) with $Q_0 = Q/2$. This measurement is independent of the time in the ring down.

For a linear material one can learn about Q_0 by studying the amplitude of the resonance curve at resonance, by examining the width of the resonance curve, or by letting the nonequilibrium steady state ring down. The latter two methods are preferred because only directly observed quantities are called for, amplitudes in Eq. (8.23) or frequencies in Eq. (8.21).

8.4.2
Measurement of Q in a Nonlinear System

In a nonlinear elastic material the basic forces at work in the material and the mechanisms of energy loss from the elastic system depend on the amplitude of the displacement field. In the context of a resonant bar this physics can be reduced to a rewriting of Eq. (8.19) as

$$A(\omega) = \frac{1}{\sqrt{(\omega^2 - \omega_0^2(A))^2 + \frac{\omega^2\omega_0^2(A)}{Q_0^2(A)}}}\frac{2F_0}{M} \, , \tag{8.24}$$

that is, the frequency ω_0 and Q_0 depend on amplitude. This means:

1. the ring down of a resonant bar from the nonequilibrium steady state may be more rapid at an early-time, large-amplitude, nonlinear energy loss than at a late-time, small-amplitude, linear energy loss;
2. the shape of a resonance curve is distorted, for example, Figures 6.11, 7.10, and 11.38 in Chapters 6, 7, and 11, respectively. Making use of the width as a measure of attenuation requires a model for the nonlinearity;

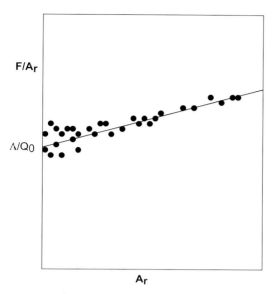

Fig. 8.1 Q^{-1} and amplitude at resonance. For $Q \gg 1$ the amplitude at resonance is proportional to Q so that Q^{-1}, the physical quantity, can be found from experiment by forming F/A_r, Eq. (8.25). The constants that turn the proportionality in Eq. (8.25) into values of Q^{-1} can be found from learning the linear (low amplitude) value of Q by suitable means.

3. the amplitude at resonance, A_r, gives Q at that amplitude, that is,

$$\frac{1}{Q(A_r)} = \frac{2}{M\omega_0^2} \frac{F}{A_r} \,. \tag{8.25}$$

This last method of measuring Q is particularly useful for extracting the nonlinear contribution to Q. The quantity F/A_r is easily formed from experimental data. If this quantity is plotted as a function of A_r, a curve like that in Figure 8.1 may result. For example, from Eq. (8.25) $F/A_r \propto 1/Q(A_r) = \Lambda/Q(A_r)$. The constant of proportionality can be found from knowing $Q(A_r \to 0) = Q(0)$, which can be reliably found by treating the low F data by the linear methods above. Application of this *amplitude at resonance* method is illustrated in Chapter 11, Section 11.3.

The discussion up till now has been somewhat loose. Below we describe a pair of data analysis procedures that resulted from the demands of nonlinear resonant bar data. These employ unconventional methods in an attempt to extract the maximum amount of information about the material of a resonant bar from data on it *without recourse* to a model for the bar material.

8.5
Resonant Bar Revisited

One of the most precise ways to learn the elastic state of a material is to do a resonant bar measurement. In the absence of complications, this method allows one to drive at a fixed frequency, achieve steady state, and measure the in-phase and out-of-phase components of the response at the frequency of the drive. A collection of such data for a range of drive amplitudes and frequencies is the basic experimental output. There are two aspects related to such experiments that we want to address. The first is the connection between the physical system, a resonant bar, and the model system that is used to ground the analysis of the data, and the second has to do with particular methods of data processing.

8.5.1
Modeling a Resonant Bar

A resonant bar, say a cylinder 2 cm in diameter and 20 cm long, has a fundamental mode resonant frequency of order 5 kHz, $f = c/(2L) = 5 \times 10^4$ Hz for $c = 2 \times 10^5$ cm/s. While this is the fundamental mode of a two-dimensional system (it is assumed that the end of the bar is driven uniformly so that the modes have at most a radial structure away from the bar axis [14]), it is typically approximated as the mode of a one-dimensional line. For compressive motion along a line one takes an equation of motion like

$$\frac{\partial^2 u}{\partial t^2} + \frac{1}{\tau}\frac{\partial u}{\partial t} = c^2 \frac{\partial^2 u}{\partial x^2} + F(t). \tag{8.26}$$

But usually, for data analysis, not even this equation is solved. It is conventional to analyze data as if each mode of a resonant bar can be mapped onto the mode of a lumped element, that is, the mode of

$$\ddot{u} + \frac{1}{\tau}\dot{u} = \omega_0^2 u + F(t), \tag{8.27}$$

where an ω_0^2 and τ are chosen for each bar mode. We saw in Chapter 6 that a lumped element equation, like Eq. (8.27), for a resonant bar could be derived from Eq. (8.26). Then, u in Eq. (8.27) is understood to be some measure of the average strain field in the bar. Nonlinearity in the response of the bar may be able to be introduced explicitly into Eq. (8.27) from a model of the bar material, as in Eq. (6.26), or it may be introduced phenomenologically,

$$\ddot{u} + \frac{1}{\tau_n(u)}\dot{u} = \omega_n^2(u)u + F(t), \tag{8.28}$$

where both the resonant frequency and the attenuation for a mode are taken to depend on the amplitude of the strain field. The u dependence of ω_0^2 is taken to represent the strain dependence of the elastic state of the bar, and u dependence

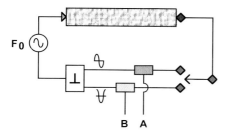

Fig. 8.2 Resonant bar data; in and out. In a resonant bar experiment in which one detects the steady-state amplitude at the drive frequency, the response has two components. The *in-phase* component is found from the time-averaged product of the drive signal and the detected signal. The *out-of-phase* component is found from the time-averaged product of the drive signal shifted by 90° and the detected signal, Eqs. (8.29)–(8.31).

of τ is taken to represent the strain dependence of the dissipative mechanisms operating in the bar. A particular choice of $\omega_0^2(u)$ could well give rise to a nonlinear attenuation that is in addition to any nonlinear attenuation that is introduced through $\tau(u)$. The result in Eq. (6.39) is an example of this.

A particular nonlinear lumped element model must be solved for the response. This is usually the detected amplitude at the frequency of the drive. When the drive is $F_0 \sin \omega_0 t$ there are three outputs, the in-phase amplitude of the response, the out-of-phase amplitude of the response, and the magnitude of the amplitude of the response. Formally these quantities are

$$a_0 = \langle S_0(t) | u(t) \rangle \,, \tag{8.29}$$

$$b_0 = \langle C_0(t) | u(t) \rangle \,, \tag{8.30}$$

$$u_0 = \sqrt{a_0^2 + b_0^2} \,, \tag{8.31}$$

where $S_0(t) = \sin \omega_0 t$ and $C_0(t) = \cos \omega_0 t$. This computational procedure, the projection procedure illustated in Chapter 6, mimics exactly the experimental detection procedure carried out by a lock-in amplifier, Figure 8.2, that is, homodyne style detection [15]. It is a, b, and u that are the model quantities that interface with experiments. Thus, between the elastic response of a resonant bar and the model it interfaces with is (1) dimensional reduction, $3 \rightarrow 2 \rightarrow 1 \rightarrow 0$, (2) the introduction of nonlinear lumped element parameters, (3) a solution to the resulting nonlinear lumped element equation, and (4) projection of the desired experimental quantities out of the solution.

8.5.2
Data Processing

A primary use of resonant bar measurements is to monitor changes in the elastic
state of the bar material. In this application it is not as important to have an exact
theory of how to go between Eqs. (8.26) and (8.27) as it is to have reason to believe
that there is a one-to-one correspondence between the elastic state of the bar material and u. When one wants to sense small changes in the elastic state, it is crucial
to carry out a data analysis that carefully fits the solution to Eq. (8.28) to resonant
bar data. The discussion here addresses this point.

In Figures 8.3 and 8.4 we show an example of resonant bar data and the results
of data analysis [16]. Figure 8.3 is a plot of the strain amplitude as a function of
frequency for 12 drive amplitudes. It is apparent from these curves that the resonance frequency, defined as the frequency of maximum amplitude, shifts to a lower

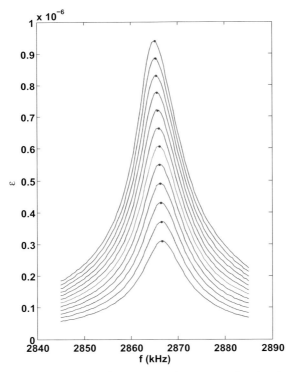

Fig. 8.3 Resonant bar data; Berea sandstone.
The magnitude of the strain, ε, as a function
of frequency, f, for 12 values of the drive (voltage to a PZT) for a Berea sandstone. There
are 81 frequency points spaced by 0.5 kHz
at each drive level. The resonance frequency
is found as the maximum of each constant
drive curve, Figure 8.4a. The value of Q^{-1} is
found from F/A_r, Eq. (8.25), and Q_0 from the
width of a resonance curve with very low drive,
Figure 8.4b [16].

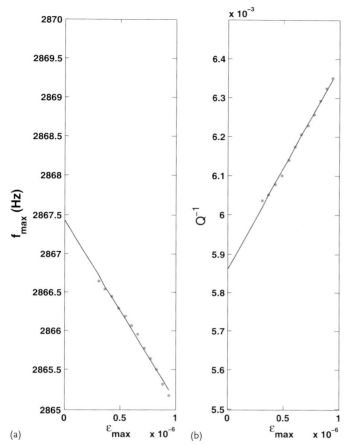

Fig. 8.4 Resonance frequency and Q^{-1}. The resonance curves
in Figure 8.3 yield (a) resonance frequencies and (b) values
of Q^{-1} as a function of the strain at resonance. The frequency
shift and shift in Q^{-1} is approximately linear in strain over the
strain range covered in the experiment [16].

frequency as the drive amplitude increases. Accompanying this frequency shift is
a distortion of the resonance curve away from symmetry about the resonance frequency. The resonance frequency as a function of the strain at resonance is plotted
in Figure 8.4a; a shift of about one part in a thousand is brought about by a strain
of order 10^{-6}. The attenuation, Q^{-1}, as a function of the strain at resonance, found
from the data using the amplitude-at-resonance method above, Eq. (8.25), is plotted
in Figure 8.4b. When data are as nice as those being discussed here or when a resonance curve is weakly distorted from those appropriate to linear systems, there is
little need for sophisticated data analysis.

Perhaps the first place one notices the need to exercise caution in getting numbers from resonance curves is in the calculation of Q^{-1}. A usual way of getting at this quantity is to make some measurement of the width of a resonance curve at half height, Eq. (8.21). It is apparent, for example, from Figure 3.7 (Chapter 3) and Figure 11.38 (Chapter 11) that distortion of the shape of a resonance curve puts this "definition" in doubt.

But we have to step back for a moment. The point of any of these measurements is to characterize the elastic state of a system. **For a nonlinear material a single resonance curve has information about many elastic states of the material.** If the parameters that characterize the elastic state have different values at different points on a resonance curve, then one wants to make pointwise measurements along the resonance curve. The focus on the resonance, the maximum in the curve, picks out only one elastic state, albeit the one of maximum strain, and disregards the others. A nonlocal measurement of the attenuation, say using the notion of line width, is some average over many different elastic states.

Finally there are 972 data points that probe the elastic state of the Berea sandstone shown in Figure 8.3 – 81 frequencies at each of 12 drive levels. Very few of these data points are used to acquire the information displayed in Figure 8.4. It would be useful to have methods of data analysis that take maximum advantage of the data available and that avoid the potential ambiguity introduced by nonlinearity. We describe two such schemes: (1) constant field analysis [16] and (2) template filtering [17].

8.5.2.1 Constant Field Analysis, CFA

We will process the data in Figure 8.5, from a hypothetical sample, using constant field analysis. The data in Figure 8.5 are the displacement, $U(i, j)$, for 80 values of the frequency, $f_i = 0.85 + 0.00253 * (i - 1)$, $i = 1, \ldots, 80$, for 40 values of the drive force, $F_j = 0.001 + 0.00254 * (j - 1)$, $j = 1, \ldots, 40$. These data are on 40 resonance curves, shown in Figure 8.6 as smooth curves, for clarity. It is apparent from the appearance of the data that the resonance frequency shifts first to a lower frequency and then to a higher frequency as the drive force is increased. The data are noisy and without the guide to the eye in Figure 8.6 not always discernible as resonance curves.

Suppose that we believe the elastic state of the sample is determined by the strain field in it. As this strain field is proportional to U, this means that the elastic state of the sample is the same at every point along the lines of constant U, for example along the line at $U = 0.432$ on Figure 8.6. If we assume that *when in a particular elastic state* the behavior of the sample can be described by a lumped element model, Eq. (8.28), then two parameters, $\tau(0.432)$ and $\omega_0^2(0.432)$, should describe all of the data on the $U = 0.432$ line.

As background we need results from the solution to Eq. (8.28). For $F(t) = F_0 \sin(2\pi f t)$ in Eq. (8.28) we have

$$u = [A \sin(2\pi f t) + B \cos(2\pi f t)]F_0, \qquad (8.32)$$

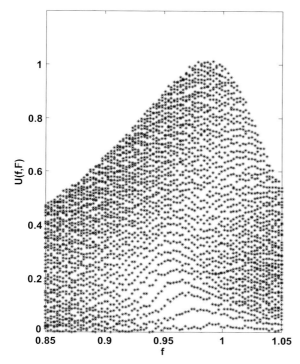

Fig. 8.5 Resonant bar data; Synthetic data Ia. Synthetic resonance curves, constructed from the self-consistent solution to Eq. (8.28), for 30 frequency values and 40 drive values. The resonance frequency depends on the magnitude of U as $f_0 = f_{00} - \gamma U + \Gamma U \tanh 2U$, Q^{-1} depends on the magnitude of U as $Q^{-1} = Q_0^{-1} + \delta U \tanh 4U$, and the data points have uniform random noise. These curves are to be used in demonstrating the CFA procedure. The solution to Eq. (8.28) yields the matrices $A(i, j)$, $B(i, j)$, and $U(i, j) = \sqrt{A(i, j)^2 + B(i, j)^2}$. The matrix $U(i, j)$ is shown here.

$$A = [f^2 - f_0^2(U)]\frac{1}{R^2} , \tag{8.33}$$

$$B = -\frac{f}{\tau(U)}\frac{1}{R^2} , \tag{8.34}$$

$$R^2 = [f^2 - f_0^2(U)]^2 + \left[\frac{f}{\tau(U)}\right]^2 = \frac{1}{A^2 + B^2} , \tag{8.35}$$

$$U = \frac{1}{R}F_0 , \tag{8.36}$$

where $u = U \sin(2\pi f t + \phi)$, and in order to feature the frequency f we have moved factors of 2π into a redefinition of F_0. A useful rearrangement of these equations yields

$$f_0^2(U) - f^2 = \frac{A}{A^2 + B^2} = AR^2 , \tag{8.37}$$

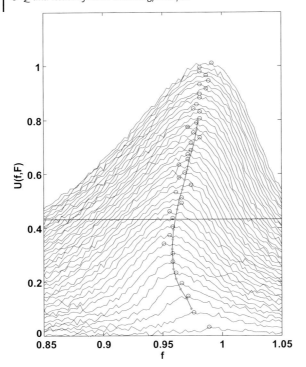

Fig. 8.6 Resonant bar data; Synthetic data Ib. The resonance curves from Figure 8.5 as smooth curves. The resonance frequency found from the maximum of each constant drive curve is shown with a star on the curve. The resonant frequency from CFA are shown as open circles. Since the constant field analysis is at constant U, the values of these resonant frequencies need not lie on a particular drive value curve. The data on the trajectory T_{10}, horizontal line at $U = 0.432$, will be discussed in detail.

$$\frac{1}{\tau(U)} = -\frac{1}{f}\frac{B}{A^2 + B^2} = -\frac{1}{f}B R^2 \,. \tag{8.38}$$

Measurement of the amplitudes $U(i, j)$ entails measurement of the in-phase component of the response to F_0, $A(i, j)$, and the out-of-phase component of the response to F_0, $B(i, j)$. Thus from experimentation we have three data matrices U, A, and B. We proceed as follows:

1. Find contours of constant U (strain) in $U(i, j)$ at values U_m, $m = 1\ldots M$. The level line U_m is a trajectory in (f, F_0) space denoted T_m along which there are (f, F_0) pairs denoted $F_0(T_m(k))$ and $f(T_m(k))$, $k = 1\ldots K_m$. Here K_m, the number of points on a level line, varies, with U_m being relatively small for U_m near the maximum, for example, 0.80.
2. Along each trajectory T_m find $A(T_m)$ and $B(T_m)$.

3. Guided by Eq. (8.37), along each trajectory form

$$L_1(U_m) = \frac{A(T_m)}{A^2(T_m) + B^2(T_m)} \,. \tag{8.39}$$

Guided by Eq. (8.38), along each trajectory form

$$L_2(U_m) = -\frac{1}{f(T_m)} \frac{B(T_m)}{A^2(T_m) + B^2(T_m)} \,. \tag{8.40}$$

4. Fit $L_1(U_m)$ to $L_1(U_m) = a_2(U_m)f^2(T_m) + a_0(U_m)$; expect $a_2(U_m) = -1.0$ and set $f_0^2(U_m) = -a_0(U_m)$ in accord with Eq. (8.37).
Fit $L_2(U_m)$ to $L_2(U_m) = b_2(U_m)f^2(T_m) + b_0(U_m)$; expect $b_2(U_m) = 0$ and set $1/\tau(U_m) = b_0(U_m)$ in accord with Eq. (8.38).

The steps in the procedure described here are illustrated in a sequence of figures. As we go through these steps we will follow a particular constant U level,

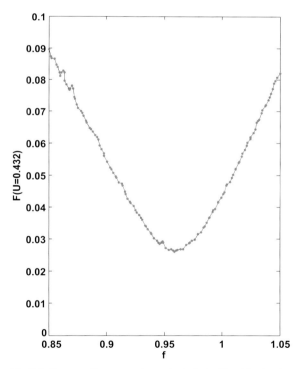

Fig. 8.7 Force vs. frequency at constant strain. The drive level (force) in Figure 8.5 necessary to maintain a constant value $U = 0.432$ (strain) is shown as a smooth curve. There are 129 force/frequency pairs on this curve, which constitute the trajectory T_{10}. See below, Eq. (8.38).

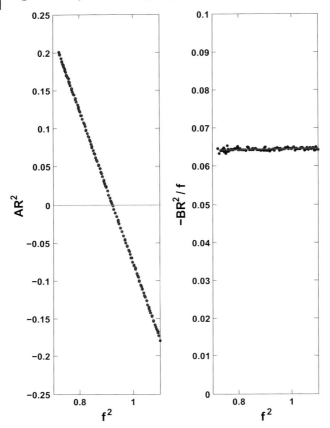

Fig. 8.8 Forming $f_0(U)^2 - f^2$ and $1/\tau(U)$. The fields A, B, and U are arranged to yield $f_0(U)^2 - f^2$ and $Q^{-1} \propto 1/\tau(U)$ as in Eqs. (8.37) and (8.38). For the trajectory T_{10}: (a) AR^2 as a function of f^2 and (b) $-BR^2/f$ as a function of f^2. The intercept at $AR^2 = 0$ from (a) yields the resonant frequency at the strain level $U = 0.432$. The horizontal line in (b) is $1/\tau(U)$, essentially Q^{-1}, at $U = 0.432$.

$U_{10} = 0.432$, shown in Figure 8.6. The trajectory T_{10} appropriate to U_{10} is a sequence of (f, F_0) pairs that are displayed in Figure 8.7, $F_0(T_{10}(k))$, as a function of $f(T_{10}(k))$, $k = 1 \ldots K_{10} = 129$. When the quantities on the RHS of Eqs. (8.37) and (8.38) are formed, they are as shown in Figure 8.8, where the independent variable is $f^2(T_{10}(k))$. It is apparent that to good approximation these quantities have the functional form sought. A fit to the polynomials above yields $(a_0(U_{10}), a_2(U_{10})) = (0.9333, -0.99999)$ and $(b_0(U_{10}), b_2(U_{10})) = (0.06434, -1.9314 \times 10^{-5})$, $f_0(U_{10}) = 0.9661$ and $1/\tau(U_{10}) = 0.06434$. The results of this type of analysis, $f_0(U_m)$ and $1/\tau(U_m)$ for 20 values of U_m that span most of the data, are plotted in Figure 8.9.

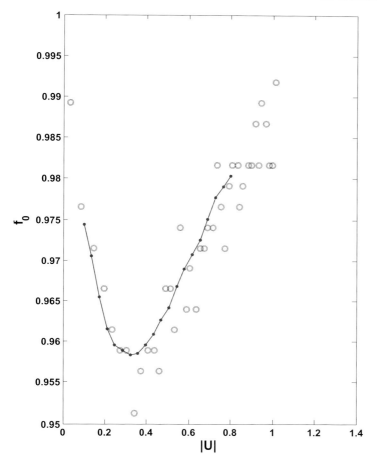

Fig. 8.9 Frequency and Q^{-1}. The results from CFA for the resonance frequency as a function of U (open circles and smooth curve). The open circles are the resonance frequency from the maximum of the constant drive curve, see Figure 8.6. The smooth curve is from CFA. See also Figure 8.10.

The results shown here prompt several observations.

1. Forming the quantities on the RHS of Eqs. (8.37) and (8.38) is simply a data manipulation with data cut at constant U instead of the conventional constant F_0.

2. The form of A/R^2 and $B/R^2/f$ in the figures confirms that U is the proper choice for the field that determines the elastic state of the system. Should it happen that some other field, for example, the velocity field or a combination of U and the velocity field, are the proper physical fields, it will be necessary to cut the data differently to search for a similar collapse.

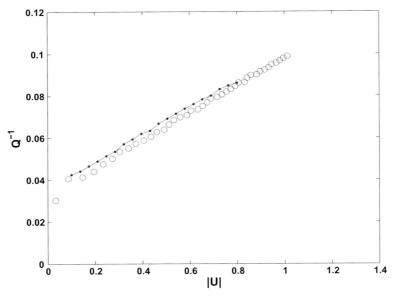

Fig. 8.10 Frequency and Q^{-1}. The result from CFA for Q^{-1} as a function of U (open circles and smooth curve). The open circles are Q^{-1} from the maximum of the constant drive curves, Fig. 8.6. The smooth curve is from CFA, see also Figure 8.9.

3. The simple form of the data on a resonance curve in these figures makes it possible to use all of the data confidently to learn the intercept, Figure 8.8a and $f_0(U)$, or the constant, Figure 8.8b and $1/\tau(U)$.

4. The scheme here stands in contrast to the constant F_0 scheme of analysis for which the quantities on the RHS of Eqs. (8.37) and (8.38) behave, as illustrated in Figure 8.11. In the constant F_0 scheme of analysis it is necessary to construct models of $f_0(U)$ and $1/\tau(U)$ to explain the kinks in these curves. In constant field analysis there are no kinks, Figure 8.8, and the functions $f_0(U)$ and $1/\tau(U)$ are the output.

8.5.2.2 Template Analysis

We will process the data in Figure 8.12, from a hypothetical sample, using template analysis [17]. The data in Figure 8.12 are the displacement, $U(i, j)$, for 80 values of the frequency, $f_i = 0.90 + 0.00253 * (i-1), i = 1 \ldots N_f$, $N_f = 80$, for 30 values of the drive force, $F_j = 0.001 + 0.00341 * (j-1), j = 1 \ldots N_F$, $N_F = 30$. These data are on 30 resonance curves, shown in Figure 8.13, in the form U/F_0 vs. f for clarity. It is not apparent from the appearance of the data that the resonance frequency shifts, but there is a marked change in the *width* of the curves as the driving force increases. The data are noisy.

Template analysis makes use of the set of equations above. Equations (8.37) and (8.38) allow one to arrange the experimental data to find f_0 and τ^{-1} at every

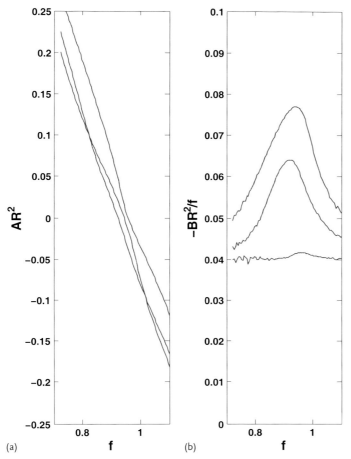

Fig. 8.11 AR^2 and $-BR^2/f$ at Constant Drive. When the arrangement of the fields A, B, and U that produces the results in Figure 8.8 is employed at constant drive, the results for three values of the drive (low, medium, and high) are as here (a) AR^2 and (b) $-BR^2/f$. Compare to Figure 8.8.

point (i, j). We do this along the curves of constant force. For three values of F_0, (0.0215, 0.0488, 0.100), the curves $f_0(F_0, f)$ and $\tau^{-1}(F_0, f)$ are plotted as a function of f, Figures 8.14 and 8.15. Although a frequency shift is not easily seen by simply looking at the data, Figure 8.12, a $\pm 2\%$ shift in frequency occurs across the resonance for $F_0 = 0.10$. A much more dramatic change in the attenuation occurs; $\tau^{-1} \approx 0.10$ for $F_0 = 0.10$ is more than twice the small-amplitude value, 0.04. To assess which physical fields might be responsible for what is seen, we form two fields from the experimental data, the in-phase field $|A\rangle$ and the out-of-phase field $|B\rangle$. This is done as follows.

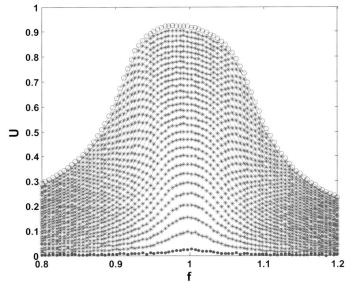

Fig. 8.12 Resonant bar data; Synthetic data IIa. Synthetic resonance curves, constructed from the self-consistent solution to Eq. (8.28), for 80 frequency values and 30 drive values. The resonance frequency depends on the strength of the A field as $f_0 = 1 - \gamma A$ and Q^{-1} depends on the B field as $Q^{-1} = Q_0^{-1} - \Gamma B$. These curves are to be used in demonstrat-ing the RTMF procedure. The solution to Eq. (8.28) yields the matrices $A(i, j)$, $B(i, j)$, and $U(i, j) = \sqrt{A(i, j)^2 + B(i, j)^2}$. The matrix $U(i, j)$ is shown here. The curve for the lowest (highest) drive level is shown as closed (open) circles. Notice these curves on Figure 8.13

1. Along a curve of constant force, say F_j, we have $A(i, j)$ and $B(i, j)$ for $i = 1 \ldots N_f$.

2. Form the vector $|A_j(i)\rangle \propto A(i, j)$ that is orthogonal to a constant and normed to 1:

$$\langle A_j | A_j \rangle = \sum_{i=1}^{N_f} \langle A_j(i) | A_j(i) \rangle = 1 , \quad j = 1 \ldots N_F , \tag{8.41}$$

$$\langle A_j | 1 \rangle = \sum_{i=1}^{N_f} \langle A_j(i) | = 0 , \quad j = 1 \ldots N_F . \tag{8.42}$$

3. Form the vector $|B_j(i)\rangle$, approximately proportional to $B(i, j)$, that is orthogonal to a constant and to $|A_j(i)\rangle$ and normed to 1:

$$\langle B_j | B_j \rangle = \sum_{i=1}^{N_f} \langle B_j(i) | B_j(i) \rangle = 1 , \quad j = 1 \ldots N_F , \tag{8.43}$$

$$\langle B_j | A_j \rangle = \sum_{i=1}^{N_f} \langle B_j(i) | A_j(i) \rangle = 0 , \quad j = 1 \ldots N_F , \tag{8.44}$$

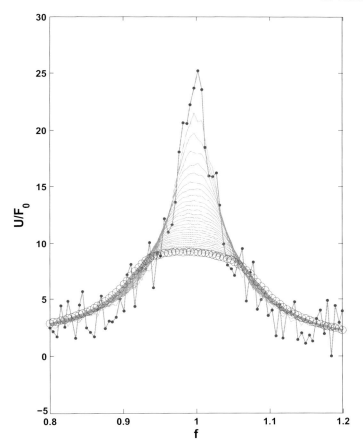

Fig. 8.13 Resonant bar data; Synthetic data IIb. The resonance curves from Figure 8.12, scaled by the drive, as smooth curves. The curve for the lowest (highest) drive level is shown as closed (open) circles. Notice these curves on Figure 8.12. While it is hard to see evidence of frequency shift, the nonlinearity of Q is manifest.

$$\langle B_j | 1 \rangle = \sum_{i=1}^{N_f} \langle B_j(i) | = 0, \ j = 1 \ldots N_F \, . \tag{8.45}$$

4. Assume that $f_j(i)$ $(\tau_j^{-1}(i))$, regarded as the vector $|f_j\rangle$ $(|\tau_j^{-1}\rangle)$, can be written in the form

$$|f_j\rangle = \gamma_j |1\rangle + \alpha_j |A_j\rangle + \beta_j |B_j\rangle \, , \tag{8.46}$$

$$|\tau_j^{-1}\rangle = g_j |1\rangle + a_j |A_j\rangle + b_j |B_j\rangle \, . \tag{8.47}$$

The amplitudes α_j, β_j, and γ_j are found from the inner products

$$\alpha_j = \langle A_j | f_j \rangle,$$ (8.48)

$$\beta_j = \langle B_j | f_j \rangle,$$ (8.49)

$$\gamma_j = \langle 1 | f_j \rangle$$ (8.50)

and an analogous set of equations for a_j, b_j, and g_j.

5. A measure of the degree to which the combination of $|1\rangle$, $|A_j\rangle$, and $|B_j\rangle$ provides an adequate description of $|f_j\rangle$ and τ_j^{-1} is provided by the magnitude of the residuals, for example, $\delta f_j^2 = \langle R_j | R_j \rangle$,

$$|R_j\rangle = |f_j\rangle - \gamma_j |1\rangle - \alpha_j |A_j\rangle - \beta_j |B_j\rangle,$$ (8.51)

for the case of $|f_j\rangle$.

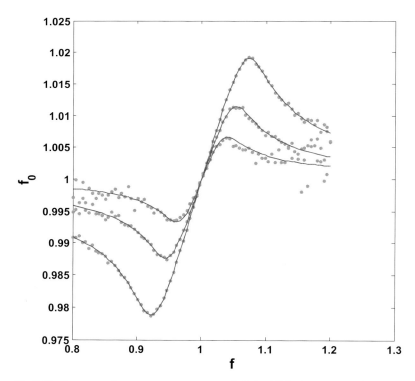

Fig. 8.14 Resonance frequency on curves of constant force. The resonance frequency, f_0, found from manipulating the data as in Eq. (8.37), as a function of f for three values of the drive, 0.021483, 0.048793, 0.100000. There is little shift in the resonance frequency (from $f_0 = 1$) near $f \approx 1$, where the strain field is large. The frequency shift is of opposite sign on the two sides of the nominal resonance frequency $f_0 = 1$. See the caption to Figure 8.12.

If we choose to norm the vector $|f_j(i)>$ to 1, we have

$$\delta f_j^2 = \langle R_j | R_j \rangle = 1 - (\alpha_j^2 + \beta_j^2 + \gamma_j^2). \tag{8.52}$$

When this procedure is carried out, an intermediate step is the formation of the templates $|A_j\rangle$ and $|B_j\rangle$ along each curve of constant F_j. An example of the templates $|A_j\rangle$ and $|B_j\rangle$ is shown in Figure 8.16 for the case $F_j = 0.0488$, cf. Fig-

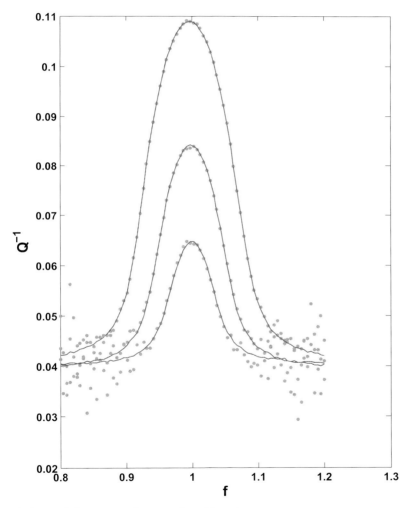

Fig. 8.15 Q^{-1} on curves of constant force. The value of Q^{-1}, found from manipulating the data as in Eq. (8.38), as a function of f for three values of the drive, 0.021483, 0.048793, 0.100000. Q^{-1} changes by almost a factor of 3 as the drive changes.

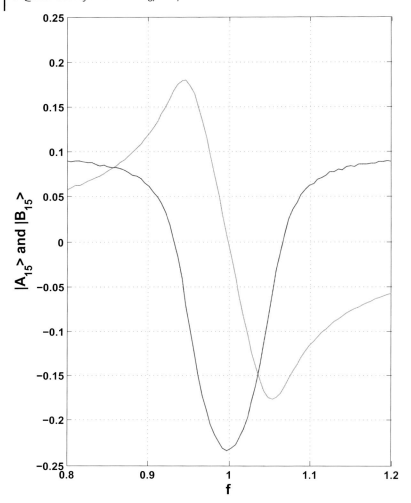

Fig. 8.16 The templates $|A\rangle$ and $|B\rangle$. The *in-phase* and *out-of-phase* templates $|A\rangle$ and $|B\rangle$, formed according to the prescriptions Eqs. (8.42) and (8.45), for the 15th drive level, $F = 0.04879$. Compare to Figures 8.14 and 8.15.

ures 8.14 and 8.15. The $|A_j\rangle$ template looks much like the difference between f_0 and a constant and the $|B_j\rangle$ template looks much like the departure of τ^{-1} from a constant. Indeed the solid curves in Figures 8.14 and 8.15 are from the fit using Eqs. (8.47). Further, the coefficient β_j is much less than α_j, and similarly the coefficient a_j is much less than b_j. These results suggest that the frequency shift is determined primarily by the in-phase strain, the $|A\rangle$ template, and the attenuation

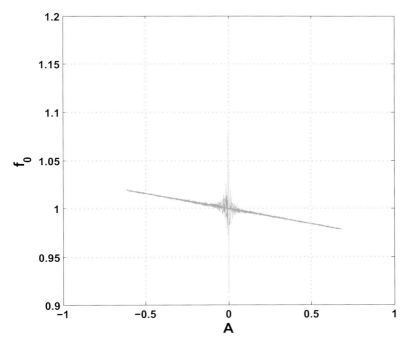

Fig. 8.17 Scaling of f_0 and Q^{-1} with A and B. The value of f_0 from Eq. (8.37) as a function of $A(i, j)/F_j$ for the 30 constant drive curves. These data collapse as a function of A/F except for $A \approx 0$, where they are very noisy. See Figure 8.18.

is determined primarily by the out-of-phase strain, the $|B\rangle$ template. To test this as a hypothesis of how the physical system behaves, we plot $f_0(i, j)$ as a function of $A(i, j)$ and $\tau^{-1}(i, j)$ as a function of $B(i, j)$ in Figure 8.16. While there is considerable noise at the low-strain field, where departure from linearity and the noise are comparable, as the strain field increases a simple relationship between the frequency shift and the strength of the A field is apparent, $f_0(i, j) = 1.00 - 0.0299A(i, j)$. Also apparent is a simple relationship between the attenuation and the strength of the B field, $\tau^{-1}(i, j) = 0.040 + 0.0750B(i, j)$.

The success of the procedure illustrated here depended on being able to identify the fields that are responsible for the observed behavior. In an experiment one makes measurements of two fields, the in-phase and out-of-phase strain fields. From these the magnitude of the strain can be formed. One expects that these three fields, A, B, and U, will suffice to determine the behavior. When they do, the procedure above or the CFA procedure can be implemented. When a more complex set of internal fields is involved, if they can be formed from the measured fields, it is in principle possible to search among possibilities and find the proper field.

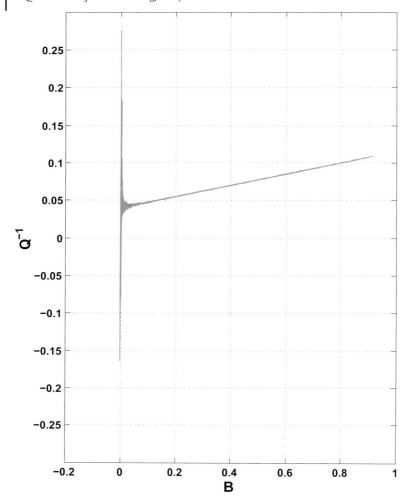

Fig. 8.18 Scaling of f_0 and Q^{-1} with A and B. The value of Q^{-1} from Eq. (8.38) as a function of $B(i, j)/F_j$ for the 30 constant drive curves. These data collapse as a function of B/F except for $B \approx 0$, where they are very noisy. See Figure 8.17.

There are two caveats that we leave to last (but one). They are by no means unimportant. Often a measurement system creates an electronic phase shift that adds to the physical phase shift and misaligns the two measured fields. One needs the means to detect that this is happening and remove it. Further, it is necessary to be able to divide out the applied force, say, to form U/F in the constant field analysis treatment of data. The information necessary to deal with these two difficulties can

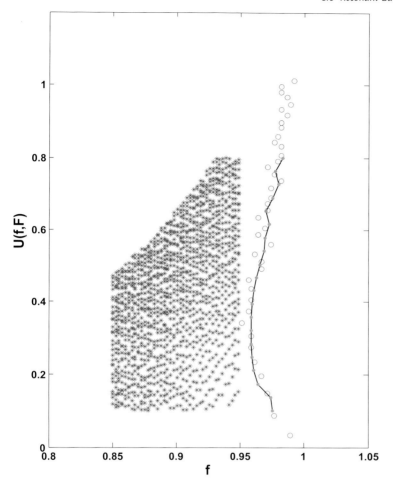

Fig. 8.19 Clipped data set. Constant field analysis was performed on the data shown here, a clipped version of the data in Figure 8.5. To the right of the data are the locations of the amplitude maxima transferred from Figure 8.6 (open circles) as well as the amplitude maximum found from constant field analysis of the data to the left. The values of $f_0(U)$ and $Q^{-1}(U)$ found in the constant field analysis are shown in Figures 8.20 and 8.21.

be gleaned from measurements on well-understood linear systems and/or from measurements on nonlinear systems at low drive amplitude.

Lastly, we show three figures, Figures 8.19, 8.20, and 8.21. In Figure 8.19 a resonant bar data set that has resonance curves found at constant F_0, none of which has a maximum, are shown. The data are a fraction of the data set in Figure 8.5. When this data set is subjected to constant field analysis, the results are those illustrated in Figures 8.20 and 8.21. To the right of the data in Figure 8.19 are the

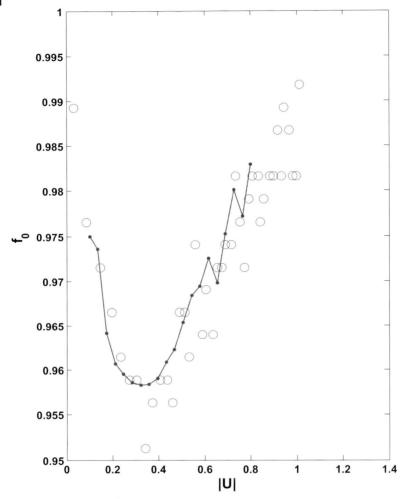

Fig. 8.20 $f_0(U)$ and $Q^{-1}(U)$ for clipped data set, Figure 8.19. (open circles) the resonance frequencies transferred from Figure 8.9 and (closed circles and line) $f_0(U)$ from analysis of the clipped data set.

maximum of each resonance curve from Figure 8.5 (full circles) and the resonance frequency found from constant field analysis of the data in the figure (solid circle and line). On the companion figures, Figures 8.20 and 8.21, the frequency $f_0(U)$ and $1/Q(U)$ are plotted as a function of U. On these curves as open circles are the results from traditional analysis that are also shown in Figures 8.9 and 8.10. Comparison of the various determinations of $f_0(U)$ and $1/Q(U)$ shows that constant field analysis can provide accurate information about a set of resonance curves,

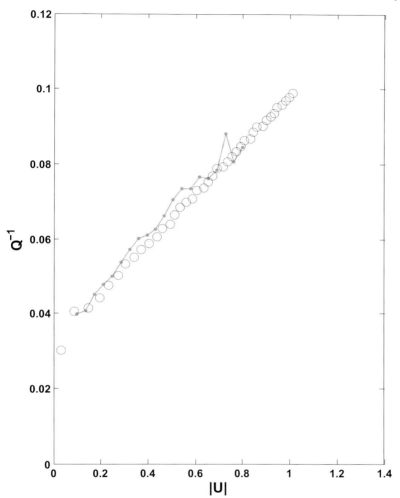

Fig. 8.21 $f_0(U)$ and $Q^{-1}(U)$ for clipped data set, Figure 8.19. (Open circles) $Q^{-1}(U)$ transferred from Figure 8.10 and (closed circles and line) $Q^{-1}(U)$ from analysis of the clipped data set.

that is, about the elastic state of a system, where the traditional methods of analysis have nothing to say.

References

1 Sato, H. and Fehler, M. (1998) *Seismic Wave Propagation and Scattering in the* *Heterogeneous Earth*, American Institute of Physics Press.

2 Landau, L.D. and Lifshitz, E.M. (1999) *Theory of Elasticity*, Butterworth and Heinemann, Oxford.

3 McCall, K.R. (1994) *J. Geophys. Res.*, **99**, 2591.

4 Huang, K. (1987) *Statistical Mechanics*, 2nd edn, John Wiley & Sons, Inc., New York.

5 O'Connell, R.J. and Budiansky, B. (1978) Measures of dissipation in viscoelastic media. *Geophysic. Res. Lett.*, **5**, 5–8.

6 Lomnitz, C. (1956) Creep measurements in igneous rock. *J. Geol.*, **64**, 473–479.

7 Murphy, W.F., Winkler, K.W., and Kleinberg, R.L. (1986) Acoustic relaxation in sedimentary rocks: dependence of grain contacts and fluid saturation. *Geophys.*, **51**, 757–766.

8 Hamilton, M.F. and Blackstock, D.T. (1997) *Nonlinear Acoustics*, Academic Press, San Diego.

9 McCall, K.R. and Guyer, R.A. (1994) Equation of state and wave propagation in hysteretic nonlinear elastic materials. *J. Geophys. Res.*, **99**, 23887–23897.

10 Capogrosso-Sansone, B. and Guyer, R.A. (2002) Dynamic model of hysteretic elastic systems. *Phys. Rev. B*, **66**, 224101.1–220101.12.

11 Puzrin, A., Frydman, S., and Talesnick, M. (1995) Normalized nondegrading behavior of soft clay under cyclic simple shear loading. *J. Geotech. Eng.*, **121**, 836–843.

12 Laurent, J. and Jia, X. Université Paris-Est Marne-la-Vallee (manuscript in preparation)

13 Xu, H., Day, S.M., and Minster, B. (2000) Hysteresis and two-dimensional nonlinear wave propagation in Berea Sandstone. *J. Geophys. Res.*, **105**, 6163–6175.

14 Graff, K. (1975) *Wave Motion in Elastic Solids*, Dover Pbl., New York.

15 http://en.wikipedia.org/wiki/Homodyne_detection, 15 May 2009.

16 Guyer, R.A., TenCate, J.A., and Johnson, P.A. (1999) Hysteresis and the dynamic elasticity of consolidated granular materials. *Phys. Rev. Lett.*, **82**, 3280–3283.

17 Smith, D.E. and TenCate, J. (2000) Sensitive determination of nonlinear properties of Berea sandstone at low strains. *Geophys. Res. Lett.*, **27**, 1985–1988.

9
Elastic State Spectroscopies and Elastic State Tomographies

In this chapter we discuss spectroscopic and tomographic methods of probing nonlinear elastic systems. By spectrosopies we mean methods of making use of the spectrum of normal modes to determine linear and nonlinear elastic constants. Implicit in this discussion is that the elastic systems are uniform, that is, K, μ, A, B, C, ..., Eq. (2.67), are independent of x. By tomographies we will mean methods of discerning the spatial structure of the elastic state of a system, linear or nonlinear, that is, $K(x)$, $\mu(x)$, $A(x)$, $B(x)$, $C(x)$, ... (This language is prompted by current practice, which uses *tomography* much more loosely than Merriam Webster [1].) We will touch on spectroscopies only briefly. Most of what we say will deal with tomographies. Linear and nonlinear methods will be discussed. The detail we provide in our discussion of linear methods is largely intended to provide a framework for a discussion of the nonlinear variant. In Section 9.1 we discuss linear spectrosopy (described well in the monograph of Migliori and Sarro [2]), Section 9.1.1, and nonlinear spectroscopy, Section 9.1.2. In Section 9.2 we discuss linear tomographies. Three methods are described: time-of-flight tomography, Section 9.2.1, normal-mode tomography, Section 9.2.2, and time-reversal tomography, Section 9.2.4. Numerical examples of normal-mode tomography (linear) and time reversal tomography (linear) are given in Section 9.2.3 and Section 9.2.5 respectively. In Section 9.3 we discuss nonlinear tomographies. The three methods, time of flight, normal mode, and time reversal, are discussed in Sections 9.3.1, 9.3.2, and 9.3.3, respectively.

9.1
Spectroscopies

Consider homogeneous elastic systems, that is, systems of uniform structure or composition throughout, K, μ, A, B, C, ... are independent of x.

Nonlinear Mesoscopic Elasticity: The Complex Behaviour of Granular Media including Rocks and Soil. Robert A. Guyer and Paul A. Johnson
Copyright © 2009 WILEY-VCH Verlag GmbH & Co. KGaA, Weinheim
ISBN: 978-3-527-40703-3

9.1.1
Linear, Homogeneous

For a linear homogeneous elastic system *the elastic tensor is the elastic state*. The normal modes of the system are a direct probe of the elastic tensor of the system. Measurement of a set of normal modes provides the means to learn this tensor with great accuracy. The monograph by Migliori and Sarro [2] describes all aspects of this problem carefully.

Information about the linear elastic state of a homogeneous system can also be learned from of a set of time of flight tone bursts having varying polarization, direction, etc. [3].

9.1.2
Nonlinear, Homogeneous

All elastic systems are nonlinear (Chapter 3). *Homogeneous* nonlinearity can be probed by doing normal-mode spectroscopy at varying probe amplitudes. (Any experiment whose outcome depends on a nonlinear elastic constant is a potential probe of that constant. See the discussion in Chapter 3, Eq. (3.30), etc.) We discuss two schemes.

1. In the case which is directly analogous to linear spectroscopy (where you only get out the frequency you put in) one drives the system in steady state at (frequency/amplitude) = (f_i, A_j) and observes the steady state response at the drive frequency, R_{ij}. Suppose the input and output signals, $A_{ij}(t) = A_j \sin(2\pi f_i t + \phi_i)$ and $R_{ij}(t)$ respectively, are digitized on time points t_m, $m = 1 \ldots N_T$. Form the two unit vectors, $U_i(t) \propto \sin 2\pi f_i t$ and $V_i(t) \propto \cos 2\pi f_i t$, and calculate

$$A_S^{ij} = \langle U_i | A_{ij} \rangle, \tag{9.1}$$

$$A_C^{ij} = \langle V_i | A_{ij} \rangle, \tag{9.2}$$

$$R_S^{ij} = \langle U_i | R_{ij} \rangle, \tag{9.3}$$

$$R_C^{ij} = \langle V_i | R_{ij} \rangle, \tag{9.4}$$

where $\langle A | B \rangle = \sum_m A(t_m) B(t_m)$. Then the input amplitude at (f_i, A_j) is $|A| = \sqrt{A_S^2 + A_C^2}$, the output amplitude at (f_i, A_j) is $|R| = \sqrt{R_S^2 + R_C^2}$, and the phase of input and output is found from A_S/A_C and R_S/R_C. Generally it is the amplitude ratio $|R|/|A|$ (usually with the zero of time chosen so that $A_S = 0$) that is studied as a function of (f_i, A_j). In circumstances where the number of time points per period is small, one should numerically construct U and V with V orthogonal to U and both orthogonal to a constant.

2. Alternatively, one drives the system at (*frequency/amplitude*) = (f_i, A_j) and observes the time train of the response $R_{ij}(t)$. The Fourier transform of the

$R_{ij}(t)$ provides a spectrum $S_{ij}(f)$. The amplitude of $S_{ij}(f)$ at f_i, a complex number, has the same information as that in the pair (R_S, R_C).

An experiment of style 1 asks the question "when the system is driven at f_i, what amplitude-dependent mechanisms are present that modify the response at f_i?" An experiment of style 2 asks the question "when the system is driven at f_i, what amplitude-dependent mechanisms are present that can create response at frequencies different from f_i?" The *4-phonon* process described in Chapter 3 does both, that is, the drive frequency is returned by a process like $\omega_i + \omega_i - \omega_i \rightarrow \omega_i$ and the third harmonic is created by $\omega_i + \omega_i + \omega_i \rightarrow 3\omega_i$.

1. $\omega_i + \omega_i - \omega_i \rightarrow \omega_i$ is seen in an experiment of type 1 as an A_j-dependent frequency shift at f_i (there is an associated small shift in the amplitude $|R|$), Chapter 3.
2. $\omega_i + \omega_i + \omega_i \rightarrow 3\omega_i$ is seen in an experiment of type 2 as an A_j-dependent Fourier component at $\pm 3f_i$, Chapter 3.

In classical nonlinear elasticity in which the stress is an analytic function of the strain, an initial frequency can produce frequencies different from itself only in multiples of itself, $\omega_i \rightarrow \pm n\omega_i$, $n = 0, 1, \ldots$ If the system is driven in two normal modes, ω_1 and ω_2, a complicated frequency spectrum will result with the amplitudes at various frequencies proportional to the weighted overlap of the strain fields associated with the normal modes. (See Chapter 3 and the discussion of the Luxemburg–Gorky effect.)

More complex frequency structures can arise in an experiment of style 2 if the elastic system has within itself elastic features with nonclassical dynamics. This is the case with an elastic system having elastic elements behaving like those of the MG model. From the discussion in Chapter 6, this could be Hertzian contacts, Masing material, etc.

Both the linear and nonlinear elastic state of homogeneous elastic systems can also be learned from a suitable set of time-of-flight measurements. When an elastic system is inhomogeneous, normal-mode spectroscopic methods detect an average elastic state that is mode dependent. As an example, consider the uv problem in Chapter 3. If the system were inhomogeneous, γ in Eq. (3.69) replaced by $\gamma(x)$, the result in Eq. (3.75) would be replaced by

$$\frac{v^{(2)}(L, \omega)}{L} = \delta(\omega - \omega_1 - \omega_2)\overline{\gamma}\varepsilon_L(\omega_1)\varepsilon_S(\omega_2)\mathcal{F}, \tag{9.5}$$

$$\overline{\gamma} = \int_0^L dx\,\gamma(x), \tag{9.6}$$

$$f_\gamma(x) = \gamma(x)/\overline{\gamma}, \tag{9.7}$$

$$\mathcal{F} = \frac{1}{L}\int_0^L dx\,f_\gamma(x)\frac{\sin(q_{\omega_3}x)}{\sin(q_{\omega_3}L)}\sin(k_{\omega_1}x)\sin(q_{\omega_2}x), \tag{9.8}$$

where $\omega_3 = \omega_1 + \omega_2$. A measurement of a side-band amplitude would provide one particular integral over $\gamma(x)$. Finding the spatial structure of $\gamma(x)$ would be an inverse problem having a familiar difficulty (e.g., invert for Preisach space or later on in this chapter). But then one is using mode amplitudes importantly (not just mode frequencies) and doing tomography. That comes next.

9.2
Tomographies, Linear, Inhomogeneous

Here we deal with $c_{ijkl}(\mathbf{x})$, which we take to be of the form $c_{ijkl}(\mathbf{x}) = c^0_{ijkl} + \delta c_{ijkl}(\mathbf{x})$, where $\delta c_{ijkl}(\mathbf{x})$ can be regarded as a perturbation in the stress-strain relation. We sketch most results as the details are found in what has gone before. For the case of time reversal we will say more. See Figures 9.1 and 9.2.

9.2.1
Time of Flight

Consider the elastic system described by

$$\delta c_{ijkl}(\mathbf{x}) = \phi(\mathbf{x})\delta_{ix}\delta_{jx}\delta_{kx}\delta_{lx},$$

(9.9)

$$\phi(\mathbf{x}) = \sum_{i=1}^{N} R_i \delta(x - x_i),$$

(9.10)

where $R_i = b\delta c_i^2/c_0^2$ and b is a length to replace the units lost to $\delta(x-x_i)$. For $u_x = u$ we have

$$\ddot{u} + \frac{1}{\tau_0}\dot{u} - c_0^2 u'' = c_0^2 \frac{\partial}{\partial x}\left(\sum_{i=1}^{N} R_i \delta(x - x_i)\frac{\partial u}{\partial x}\right) + F(t)\delta(x),$$

(9.11)

$-\infty < x < +\infty$. It is convenient to define the amplitude Λ and the dimensionless form factor $f(x)$

$$\Lambda = \frac{1}{b}\sum_{i=1}^{N} R_i = \sum_{i=1}^{N} \frac{\delta c_i^2}{c_0^2},$$

(9.12)

$$f(x) = \sum_{i=1}^{N} R_i \delta(x - x_i)/\Lambda,$$

(9.13)

$$\frac{1}{b}\int dx\, f(x) = 1.$$

(9.14)

The *average strength* of the inhomogeneity is measured by Λ, whereas its *shape* is described by $f(x)$. Then

$$\ddot{u} + \frac{1}{\tau_0}\dot{u} - c_0^2 u'' = c_0^2 \Lambda \frac{\partial}{\partial x}\left(f(x)\frac{\partial u}{\partial x}\right) + F(t)\delta(x).$$

(9.15)

We are concerned with the effect of $\delta c(x)$ on the time of flight of a wave, $u(x,t) = u_0 \exp i(kx - \Omega t)$, that crosses over it. We examine this question using the perturbative methods illustrated above. Write $u = u^{(1)} + \Lambda u^{(2)} + \dots$ Then $F(t)\delta(x)$ creates

$u^{(1)} = u_0 \exp i(kx - \Omega t)$, or in Fourier space

$$u^{(1)}(x; \omega) = G(x|0; \omega) F(\omega) \tag{9.16}$$

where $F(\omega) = F_0 \delta(\omega - \Omega)$ and

$$u^{(2)}(x; \omega) = -c_0^2 \left(\int dx' \frac{\partial G(x|x'; \omega)}{\partial x'} f(x') \frac{\partial G(x'|0; \omega)}{\partial x'} F(\omega) \right) . \tag{9.17}$$

Using the equation for G we have

$$u^{(2)}(x; \omega) = \frac{b}{4c_0^2} e^{ik_\omega x} \tag{9.18}$$

$$u = u^{(1)}(x; \omega) + \Lambda u^{(2)}(x; \omega) = \frac{iF_0}{2c_0^2 k_\omega} (1 - ik_\omega b\Lambda) e^{ik_\omega x} \delta(\omega - \Omega) , \tag{9.19}$$

and on rearrangement

$$u(x, t) = u_0 (1 - ik_\Omega b\Lambda) e^{i(k_\Omega x - \Omega t)} , \tag{9.20}$$

$$u(x, t) = u_0 e^{i(k_\Omega (x-a) - \Omega t)} , \tag{9.21}$$

where $k_\Omega b\Lambda \ll 1$. An encounter with the perturbation $\delta c_{ijkl}(x)$ appears as a phase shift. At $x = \lambda$ the phase is zero at $T_\lambda = k(\lambda - a)/\Omega$. At $x = \lambda$, for $a = 0$, the phase is zero at $T_0 = k\lambda/\Omega = 1/f$. So $T_\lambda = T_0(1 - a/\lambda)$. For $\Lambda > 0$ the moment of zero phase arrives sooner, $T_\lambda < T_0$, and for $\Lambda < 0$ the moment of zero phase arrives later, $T_\lambda > T_0$. This is equivalent to an effective velocity shift that goes as $(a \ll \lambda)$

$$\frac{c_{\text{eff}}}{c_0} = 1 + \frac{a}{\lambda} = 1 + \Lambda \frac{b}{\lambda} . \tag{9.22}$$

Thus the phase shifts of a suitable set of "beams" can provide the means to find $f(x)$ and determine Λ.

9.2.2
Normal Mode

Consider a resonant bar, $0 \leq x \leq L$, for which the displacement obeys the equation

$$\ddot{u} + \frac{1}{\tau_0} \dot{u} - c_0^2 u'' = c_0^2 \Lambda \frac{\partial}{\partial x} \left(f(x) \frac{\partial u}{\partial x} \right) + F(t)\delta(x) \tag{9.23}$$

using the notation in Eq. (9.15) with $u'(0) = u'(L) = 0$. We seek the frequencies of the normal modes of this system. To find these we employ the methods of conventional time-independent perturbation theory [4] using the complete set of spatial states associated with the unperturbed problem set by the leading terms on the LHS of the equation,

$$c_0^2 u'' = \ddot{u} \rightarrow c_0^2 \phi_n'' = \lambda_n \phi_n = -\omega_n^2 \phi_n . \tag{9.24}$$

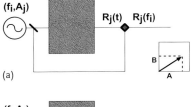

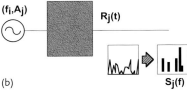

Fig. 9.1 Nonlinear spectroscopy. For a homogeneous non-linear elastic system the presence of nonlinearity can make itself known in spectroscopy through (a) amplitude-dependent modification of the response at a drive frequency or (b) the generation of response frequencies that are different from the drive frequency.

These states are $|n\rangle = \phi_n(x) \propto \cos k_n x$, $k_n = n\pi/L$, $n = 1 \ldots$, $\omega_n^0 = k_n c_0$, for which we have the notation and algebra

$$\psi_n = |\hat{n}\rangle = \frac{1}{k_n}|n'\rangle , \tag{9.25}$$

$$\langle n|m\rangle = \int_0^L dx\, \phi_n(x)\phi_m(x) = \delta_{nm} , \tag{9.26}$$

$$\langle \hat{n}|\hat{m}\rangle = \int_0^L dx\, \psi_n(x)\psi_m(x) = \delta_{nm} , \tag{9.27}$$

$$\langle n'|m'\rangle = \int_0^L dx\, \phi_n'(x)\phi_m'(x) = k_n^2\delta_{nm} , \tag{9.28}$$

where for convenience we have introduced the second set of functions ψ_n. Use the complete set of functions ϕ_n to express $u(x,t)$ in the form

$$u = \sum_{m=1}^M a_m(t)\phi_m(x) , \tag{9.29}$$

where M is a practical limit on m. Substitute into Eq. (9.23) and take the inner product of the resulting equation with $\langle n|$. Use one integration by parts, Eq. (9.28), and rearrange to bring the resulting equation to the form

$$\ddot{a}_n + \frac{1}{\tau_0}\dot{a}_n + c_0^2 k_n^2 a_n = -c_0^2 \Lambda \sum_{m=1}^M k_n k_m \langle \hat{n}|\, f(x)\,|\hat{m}\rangle a_m(t) + F(t)\langle n|\delta(x)\rangle ,$$

$$\tag{9.30}$$

$$= -c_0^2 \Lambda \sum_{m=1}^{M} k_n k_m G_{nm} a_m(t) + F(t)\langle n|\delta(x)\rangle, \tag{9.31}$$

where $G_{nm} = \langle \hat{n}|f(x)|\hat{m}\rangle$ carries the information about the geometry of the perturbation. Use the usual Fourier representation of $F(t)$ and $a_n(t)$, $F(\omega)$ and $a_n(\omega)$, to write

$$\left(-\omega^2 - i\frac{\omega}{\tau_0} + c_0^2 k_n^2\right) a_n(\omega) = -c_0^2 \Lambda \sum_{m=1}^{M} k_n k_m G_{nm} a_m(\omega) + F(\omega)\langle n|\delta(x)\rangle. \tag{9.32}$$

Take $F(t)$ to drive the system near a normal mode, that is, $F(\omega) \propto \delta(\omega \pm \Omega)$, $\Omega \approx c_0 k_n$. Then the amplitude a_n is relatively large and we can write

$$a_m(\omega) = a_n^0(\omega)\delta_{mn} + \Lambda g_m(\omega) + \Lambda^2 h_m(\omega) + \ldots \tag{9.33}$$

Substitution of this form in Eq. (9.32) and arrangement according to powers of Λ leads to (for $m = n$)

$$D_n(\omega)a_n^0(\omega) = F(\omega)\langle n|\delta(x)\rangle = F(\omega)X_n, \tag{9.34}$$

$$D_n(\omega)g_n(\omega) = -c_0^2 k_n^2 G_{nn} a_n^0(\omega), \tag{9.35}$$

$$\vdots \tag{9.36}$$

where

$$D_n(\omega) = \left(-\omega^2 - i\frac{\omega}{\tau_0} + c_0^2 k_n^2\right). \tag{9.37}$$

Solve for a_n^0 and g_n and write

$$a_n \approx a_n^0 + \Lambda g_n = \left(1 - \frac{1}{D_n}\Lambda c_0^2 k_n^2 G_{nn}\right) a_n^0, \tag{9.38}$$

$$= \frac{1}{D_n + \Lambda c_0^2 k_n^2 G_{nn}} F(\omega)X_n, \tag{9.39}$$

$$= \frac{1}{-\omega^2 - i\frac{\omega}{\tau_0} + c_0^2 k_n^2(1 + \Lambda G_{nn})} F(\omega)X_n. \tag{9.40}$$

From the denominator we read off the resonance frequency

$$\omega_n^2 - \left(\omega_n^0\right)^2 \approx \left(\omega_n^0\right)^2 \Lambda G_{nn} \tag{9.41}$$

or

$$\frac{\omega_n - \omega_n^0}{\omega_n^0} \approx \frac{1}{2}\Lambda G_{nn}. \tag{9.42}$$

The magnitude of the frequency shift is in Λ, and the geometry of the perturbation causing this shift is in G_{nn}. We can learn the geometrical arrangement of the inhomogeneity by studying a set of the G_{nn}, which are integrals of the unknown function $f(x)$ over the known functions $k_n^{-1}d\varphi_n(x)/dx \propto \sin k_n x$. Again we have an inverse problem of a familiar kind.

(a)

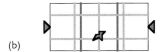

(b)

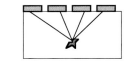

(c)

Fig. 9.2 Tomographies. The position of a localized elastic inhomogeneity, $\delta c_{ijkl}(x)$, can be learned using at least three methods. (a) Time-of-flight tomography, in which the travel time over a multiplicity of paths is used to find the involvement of $\delta c_{ijkl}(x)$. (b) Normal-mode tomography, in which alteration of the normal-mode spectrum due to $\delta c_{ijkl}(x)$ is used to locate it. (c) Time-reversal tomography, in which a set of sources, mirrors, are trained to have an amplitude/phase that focuses on a scatterer, that is, $\delta c_{ijkl}(x)$.

9.2.3
Normal Mode, a Numerical Example

Let us look at an example of the principle enunciated above to see some of the features that appear in a problem of the type set. For an inhomogeneity with $f(x)$ as shown in Figure 9.3 of strength $\Lambda = -0.02$ we find the frequencies $\{\omega_n\}$ from an *exact solution* to Eq. (9.31). These are shown as frequency shifts in Figure 9.4 using $c_0 = L = 1$ and $\omega_n^0/\pi = n$ ($\omega_n^0/\pi = n$ is the horizontal axis in Figure 9.4). Error bars have been attached to the $\{\omega_n\}$ (relatively larger as the frequency increases) that transfer directly to the frequency shifts. We take the frequencies $\{\omega_n\}$ (or the frequency shifts) as *the data*.

1. To have data to display as in Figure 9.4 from an experiment one needs measurements of the frequency of M normal modes for a homogeneous sample as well as measurements of the frequency of the same M normal modes for a sample identical in every regard except for the inhomogeneity. "Differing unimportantly compared to the inhomogeneity" also works. But how would you know? The need to compare measurements on two different samples is troublesome.

2. The frequency shift, $\Delta\omega_n/\pi = (\omega_n - \omega_n^0)/\pi$, scales approximately as the frequency ω_n^0. A fit of $\Delta\omega_n/\pi$ to a line yields

$$\Delta\omega_n/\pi \approx -0.00337 - 0.00981n , \tag{9.43}$$

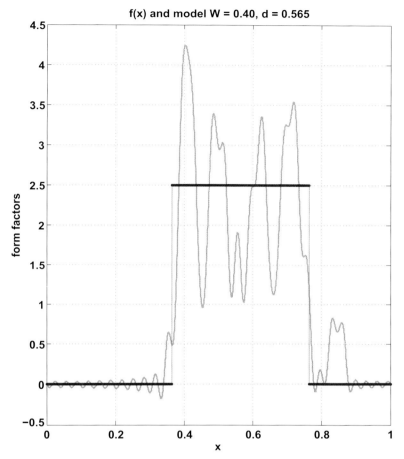

Fig. 9.3 Form Factor for $\delta c_{ijkl}(x)$. The spatial structure of an elastic inhomogeneity is described by a form factor, for example, Eq. (9.14). For the numerical example in Section 9.2.2 $f(x)$ is the light smooth curve with approximately four oscillations between $0.4 < x < 0.75$. The model form factor, $f_M(x) = 1/W, -W/2 + d \leq x \leq d + W/2$, is the heavy line "box" with $d = 0.565$ and $W = 0.40$ from the minimum of the energy $\mathcal{E}$ in Eq. (9.44).

in qualitative accord with Eq. (9.42). If we assume that the integrals G_{nn} are of order 1, then the coefficient of $\omega_n^0/\pi = n$ provides an estimate of Λ, $\Lambda \approx -0.0196$.

3. If $f(x) = 1, 0 \leq x \leq L$, then $G_{nn} = 1, \forall n$. The error bars in Figure 9.4 are to be taken seriously. It is modes $n = 1, 7 \ldots 10$ that locate $f(x) \neq 1$ and contain the information that will pin it down. To see this, remove the trivial n dependence from $\Delta \omega_n$ by dividing by ω_n^0, see Figure 9.6 (circles).

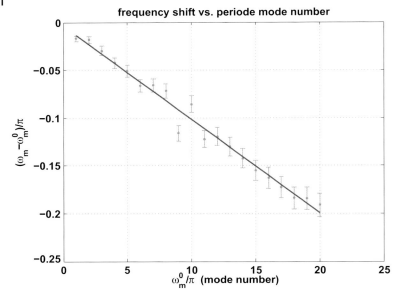

Fig. 9.4 Frequency Shift. The frequency shift as a function of mode number for an inhomogeneity of strength $\Lambda = -0.02$ having the form factor shown in Figure 9.3. For illustrative purposes an error bar was assigned to each *experimental* data point. The frequency shift is proportional to the mode number as it must be from simple scaling, Eq. (9.42) and from Eq. (9.43). Since the G_{nn} are of order 1, the approximate amplitude of the inhomogeneity is found from the slope in Eq. (9.43). The shape of the form factor comes from notable departures of the frequency shift from $f(x) = 1/L$. See Figure 9.6.

4. To estimate $f(x)$ in order to locate the inhomogeneity, we use a model with 2 degrees of freedom (W, d), $f_M(x) = 1/W$, $-W/2 + d \le x \le d + W/2$. For $f_M(x)$ we calculate the integrals G_{nn}^M, $n = 1 \dots M$, and form them into the M component unit vector, e_M ($e_M \cdot e_M = 1$). We form the corresponding unit vector, e_X ($e_X \cdot e_X = 1$), from the experimental data, $\Delta\omega_n/\omega_n^0$. Forming the unit vectors takes all of the issues of magnitude out of the comparison of $f(x)$ with $f_M(x)$. We seek $e_M \approx e_X$. Find this as the minimum in the energy, $\mathcal{E}(d, W)$, given by

$$\mathcal{E}(d, W) = \sum_{n=1}^{M} (e_X(n) - e_M(n)) K(n) (e_X(n) - e_M(n)), \qquad (9.44)$$

where $K(n)^{-1} \propto \sigma(n)$, the uncertainty in data point n, that is, the uncertainty in $\Delta\omega_n/\omega_n^0$.

5. In Figure 9.5 we show $\mathcal{E}(d, W)$ as a function of (W, d). There is an absolute minimum at $(W, d) = (0.40, 0.565)$. On Figure 9.3 we show this $f_M(x)$ along with $f(x)$. It is apparent that the very crude $f_M(x)$ we have used captures the gross properties of $f(x)$ quite well, but it certainly cannot get the details. We see this in Figure 9.6, where we show the $M = 20$ components of e_X and e_M

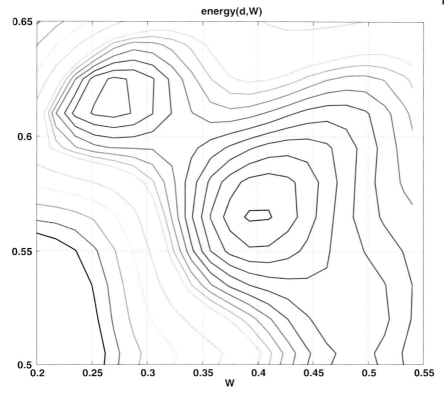

Fig. 9.5 Energy Landscape. The energy in Eq. (9.44), due to the difference between the data vector e_X and the model vector e_M, as a function of d and W, the parameters of the model form factor $f_M(x)$. The energy minimum is at $d = 0.565$ and $W = 0.40$. See Figure 9.3. (Please find a color version of this figure on the color plates)

(0.40,0.565). The long-wavelength amplitudes are in good agreement, say out to $n = 7$. The amplitudes $n = 8\ldots11$, which must pick up the short-wavelength structure in $f(x)$, do not do particularly well. On the other hand, these observations point toward the direction one would go in to seek improvement.

9.2.4
Time Reversal

For this discussion we are more careful with details than we were above, where they are relatively easily provided from what has gone before. So be prepared to

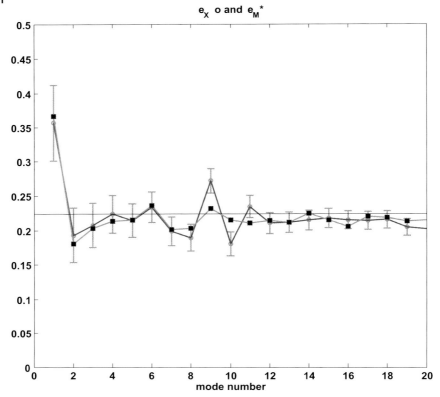

Fig. 9.6 Frequency-Shift Vectors. The experimental data points, in the form $(\omega_n - \omega_n^0)/\omega_n^0$, are formed into a unit vector, e_X (open circles with error bars), as are the corresponding points in the best model of the data, e_M ($d = 0.565$ and $W = 0.40$) (squares). These vectors have no information about the amplitude of the inhomogeneity, but they do carry the information about the shape, that is, $f(x)$. Further, since the trivial factor ω_n^0 in Eq. (9.42) has been divided out, the importance of the error bars is apparent. If $f(x) = 1/L$, then $e(n) = 1/\sqrt{M}\,\forall n$ (the line at $1/\sqrt{20}$ on the figure). Departure of $e(n)$ from $1/\sqrt{20}$ for modes $n = 1, 7, 8, 9, 10$ by more than the error bars means that these modes are particularly sensitive to $f(x)$. The very simple model form factor, $f_M(x)$, is adequate to get $e(n)$ about right for $n = 1...7$. A model form factor with more structure is necessary to do a better job with $n = 8...11$.

see lots of indices. We begin with the classical continuum equation for an isotropic material:

$$\varrho \ddot{u}_i = \sum_{jkl} \frac{\partial}{\partial x_j} c_{ijkl}(x) \frac{\partial u_k}{\partial x_l} , \qquad (9.45)$$

where

$$c_{ijkl}(x) = \lambda(x)\delta_{ij}\delta_{kl} + \mu(x)(\delta_{ik}\delta_{jl} + \delta_{il}\delta_{jk}) \qquad (9.46)$$

and λ and μ are the Lamê constants. In detail, Eq. (9.45) is

$$\varrho \ddot{u}_i = \frac{\partial}{\partial x_i} \left(\lambda(\boldsymbol{x}) \sum_k \frac{\partial u_k}{\partial x_k} \right) + \sum_k \frac{\partial}{\partial x_k} \left(\mu(\boldsymbol{x}) \frac{\partial u_i}{\partial x_k} \right) + \sum_k \frac{\partial}{\partial x_k} \left(\mu(\boldsymbol{x}) \frac{\partial u_k}{\partial x_i} \right).$$

(9.47)

As above, we examine the case where the elastic system is uniformly isotropic except for a spatially local perturbation. We write

$$\varrho \ddot{u}_i + \varrho \frac{\dot{u}}{\tau_0} = [Bu]_i + \sum_{jkl} \frac{\partial}{\partial x_j} \left(\delta c_{ijkl}(\boldsymbol{x}) \frac{\partial u_k}{\partial x_l} \right) + \varrho f_i(\boldsymbol{x}, t),$$

(9.48)

where

$$[Bu]_i = \lambda_0 \frac{\partial}{\partial x_i} \left(\sum_k \frac{\partial u_k}{\partial x_k} \right) + \mu_0 \sum_k \frac{\partial}{\partial x_k} \left(\frac{\partial u_i}{\partial x_k} \right) + \mu_0 \sum_k \frac{\partial}{\partial x_k} \left(\frac{\partial u_k}{\partial x_i} \right).$$

(9.49)

In the usual way, write u_i in the form $u_i = u_i^{(1)} + u_i^{(2)} + \dots$, ($u^{(n+1)}$ of order $(\delta c)^n$), and find

$$\varrho \ddot{u}^{(1)} = [Bu^{(1)}]_i + \varrho f_i(\boldsymbol{x}, t),$$

(9.50)

$$\varrho \ddot{u}^{(2)} = [Bu^{(2)}]_i + \sum_{jkl} \frac{\partial}{\partial x_j} \left(\delta c_{ijkl}(\boldsymbol{x}) \frac{\partial u_k^{(1)}}{\partial x_l} \right),$$

(9.51)

$$\vdots$$

(9.52)

Choose the geometry shown in Figure 9.7. Place $m = 1 \dots M$ mirrors (displacement source/receivers) [5, 6] at $\boldsymbol{x}_m = (0, y_m, 0)$. These are the locations of f,

$$f_i(\boldsymbol{x}, t) = \sum_{m=1}^{M} f_i(\boldsymbol{x}_m, t) \delta(\boldsymbol{x} - \boldsymbol{x}_m).$$

(9.53)

Use the usual Fourier representation of u and f and find

$$u_i^{(1)} = U_i^{(1)} + U_i^{(2)} + \dots U_i^{(M)},$$

(9.54)

$$U_i^{(m)}(\boldsymbol{x}; \omega) = G_{i2}(\boldsymbol{x}, \boldsymbol{m}; \omega) f_2(\boldsymbol{m}; \omega),$$

(9.55)

$$u_1^{(2)}(\boldsymbol{x}; \omega) = \sum_{m=1}^{M} \sum_{ijkl} \int d\boldsymbol{x}' G_{1i}(\boldsymbol{x}, \boldsymbol{x}'; \omega) \frac{\partial}{\partial x_j'} \left(\delta c_{ijkl}(\boldsymbol{x}') \frac{\partial U_k^{(m)}}{\partial x_l'} \right),$$

(9.56)

where $\boldsymbol{m}$ stands for $\boldsymbol{x}_m$ and $G_{ij}(\boldsymbol{x}, \boldsymbol{m})$ is the Green function for the isotropic elastic system. The component of f and displacement $u^{(2)}$ have been chosen to be 2 and 1, respectively, for clarity.

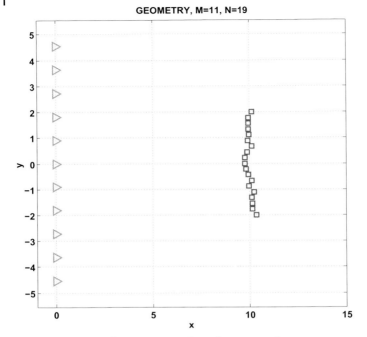

Fig. 9.7 Time-Reversal Geometry. A uniform elastic system has $M = 11$ mirrors at $x_m = 0, -5 \le y_m \le +5$ and $N = 19$ scatterers at $x_n \approx L = 10, -2 \le y_m \le +2$.

The first-order field is built up as a sum of the fields $U_i^{(m)}$ due to $f_i(\boldsymbol{m})$. The superposition of the scattering of these fields from $\delta c(\boldsymbol{x})$ gives $u_i^{(2)}$. Using the equations for $U_i^{(m)}$ in the equation for $u_i^{(2)}$ and doing one integration by parts leads to

$$u_1^{(2)}(\boldsymbol{x}; \omega) = -\sum_{m=1}^{M} \sum_{ijkl} \int d\boldsymbol{x}' \frac{\partial G_{1i}(\boldsymbol{x}, \boldsymbol{x}'; \omega)}{\partial x'_j} \delta c_{ijkl}(\boldsymbol{x}') \frac{\partial G_{k2}(\boldsymbol{x}', \boldsymbol{m}; \omega)}{\partial x'_l} f_2(\boldsymbol{m}; \omega).$$

(9.57)

For the scattered field at $\boldsymbol{m}'$ we have

$$u_1^{(2)}(\boldsymbol{m}'; \omega) = -\sum_{m=1}^{M} \sum_{ijkl} \int d\boldsymbol{x}' \frac{\partial G_{1i}(\boldsymbol{m}', \boldsymbol{x}'; \omega)}{\partial x'_j}$$

$$\delta c_{ijkl}(\boldsymbol{x}') \frac{\partial G_{k2}(\boldsymbol{x}', \boldsymbol{m}; \omega)}{\partial x'_l} f_2(\boldsymbol{m}; \omega),$$

(9.58)

$$= -\sum_{m=1}^{M} \mathbf{t}_{12}(\boldsymbol{m}', \boldsymbol{m}; \omega) f_2(\boldsymbol{m}; \omega).$$

(9.59)

Matrix $\mathbf{t}$ is called the *transfer matrix*. Under certain conditions it obeys reciprocity. To show this we use the sequence of steps

$$\mathbf{t}_{12}(\boldsymbol{m'},\boldsymbol{m};\omega) = \sum_{ijkl} \int d\boldsymbol{x'} \frac{\partial G_{1i}(\boldsymbol{m'},\boldsymbol{x'};\omega)}{\partial x'_j} \delta c_{ijkl}(\boldsymbol{x'}) \frac{\partial G_{k2}(\boldsymbol{x'},\boldsymbol{m};\omega)}{\partial x'_l} , \quad (9.60)$$

$$= \sum_{ijkl} \int d\boldsymbol{x'} \frac{\partial G_{i1}(\boldsymbol{x'},\boldsymbol{m'};\omega)}{\partial x'_j} \delta c_{ijkl}(\boldsymbol{x'}) \frac{\partial G_{2k}(\boldsymbol{m},\boldsymbol{x'};\omega)}{\partial x'_l} , \quad (9.61)$$

$$\mathbf{t}_{21}(\boldsymbol{m},\boldsymbol{m'};\omega) = \sum_{ijkl} \int d\boldsymbol{x'} \frac{\partial G_{i2}(\boldsymbol{x'},\boldsymbol{m};\omega)}{\partial x'_j} \delta c_{ijkl}(\boldsymbol{x'}) \frac{\partial G_{1k}(\boldsymbol{m'},\boldsymbol{x'};\omega)}{\partial x'_l} , \quad (9.62)$$

$$= \sum_{ijkl} \int d\boldsymbol{x'} \frac{\partial G_{k2}(\boldsymbol{x'},\boldsymbol{m};\omega)}{\partial x'_l} \delta c_{klij}(\boldsymbol{x'}) \frac{\partial G_{1i}(\boldsymbol{m'},\boldsymbol{x'};\omega)}{\partial x'_j} . \quad (9.63)$$

1. The second line follows from the reciprocity of G [7], $G_{ij}(1, 2; \omega) = G_{ji}(2, 1; \omega)$.
2. The third line follows from the second upon interchanging $1 \leftrightarrow 2$ and $\boldsymbol{m} \leftrightarrow \boldsymbol{m'}$.
3. The fourth line follows from the third upon interchange of the dummy indices $i \leftrightarrow k$ and $j \leftrightarrow l$.

Then, if the perturbation is such that $\delta c_{klij}(\boldsymbol{x}) = \delta c_{ijkl}(\boldsymbol{x})$,

$$\mathbf{t}_{21}(\boldsymbol{m},\boldsymbol{m'};\omega) = \mathbf{t}_{12}(\boldsymbol{m'},\boldsymbol{m};\omega) , \quad (9.64)$$

the transfer matrix obeys reciprocity. For a linear elastic material with elastic energy density that is pointwise an analytic function of the strain field one necessarily has $\delta c_{ijkl}(\boldsymbol{x}) = \delta c_{klij}(\boldsymbol{x})$ [8].

Now let us look at the iterative time reversal experimental procedure [5, 6].
1. The initial broadcast is a tone burst (time train) with central frequency Ω from each of the mirrors $m = 1\ldots M$.
2. The resulting time trains $u_1(\boldsymbol{m'};t)$ are detected at mirrors $m' = 1\ldots M$.
3. The time trains $u_1(\boldsymbol{m'};t)$ are time reversed and rebroadcast from mirrors $m' = 1\ldots M$.
4. The resulting time trains $u_1(\boldsymbol{m''};t)$ are detected at mirrors $m'' = 1\ldots M$.
5. The time trains $u_1(\boldsymbol{m''};t)$ are time reversed and used as the input to step 1.

In the Fourier domain time reversal is equivalent to complex conjugation:

$$f(t) = \int \frac{d\omega}{2\pi} G(\omega) e^{-i\omega t} , \quad (9.65)$$

$$f(-t) = \int \frac{d\omega}{2\pi} G(\omega) e^{i\omega t} = \int \frac{d\omega}{2\pi} G(-\omega) e^{-i\omega t} , \quad (9.66)$$

$$= \int \frac{d\omega}{2\pi} G^*(\omega) e^{-i\omega t} , \quad (9.67)$$

where the last line follows from $G^*(-\omega) = G(\omega)$ for a real function. Thus in the Fourier domain the description of the basic step in the time reversal experimental procedure (1 to 4 above) is

$$u_1(m'', \omega) = \sum_k \sum_l w_k \mathbf{t}_{1k}(m'', l; \omega)\mathbf{t}_{k2}^*(l, m; \omega) f_2^*(m, \omega), \tag{9.68}$$

$$= \mathbf{K}_{12}(m'', m; \omega) f_2^*(m, \omega), \tag{9.69}$$

where $w_k = 1$ if the mirrors receive/broadcast the k component of the displacement ($w_k = 0$ otherwise) and $\mathbf{K}_{12}(m'', m; \omega)$ is the *time-reversal* matrix. Employ the notation $\nu = (i, m)$ to write K in the form

$$\mathbf{K}_{\mu\nu} = \sum_\eta \mathbf{t}_{\mu\eta}\mathbf{t}_{\eta\nu}^*, \tag{9.70}$$

so that it is apparent that

$$\mathbf{K}_{\mu\nu} = \mathbf{K}_{\nu\mu}^*, \tag{9.71}$$

that is, $\mathbf{K}$ is a Hermitian (or self-adjoint) matrix. Iteration of the basic step in the experimental procedure (add 5 above) corresponds to operating with K on what has gone before. Schematically,

$$r_0 = t e_0, \tag{9.72}$$

$$e_1 = t r_0^* = t t^* e_0^* = K e_0^*, \tag{9.73}$$

$$r_1 = t e_1^* = t t^* t e_0, \tag{9.74}$$

$$e_2 = t r_1^* = t t^* t t^* e_0^* = K^2 e_0^*, \tag{9.75}$$

$$\vdots \quad . \tag{9.76}$$

Thus $e_n = K^n e_0^*$ for $n = 1, 2, \ldots$ That is, the analytic description of repeated experimental cycling through steps 1 to 5 is a repeated application of the time-reversal matrix to the amplitudes of the initial broadcast. To see what this analytic description leads to, let us spend a few moments with Hermitian matrices [4]. For an $M \times M$ Hermitian matrix there is an eigenvalue problem

$$\sum_{m'=1}^M K(m, m')\phi_n(m') = \lambda_n \phi_n(m), \quad n = 1 \ldots M, \tag{9.77}$$

for which the eigenvalues λ_n are real and the eigenvectors are orthogonal

$$\sum_{m=1}^M \chi_{n'}(m)\phi_n(m) = \delta_{n,n'}, \tag{9.78}$$

where $\chi_n = \phi_n^\dagger$. If we expand e_0 in terms of the eigenvectors ϕ_n as here

$$e_0 = \sum_{n=1}^M a_n \phi_n, \tag{9.79}$$

then repeated application of K will pull out the eigenvector with the largest eigenvalue:

$$K^N e_0 = \sum_{n=1}^{M} a_n K^N \phi_n \longrightarrow \lambda_{n_{\max}}^N \phi_{n_{\max}} + \dots \tag{9.80}$$

This eigenvector $\phi_{n_{\max}}$ has M components each of which is a complex number or a magnitude/phase, $\phi_n(m) \leftrightarrow (R(m), \theta(m))$. It can be shown that $(R(m), \theta(m))$ are the magnitude/phase that, when used in the broadcast of tone bursts, will focus on the strongest scattering structure. Thus repeated use of steps 1 to 5 is a *training procedure* that finds these amplitudes/phases. If one has a velocity model of the homogeneous elastic system in which the inhomogeneity resides, then standard seismiclike procedures will make it possible to find the location of the strongest scattering structure in the inhomogeneity.

9.2.5
Time Reversal, a Numerical Example

Let us look at some features in a numerical example of time reversal. We consider the geometry shown in Figure 9.7; $M = 11$ mirrors at $x = 0, -5 \leq y \leq +5$ and $N = 19$ scatterers near $x = L = 10, -2 \leq y \leq +2$. The scatterers have strength $S(\mathbf{n}) = \exp(-r)$, where r is a random number from a uniform distribution $0 < r < 1$, Figure 9.8a. We form K for the case $\omega = ck = 7.8540$ $(c = 1)$ from t and the scattering strengths

$$\delta c(\mathbf{x})_{ijkl} \longrightarrow \sum_{n=1}^{N} S(\mathbf{n}) \delta(\mathbf{x} - \mathbf{x}_n) \delta_{ix} \delta_{jx} \delta_{kx} \delta_{lx} \tag{9.81}$$

using Eqs. (9.63) and (9.69) and the far-field approximation to the P-wave elastic Green function. The result is an 11×11 Hermitian matrix.

1. The initial set of displacement amplitudes $e_0(m)$ at the mirrors, $m = 1 \dots 11$, are 11 random numbers from a uniform distribution $0 < r < 1$. These produce an amplitude in the plane $x = L$

$$u_0(L, y; \omega) = \sum_{m=1}^{M} G(L, y | m; \omega) e_0(m) \tag{9.82}$$

   whose magnitude (before scattering) is shown in Figure 9.8b, open circles.

2. The amplitudes $u_0(\mathbf{n}, \omega)$ are scattered back to the mirrors where they become the amplitudes $e_1(m) = \sum K(m, m') e_0(m')$ for the first rebroadcast from the mirrors. This rebroadcast produces an amplitude in the plane $x = L$

$$u_1(L, y; \omega) = \sum_{m=1}^{M} G(L, y | m; \omega) e_1^*(m) \tag{9.83}$$

   whose magnitude (before scattering) is shown in Figure 9.8b, solid line.

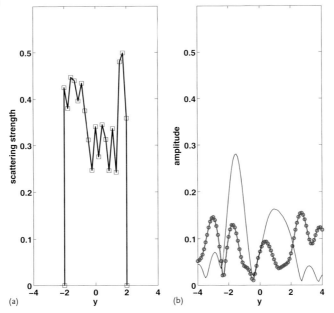

Fig. 9.8 Scattering Strength and Initial Amplitudes. (a) Scattering strength $\mathcal{S}(\boldsymbol{n})$ as a function of y_n. (b) Amplitude of first broadcast to the plane $x = L = 10$, $u_0(L, y; \omega)$, from Eq. (9.82) (circles) and $u_1(L, y; \omega)$ from Eq. (9.83) (solid line).

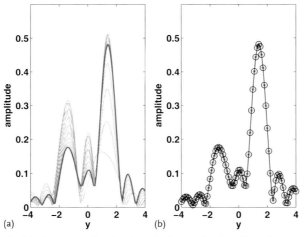

Fig. 9.9 Successive Amplitudes. (a) The amplitudes of broadcasts $n = 1 \ldots 12$, $u_n(L, y; \omega)$, as a function of y. (b) The amplitude of the final broadcast, $u_{12}(L, y; \omega)$, as a function of y. Compare $u_0(L, y; \omega)$ in Figure 9.8b to $u_{12}(L, y; \omega)$.

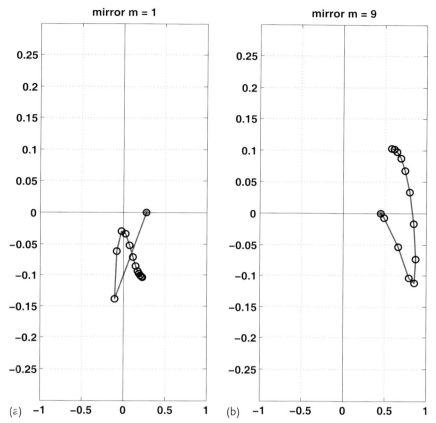

Fig. 9.10 Phase Evolution. The amplitudes $e_n(m)$ are complex numbers that evolve with successive broadcasts. (a) The amplitude $e_n(1)$ starts on the real axis at $(0.26, 0.0)$, takes a large step into the third quadrant, and evolves slowly to a location in the fourth quadrant, $(0.24, -0.62)$. (b) The amplitude $e_n(9)$ evolves qualitatively differently, from $(0.47, 0)$ to a location in the first quadrant, $(0.56, 0.62)$.

3. Repetition of the procedure produces a sequence of amplitudes in the $x = L$ plane, $u_n(L, y; \omega) = \sum G(L, y|m; \omega)e_n^*(m)$ $[e_n(m) = \sum K(m, m')e_{n-1}^*(m')]$, that are shown in Figure 9.9 a for $n = 1\ldots12$, that is, for 12 rebroadcasts.

4. The broadcast amplitude sequence $e_0, e_1, \ldots$ approaches a limit, that is, is *trained* by repetition of the time-reversal steps. This amplitude sequence produces a limiting displacement in the $x = L$ plane, shown in Figure 9.9b. The amplitudes $e(m)$ are complex numbers. In Figure 9.10 we show the amplitudes $e_n(1)$ and $e_n(9)$ for the 12 broadcasts. Each amplitude is initially a real number, see 1 above, that evolves to a location in the complex plane.

To keep the magnitude of $e(m)$ approximately constant, we norm the vector e to 1 after each TR step. (So would an experimental procedure.)

5. The trained amplitudes focus the broadcast onto the strongest scatterer if the scatterers are well separated and on the strongest scattering structure when not.

The discussion here, primarily pedagogic, is complemented by an example of current implementation, for example, [9].

9.3
Tomographies, Nonlinear, Inhomogeneous

9.3.1
Time of Flight

Consider the elastic system described by

$$\ddot{u} + \frac{1}{\tau_0}\dot{u} - c_0^2 u'' = c_0^2 \frac{\partial}{\partial x} \left(\sum_{i=1}^{N} R_i \delta(x - x_i) \left(\frac{\partial u}{\partial x} \right)^2 \frac{\partial u}{\partial x} \right) + F(t)\delta(x). \quad (9.84)$$

The localized inhomogeneity is modeled after the quartic anharmonicity examined in Chapter 3. This inhomogeneity differs from that in Eq. (9.11) in that strains at $x = x_i$ appear squared as part of $\delta c(x)$ and are time dependent. We use the definitions in Eq. (9.14) and the usual Fourier analysis with $F(x; \omega)$ and $u(x; \omega)$ to write

$$D(x; \omega)u(x; \omega) = c_0^2 \Lambda \frac{\partial}{\partial x} \left(f(x)S(x; \omega) \right) + F(\omega)\delta(x), \quad (9.85)$$

where

$$D(x; \omega) = -\omega^2 - i\frac{\omega}{\tau_0} - c_0^2 \frac{\partial^2}{\partial x^2} \quad (9.86)$$

and

$$S(x; \omega) = \int d\omega' \int d\omega'' \frac{\partial u(x; \omega - \omega' - \omega'')}{\partial x} \frac{\partial u(x; \omega')}{\partial x} \frac{\partial u(x; \omega'')}{\partial x}. \quad (9.87)$$

Represent $u(x; \omega)$ in the form $u(x; \omega) = u_1(x; \omega) + \Lambda u_2(x; \omega) + \ldots$ and find the equations

$$D(x; \omega)u_1(x; \omega) = F(\omega)\delta(x), \quad (9.88)$$

$$D(x; \omega)u_2(x; \omega) = c_0^2 \frac{\partial}{\partial x} \left(f(x)S_1(x; \omega) \right), \quad (9.89)$$

$$\vdots \quad (9.90)$$

where

$$S_1(x;\omega) = \int d\omega' \int d\omega'' \frac{\partial u_1(x;\omega-\omega'-\omega'')}{\partial x} \frac{\partial u_1(x;\omega')}{\partial x} \frac{\partial u_1(x;\omega'')}{\partial x}. \quad (9.91)$$

Solve these using the definition of the Green function $D(x;\omega)G(x|x';\omega) = \delta(x-x')$ with the result

$$u_1(x;\omega) = G(x|0;\omega)F(\omega), \quad (9.92)$$

$$u_2(x;\omega) = -c_0^2 \left(\int dx' \frac{\partial G(x|x';\omega)}{\partial x'} f(x)S_1(x';\omega) \right) = -c_0^2 T(x;\omega). \quad (9.93)$$

Use $u_1(x,\omega)$ in S_1 in the second line

$$S_1(x;\omega) = \int d\omega' \int d\omega'' H(\omega,\omega',\omega'') \frac{\partial G(x|0;\omega-\omega'-\omega'')}{\partial x} \frac{\partial G(x|0;\omega')}{\partial x}$$
$$\times \frac{\partial G(x|0;\omega'')}{\partial x}, \quad (9.94)$$

where

$$H(\omega,\omega',\omega'') = F(\omega-\omega'-\omega'')F(\omega')F(\omega''). \quad (9.95)$$

Then with the explicit form of the Green function $G(x|x';\omega) = (i/2c_0^2 k_\omega)\exp(-ik_\omega|x-x'|)$ we have

$$T(x;\omega) = -\left(\frac{1}{2c_0^2}\right)^4 \int dx' e^{ik_\omega(x-x')} f(x') \int d\omega' \int d\omega'' H(\omega,\omega',\omega'')e^{ik_\omega x'} \quad (9.96)$$

$$= -\left(\frac{1}{2c_0^2}\right)^4 b e^{ik_\omega x} \int d\omega' \int d\omega'' H(\omega,\omega',\omega''). \quad (9.97)$$

For a monochromatic source $F(\omega) = F_0[\delta(\omega-\Omega)+\delta(\omega+\Omega)]$ there are two types of terms,

$$\int d\omega' \int d\omega'' H(\omega,\omega',\omega'') = F_0^3 \big[3\delta(\omega-\Omega) + 3\delta(\omega+\Omega) + \delta(\omega-3\Omega)$$
$$+ \delta(\omega+3\Omega) \big], \quad (9.98)$$

those that return the frequencies in $F(\omega)$ and those that shift the frequency, $\delta(\omega-3\Omega) + \delta(\omega+3\Omega)$. For the terms in $\delta(\omega-\Omega)+\delta(\omega+\Omega)$ we have

$$u_2(x;\omega) = \frac{3b}{16}\frac{F_0^3}{c_0^6}[\delta(\omega-\Omega)+\delta(\omega+\Omega)]. \quad (9.99)$$

Combine this with u_1

$$u(x; \omega) \approx u_1(x; \omega) + \Lambda u_2(x; \omega) = \frac{iF_0}{2c_0^2 k_\omega} \left(1 - i\frac{3}{8}k_\omega b\Lambda \frac{F_0^2}{c_0^4}\right)$$
$$\times e^{ik_\omega x}[\delta(\omega - \Omega) + \delta(\omega + \Omega)]. \qquad (9.100)$$

Then following the discussion below Eq. (9.19) and using $F_0/c_0^2 = -ik_\Omega u_0$ and $\varepsilon_0 = ik_\Omega u_0$ we have

$$\frac{c_{\text{eff}}}{c_0} = 1 + \frac{3}{8}\frac{b}{\lambda}\Lambda\varepsilon_0^2. \qquad (9.101)$$

Thus there is a velocity shift that scales with strain squared.

For the terms that produce a frequency shift

$$\Lambda u_2(x; \omega) = \frac{b}{16}\Lambda\frac{F_0^3}{c_0^6}[\delta(\omega - 3\Omega) + \delta(\omega + 3\Omega)]. \qquad (9.102)$$

If we follow one of these terms, say $\delta(\omega - 3\Omega)$, back to x and t, we have

$$u_{3\Omega}(x, t) = \frac{b}{16}\Lambda\frac{F_0^3}{c_0^6}e^{i(k_{3\Omega} - 3\Omega t)} \qquad (9.103)$$

$$= -\frac{b}{16}\Lambda\varepsilon_0^3 e^{i(k_{3\Omega} - 3\Omega t)}. \qquad (9.104)$$

We can learn about a localized nonlinear scatterer in the path of a tone burst by (1) noticing a shift in the velocity of the center frequency proportional to the strain squared or by (2) noticing a Fourier component at 3 times the center frequency in the tone burst. The generalization is to a velocity shift proportional to ε_0^n and a Fourier component at $n + 1$ times the center frequency in the tone burst, where ε_0 is the strain amplitude at the center frequency in the tone burst. As with linear time-of-flight tomography, encounters with a localized nonlinear inhomogeneity by a suitable number of broadcasts will allow determination of its size, location, and nonlinear character.

9.3.2
Nonlinear Normal-Mode Tomography

Consider a resonant bar, $0 \leq x \leq L$, for which the displacement obeys the equation

$$\ddot{u} + \frac{1}{\tau_0}\dot{u} - c_0^2 u'' = c_0^2 \Lambda \frac{\partial}{\partial x}\left(f(x)\left(\frac{\partial u}{\partial x}\right)^2 \frac{\partial u}{\partial x}\right) + F(t)\delta(x). \qquad (9.105)$$

Following the scheme used above for the linear resonant bar problem, see below Eq. (9.24), we use the compete set of orthonormal states associated with $\ddot{u} - c_0^2 u'' = 0$ to write $u(x, t)$ in the form

$$u = \sum_{m=1}^{M} a_m(t)\phi_m(x). \qquad (9.106)$$

Substitute into Eq. (9.105) and take the inner product of the resulting equation with $< n|$. Find using one integration by parts and rearrangement

$$\ddot{a}_n + \frac{1}{\tau_0}\dot{a}_n + c_0^2 k_n^2 a_n = -c_0^2 \sum_{m=1}^{M} \sum_{m'=1}^{M} \sum_{m''=1}^{M} k_n k_m k_{m'} k_{m''} \tag{9.107}$$

$$\langle \hat{n} | f(x) | \hat{m} \hat{m}' \hat{m}'' \rangle a_m(t) a_{m'}(t) a_{m''}(t) \tag{9.108}$$

$$+ F(t)\langle n|\delta(x)\rangle . \tag{9.109}$$

Define

$$G_{nmm'm''} = \frac{2L}{3} \langle \hat{n} | f(x) | \hat{m}\hat{m}'\hat{m}'' \rangle , \tag{9.110}$$

where the factor of $2L/3$ in this definition makes G dimensionless and equal to 1 for $f(x)$ constant, a balancing factor of $2/3$ is absorbed in a redefinition of Λ and $X_n = \langle n|\delta(x)\rangle$. Write

$$\ddot{a}_n + \frac{1}{\tau_0}\dot{a}_n + c_0^2 k_n^2 a_n = -c_0^2 \Lambda L^{-1} \sum_{m=1}^{M} \sum_{m'=1}^{M} \sum_{m''=1}^{M} k_n k_m k_{m'} k_{m''}$$

$$\times G_{nmm'm''} a_m(t) a_{m'}(t) a_{m''}(t) + F(t)X_n . \tag{9.111}$$

Employ a Fourier description of $a(t)$ and $F(t)$ to find

$$D_n(\omega)a_n(\omega) = -c_0^2 \Lambda L^{-1} \sum_{m=1}^{M} \sum_{m'=1}^{M} \sum_{m''=1}^{M} k_n k_m k_{m'} k_{m''} G_{nmm'm''} J(\omega) + F(\omega)X_n ,$$

$$\tag{9.112}$$

where

$$J(\omega) = \int d\omega' \int d\omega'' a_m(\omega') a_{m'}(\omega'') a_{m''}(\omega - \omega' - \omega'') . \tag{9.113}$$

Up to this point we have just been arranging Eq. (9.105) in preparation for doing perturbation theory. For $F(\omega) \propto \delta(\omega \pm \Omega)$, $\Omega \approx c_0 k_n$, the amplitude a_n is relatively large and we can write

$$a_m(\omega) = a_n^0(\omega)\delta_{mn} + \Lambda g_m(\omega) + \Lambda^2 h_m(\omega) + \dots \tag{9.114}$$

Substitution of this form into Eq. (9.111) and arrangement according to powers of Λ lead to $(m = n)$

$$D_n(\omega)a_n^0(\omega) = F(\omega)X_n , \tag{9.115}$$

$$D_n(\omega)g_n(\omega) = -c_0^2 \Lambda L^{-1} k_n^4 G_{nnnn} J_0(\omega) , \tag{9.116}$$

$$\vdots \tag{9.117}$$

where

$$J_0(\omega) = \int d\omega' \int d\omega'' a_n^0(\omega') a_n^0(\omega'') a_n^0(\omega - \omega' - \omega'') . \tag{9.118}$$

Solve for a_n^0 and use the result to find J_0:

$$J_0 = X_n^3 \int d\omega' \int d\omega'' H(\omega, \omega', \omega'') F(\omega') F(\omega'') F(\omega - \omega' - \omega'') , \tag{9.119}$$

where

$$H(\omega, \omega', \omega'') = \frac{1}{D_n(\omega)} \frac{1}{D_n(\omega')} \frac{1}{D_n(\omega'')} . \tag{9.120}$$

In order to work this out use $F(\omega) = F_0[\delta(\omega - \Omega) + \delta(\omega + \Omega)]$ and find the terms in J_0 proportional to $\delta(\omega \pm \Omega)$. These are

$$J_0 = 3F_0^3 X_n^3 \left(\frac{1}{D_n(\Omega)^2 D_n(-\Omega)} \delta(\omega - \Omega) + \frac{1}{D_n(\Omega) D_n(-\Omega)^2} \delta(\omega + \Omega) \right) . \tag{9.121}$$

For the terms in $a_n \approx a_n^0 + \Lambda g_n$ proportional to $\delta(\omega - \Omega)$ we have (see Eq. (9.40))

$$a_n(\Omega) = \frac{F_0 X_n}{D_n(\Omega)} \delta(\omega - \Omega) - 3c_0^2 \Lambda k_n^4 G_{nnnn} \frac{F_0^3 X_n^3}{L D_n(\Omega)^3 D_n(-\Omega)} \delta(\omega - \Omega) \tag{9.122}$$

$$= \frac{1}{D_n(\Omega) + 3c_0^2 k_n^2 \Lambda \frac{k_n^2 F_0^2 X_n^2}{L |D_n(\Omega)|^2} G_{nnnn}} F_0 X_n \delta(\omega - \Omega) , \tag{9.123}$$

where $D_n(-\Omega) = D_n(\Omega)^*$. The denominator can be rearranged as

$$-\omega^2 - i\frac{\omega}{\tau_0} + c_0^2 k_n^2 \left(1 + 3\Lambda \frac{k_n^2 F_0^2 X_n^2}{L |D_n(\Omega)|^2} G_{nnnn} \right) , \tag{9.124}$$

so that the frequency shift can be read off as

$$\omega_n^2 \approx c_0^2 k_n^2 \left(1 + 3\Lambda \frac{k_n^2 F_0^2 X_n^2}{L |D_n(\Omega)|^2} G_{nnnn} \right) \tag{9.125}$$

and

$$\frac{\omega_n - \omega_n^0}{\omega_n^0} \approx \frac{3}{4} \Lambda G_{nnnn} (\varepsilon_n^r)^2 , \tag{9.126}$$

where $\varepsilon_n^r = k_n F_0 X_n / (\sqrt{L/2} D_n(\omega_n))$, and $u_n(0) \approx (F_0 X_n / D(\omega_n))$.

1. The terms in $\delta(\omega \pm 3\Omega)$ have an amplitude proportional to $(\varepsilon_n^r)^3$ and have the same information about the localized inhomogeneity as the frequency shift, that is, are proportional to ΛG_{nnnn}.

2. One could launch two first-order strain fields of different frequency, $F(\omega) = F_1\delta(\omega \pm \Omega_1) + F_2\delta(\omega \pm \Omega_2)$, where Ω_1 and Ω_2 are near resonance frequencies, and find frequency shifts in each resonance frequency proportional to the amplitude of the strain field of the other. Sweeping frequency ω near Ω_1 with Ω_2 fixed would lead to

$$\frac{\omega_1 - \omega_1^0}{\omega_1^0} \approx \Lambda G_{1221}\left(\varepsilon_2^r\right)^2 ; \qquad (9.127)$$

sweeping frequency ω near Ω_2 with Ω_1 fixed would lead to

$$\frac{\omega_2 - \omega_2^0}{\omega_1^0} \approx \Lambda G_{2112}\left(\varepsilon_1^r\right)^2 , \qquad (9.128)$$

etc. In this way one could build up a set of integrals G_{ijkl} to use in tackling the problem of finding $f(x)$.

Nonlinear normal-mode tomography as a practical scheme is subject to many of the considerations described in Section 9.2.3. It has one great advantage: two samples are not required. The normal-mode frequency shifts that are the basic data are built up in a sample by change in drive amplitude. One makes a study of the amplitude dependence of M normal modes. Then the M coefficients of

$$\frac{\omega_n - \omega_n^0}{\omega_n^0} = C_n\left(\varepsilon_n^r\right)^2 , \quad n = 1\ldots M , \qquad (9.129)$$

$C_1 \ldots C_M$, are formed into the vector $\boldsymbol{e}_X$ and treated as described above. From that point forward the only detail that is different is the integrand of the integral that determines $\boldsymbol{e}_M$, G_{nnnn} vs. G_{nn}.

9.3.3
Nonlinear Time-Reversal Tomography

To discuss nonlinear time reversal we adopt a somewhat general approach by considering a localized inhomogeneity that is time dependent, that is, $\delta c_{ijkl}(\boldsymbol{x}) \rightarrow \delta c_{ijkl}(\boldsymbol{x}, t)$. Then in place of the last line in Eq. (9.56) find

$$u_1^{(2)}(\boldsymbol{x}; \omega) = \sum_{m=1}^{M}\sum_{ijkl}\int d\boldsymbol{x}' G_{1i}(\boldsymbol{x}, \boldsymbol{x}'; \omega)\int d\omega' \frac{\partial}{\partial x_j'}$$

$$\times \left(\delta c_{ijkl}(\boldsymbol{x}', \omega')\frac{\partial U_k^{(m)}(\boldsymbol{x}', \omega - \omega')}{\partial x_l'}\right) , \qquad (9.130)$$

and in place of Eq. (9.57) find

$$u_1^{(2)}(\boldsymbol{x}; \omega) = -\sum_{m=1}^{M}\sum_{ijkl}\int d\boldsymbol{x}' \frac{\partial G_{1i}(\boldsymbol{x}, \boldsymbol{x}'; \omega)}{\partial x_j'}$$

$$\times \int d\omega' \delta c_{ijkl}(\boldsymbol{x}', \omega')\frac{\partial G_{k2}(\boldsymbol{x}', \boldsymbol{m}; \omega - \omega')}{\partial x_l'} f_2(\boldsymbol{m}; \omega - \omega') . \qquad (9.131)$$

1. Take the time dependence of $\delta c_{ijkl}(\mathbf{x}', t)$ to be such that

$$\delta c_{ijkl}(\mathbf{x}', \omega) = \sum_\alpha \delta c_{ijkl}(\mathbf{x}', \omega) \left[\delta(\omega - \omega_\alpha) + \delta(\omega + \omega_\alpha) \right] . \qquad (9.132)$$

2. Take the time dependence of $f_2(\mathbf{m}, t)$ to be such that

$$f_2(\mathbf{m}, \omega) = f_2(\mathbf{m}) \delta(\omega - \Omega). \qquad (9.133)$$

Then from Eq. (9.131) for the upper side band at $\Omega + \omega_\alpha$, $\omega' = \omega_\alpha$, we have

$$u_1^{(2)}(\mathbf{x}; \omega) = -\delta(\omega - \Omega - \omega_\alpha) \sum_{m=1}^{M} \sum_{ijkl} \int d\mathbf{x}' \delta c_{ijkl}(\mathbf{x}', \omega_\alpha)$$

$$\times \frac{\partial G_{1i}(\mathbf{x}, \mathbf{x}'; \omega)}{\partial x'_j} \frac{\partial G_{k2}(\mathbf{x}', \mathbf{m}; \Omega)}{\partial x'_l} f_2(\mathbf{m}) , \qquad (9.134)$$

and analogously to Eq. (9.59)

$$u_1^{(2)}(\mathbf{m}'; \omega) = -\delta(\omega - \Omega - \omega_\alpha) \sum_{m=1}^{M} \sum_{ijkl} \int d\mathbf{x}' \delta c_{ijkl}(\mathbf{x}', \omega_\alpha)$$

$$\times \frac{\partial G_{1i}(\mathbf{m}', \mathbf{x}'; \omega)}{\partial x'_j} \frac{\partial G_{k2}(\mathbf{x}', \mathbf{m}; \Omega)}{\partial x'_l} f_2(\mathbf{m}) \qquad (9.135)$$

$$= -\delta(\omega - \Omega - \omega_\alpha) \sum_{m=1}^{M} \mathbf{t}_{12}(\mathbf{m}', \mathbf{m}; \Omega + \omega_\alpha, \Omega) f_2(\mathbf{m}) , \qquad (9.136)$$

where $\mathbf{t}_{12}(\mathbf{m}', \mathbf{m}; \Omega + \omega_\alpha, \Omega)$ is the transfer matrix, cf. Eq. (9.59). For $\delta c_{klij}(\mathbf{x}', \omega_\alpha) = \delta c_{ijkl}(\mathbf{x}', \omega_\alpha)$, using the arguments below Eq.(9.63), we have

$$\mathbf{t}_{12}(\mathbf{m}', \mathbf{m}; \Omega + \omega_\alpha, \Omega) = \mathbf{t}_{21}(\mathbf{m}, \mathbf{m}'; \Omega, \Omega + \omega_\alpha) , \qquad (9.137)$$

a form of reciprocity. The basic thing going on here that is different from linear time reversal is that the broadcast, mirror to scatterer to mirror, involves a frequency shift, $\Omega \to \Omega \pm \omega_\alpha$.

We introduce a frequency protocol into the time-reversal procedure:

1. The initial broadcast is a tone burst (time train) with central frequency Ω from each of the mirrors $m = 1 \ldots M$.
2. The resulting time trains $u_1(\mathbf{m}'; t)$ detected at mirrors $m' = 1 \ldots M$ are Fourier analyzed and the component of each time train with center frequency $\omega = \Omega + \omega_\alpha$, $u_1^{(+)}(\mathbf{m}'; t)$, is constructed.
3. The time trains $u_1^{(+)}(\mathbf{m}'; t)$ are time reversed and rebroadcast from mirrors $m' = 1 \ldots M$.
4. The resulting time trains $u_1(\mathbf{m}''; t)$ detected at mirrors $m'' = 1 \ldots M$ are Fourier analyzed and the Fourier component of each time train at Ω, $u_1^{(+-)}(\mathbf{m}''; \Omega)$, is found.

5. From $u_1^{(+-)}(\boldsymbol{m}''; \Omega)$ the component of the time train with central frequency Ω, $u_1^{(+-)}(\boldsymbol{m}''; t)$, is constructed, possibly to be used iteratively in steps 1 to 4.

Steps 1 to 4 are described by

$$u_1^{+-}(\boldsymbol{m}''; \Omega) = \sum_m \mathbf{K}_{12}(\boldsymbol{m}'', \boldsymbol{m}, \Omega, \Omega; \omega_\alpha) f_2^*(\boldsymbol{m}; \Omega) , \qquad (9.138)$$

where

$$\mathbf{K}_{12}(\boldsymbol{m}'', \boldsymbol{m}, \Omega, \Omega; \omega_\alpha) = \sum_k \sum_l w_k \mathbf{t}_{1k}(\boldsymbol{m}'', l, \Omega, \Omega + \omega_\alpha) \mathbf{t}_{k2}^*(l, \boldsymbol{m}, \Omega + \omega_\alpha, \Omega) ,$$

$$(9.139)$$

where $w_k = 1$ for each displacement component received/broadcast by the mirrors. Using Eq. (9.137) it follows that the time-reversal operator $\mathbf{K}_{12}(\boldsymbol{m}'', \boldsymbol{m}, \Omega, \Omega; \omega_\alpha)$ is Hermitian.

Thus the experimental iteration of steps 1 to 5 or a DORT treatment [6] of the matrix $\mathbf{K}_{12}(\boldsymbol{m}'', \boldsymbol{m}, \Omega, \Omega; \omega_\alpha)$ will provide a set of time trains that, when broadcast from the mirrors, will focus on the "largest" nonlinear scattering structure in the system. Further manipulation will allow examination of a hierarchy of nonlinear scattering structures ordered by strength.

References

1 http://www.merriam-webster.com/, 15 May 2009.

2 Migliori, A. and Sarrao, J.L. (1997) *Resonant Ultrasound Spectroscopy*, John Wiley & Sons, Inc., New York.

3 Bucur, V.I. and Rasolofosaon, P.J.N. (1998) *Ultrasonics*, **36**, 813.

4 Merzbacher, E. (1998) *Quantum Mechanics*, John Wiley & Sons, Inc New York.

5 Prada, C. Thomas, J.L., and Fink, M. (1995) The iterative time reversal process: analysis of the convergence. *J. Acoust. Soc. Am.*, **97**, 62–71.

6 Prada, C., Manneville, S., Spoliansky, D., and Fink, M. (1996) Decomposition of the time reversal operator. Detection and selective focusing on two scatterers. *J. Acoust. Soc. Am.*, **99**, 2067–2076.

7 Aki, K. and Richards, P.G. (1980) *Quantitative Seismology*, W.H. Freeman, San Francisco.

8 Landau, L.D. and Lifshitz, E.M. (1999) *Theory of Elasticity*, Butterworth and Heinemann, Oxford.

9 Prada, C., de Rosny, J., Clorennec, D., Minonzio, J.G., Bernier, L., Billand, P., Hibral, S., and Folegot, T. (2007) Experimental detection and focusing in shallow water by decomposition of the Time Reversal Operator. *J. Acoust. Soc. Am.*, (**122**), 761–768.

10
Quasistatic Measurements

This chapter is about quasistatic measurements. We examine a collection of quasistatic measurements that establish the physical picture that underlies our understanding of mesoscopic elastic material and we examine the result of a further series of measurements that flesh out that understanding. We begin in Section 10.1 with a discussion of numbers, epoxy, and neutrons that suggests/supports the physical picture we advocate. In Section 10.2 we look at quasistatic stress-strain data on a variety of systems, several sandstones, a ceramic (TBC), and a soil. We give particular attention to the notion of modulus. Section 10.3 is devoted to the description of the influence of auxiliary fields on quasistatic behavior; Section 10.3.1 discusses vycor glass, sandstone, saturation, and Section 10.3.2 sandstone and temperature. The inverse problem associated with quasistatic data, that is, learning about the characteristics of Preisach space, is discussed in Section 10.4; Section 10.4.1 concerns simple stress protocols, and Section 10.4.2 discusses elaborate stress protocols. In Section 10.4.3 a simple, approximate "solution" to the inverse problem is introduced to allow comparison among quasistatic data sets and to illustrate the quantitative connection between quasistatic behavior and nonlinear dynamics.

10.1
Some Basic Observations

The simplest measurement that points to an important quality of mesoscopic elastic materials is the velocity of sound. A Berea sandstone is approximately a bonded set of quartz grains. A lightly packed system of glass beads, for example, under a pressure of order 10 atmospheres, is an aggregate of quartz grains that encounter one another through a contact network. The nominal velocities of sound in quartz, Berea sandstone, and a silica contact network are 6, 2, and 1 km/s [1–3]. These three systems have about the same density, ≈ 2–$3\,\mathrm{gm/cm^3}$. Consequently the relevant physical variables, the compressibilities, are approximately $K_Q \approx 8 \cdot 10^{11}\,\mathrm{dyne/cm^2}$ (80 GPa), $K_{SS} \approx 8 \cdot 10^{10}\,\mathrm{dyne/cm^2}$ (8 GPa), and $K_{CN} \approx 2 \cdot 10^{10}\,\mathrm{dyne/cm^2}$ (2 GPa). The simplest meaning of K is that a pressure δP produces a macroscopic strain

Nonlinear Mesoscopic Elasticity: The Complex Behaviour of Granular Media including Rocks and Soil. Robert A. Guyer and Paul A. Johnson
Copyright © 2009 WILEY-VCH Verlag GmbH & Co. KGaA, Weinheim
ISBN: 978-3-527-40703-3

given by

$$\varepsilon \approx \frac{1}{K} \delta P \, . \tag{10.1}$$

Thus for fixed δP, $\varepsilon_{CN} \approx 4 \times \varepsilon_{SS} \approx 40 \times \varepsilon_Q$. The silica component of all three of these systems is essentially the same since $\varrho_Q \approx 2.7 \, \mathrm{gm/cm^3}$, $\varrho_{SS} \approx 2.3 \, \mathrm{gm/cm^3}$, and $\varrho_{CN} \approx 2 \, \mathrm{gm/cm^3}$. The extra strain beyond that which would be present if it were taken up by the quartz grains resides in a small volume of the system, for example, in the mortar of a bricks-and-mortar scenario or in the contact area of a contact network picture of glass beads. As these volumes are small, the strains in them are large. Let us make two estimates.

1. Bricks and mortar, Section 4.4.3 the bricks-and-mortar model of Figure 10.1 we assign elastic constants K_Q to the bricks and K_m to the mortar and contact. Upon application of a uniform compressive stress, σ, we find strain across the system (as the system is uniform, we need only study the unit cell) given by

$$
\begin{aligned}
\bar{\varepsilon} = \frac{\delta(a+b)}{a+b} &= \left(\frac{1}{K_Q + \frac{b}{a} K_m} + \frac{1}{\frac{a}{b} K_m + K_c} \right) \sigma \, , \\
&= \left(\frac{1}{1 + \frac{b}{a} \frac{K_m}{K_Q}} + \frac{1}{\left(1 + \frac{a}{b}\right) \frac{K_m}{K_Q}} \right) \varepsilon_Q \, , \\
&= \chi \varepsilon_Q \, , \tag{10.2}
\end{aligned}
$$

where $K_c = K_m$ is the elastic constant of the contact and $\varepsilon_Q = \sigma/K_Q$, the strain appropriate to a system of uniform quartz. Suppose we choose $b/a \approx 0.1$; the mortar is about 10% of the volume of the sample. To have $\bar{\varepsilon} \approx 10 \times \varepsilon_Q$ requires $K_m \approx 0.01 K_Q$. Necessarily the strain in the mortar is approximately a/b times the observed macroscopic strain $\bar{\varepsilon}$. A bricks-and-mortar picture of a Berea sandstone suggests that the mortar is two orders of magnitude softer than the grains and that the strains in the mortar are an order of magnitude greater than the observed macroscopic strain.

2. Hertzian contact, Section 4.2. Suppose a Hertzian contact is prestrained to h_0, where $a_0^2 \approx R h_0$ [4]. A pressure increase on the system of δP will require each contact to support an additional force of order $\delta F = \delta P R^2$. The contact does this by crushing the additional volume $a_0^2 \delta h$ into its interior, the volume a_0^3, and producing the internal pressure $\delta P_{\mathrm{in}} = K_Q \delta h / a_0$. This internal pressure pushes on a_0^2 with force $\delta P_{\mathrm{in}} a_0^2$ that must balance $\delta P R^2$. Thus

$$\bar{\varepsilon} \approx \frac{\delta h}{R} \approx \sqrt{\frac{R}{h_0} \frac{\delta P}{K_Q}} = \sqrt{\frac{R}{h_0}} \varepsilon_Q \, , \tag{10.3}$$

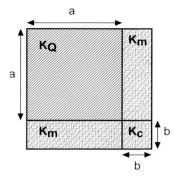

Fig. 10.1 Bricks and mortar. To illustrate the effect of mortar
on a bricks-and-mortar system, the bricks are assigned elastic
constant K_Q, the two slabs of mortar K_m, and the contact $K_c = K_m$. Then, under compression, K_Q and K_m act in parallel, as
do K_m and K_c. The (K_Q, K_m) unit is in series with the (K_m, K_c)
unit, giving the result in Eq. (10.2).

that is, the effective elastic constant is $K_{CN} \approx \sqrt{(h_0/R)} \times K_Q$. The strain in
the contact volume, a_0^3, is

$$\varepsilon_{a_0} \approx \delta h / a_0 \approx \sqrt{\frac{R}{h_0}} \bar{\varepsilon}. \qquad (10.4)$$

For $\bar{\varepsilon} \approx 10 \times \varepsilon_Q$ we need $h_0 \approx 0.01R$. Then, the strain in the contact volume
is about 10 times the observed macroscopic strain.

The characteristic of these two models is that the grains are moderated in their
interaction with one another by a physically small system that is elastically soft.
This system carries strains much larger than those in the grains and much larger than the apparent strain on the sample, the macroscopic strain. For the case of
a Berea sandstone the picture, deduced from numbers above, was confirmed in
an ingenious experiment by Gist [5, 6]. The P-wave velocity of a *dry* Berea sandstone was measured as a function of pressure, 0.1 MPa < P < 60 MPa, Figure 10.2.
The velocity (Young's modulus) varied by a factor of 2 (4) over this pressure range
with a mild hysteresis in which the velocity (modulus) at a given pressure was
greater on achieving the pressure from above than from below. The pore space
was then filled with a low-viscosity epoxy the bulk of which was centrifuged out
to a saturation of 35%. The P-wave velocity was measured a second time as the
sample was again taken through the same pressure range. The velocity was found
to be independent of pressure with a value somewhat greater than that at the
highest pressure in the initial set of measurements. The pore space geometry per-
mits the epoxy to reach those places of soft elastic material where it stiffens that
material or shunts forces around that material. In the bricks-and-mortar picture
the epoxy either stiffens the mortar or makes the mortar irrelevant. For exam-

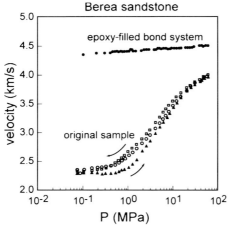

Fig. 10.2 Epoxy in the Pore Space. Gist measured the velocity of sound as a function of pressure to detect the elastic state of a sample of dry Berea sandstone. On the first pressure increase the velocity, triangles, was lower than on all subsequent pressure loops, open squares ($\downarrow$) and open circles ($\uparrow$), which show very modest hysteresis in which *the velocity stays in*. After coating the walls of the pore space with epoxy, the velocity (read: elastic state) is almost unchanged by pressure, filled squares.

ple, if K_c in the first of Eqs. (10.2) is $K_{epoxy} \geq K_Q$, then the mortar is irrelevant. For the case of sandstones the *coup de grace* was provided by the neutron scattering measurements of Frischbutter *et al.* [7]. This work was followed up by Darling and TenCate [8], and we show results from these authors for a Fontainebleau sandstone in Figure 10.3. The experimental situation is shown in Figure 10.4. A sample in a uniaxial stress apparatus is placed in a neutron beam so that the Bragg scattering from lattice planes can be monitored. The Bragg scattering yields the spacing between lattice planes. Change in that spacing as the applied stress is changed yields the *internal* strain as a function of applied stress. An external-strain gauge attached to the sample measures the macroscopic strain. This type of measurement is a two-strain gauge or an internal/external-strain gauge measurement. As the grains constitute typically more than 80% of the sample volume, the internal-strain gauge sees the strain in the grains (bricks); the external-strain gauge sees the sum of bricks and mortar. From Figure 10.3 we note the following:

1. An applied stress of 20 MPa causes a strain of 0.3 millistrain in the quartz grains (thus a Young's modulus of order 60 GPa) and a strain of 1.5 millistrain in the sample as a whole (thus a Young's modulus of order 10 MPa), about a factor of 5. See above.

2. The macroscopic strain is nonlinear and hysteretic; the internal strain is linear and shows no evidence of hysteresis.

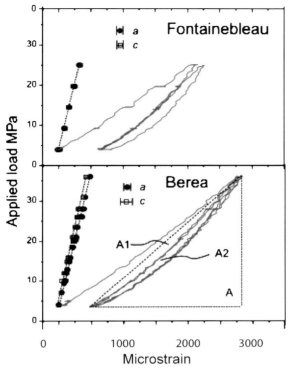

Fig. 10.3 Neutron Measurement on Fontainebleau Sandstone. The strain (microstrain) as a function of stress (MPa) for a pressure protocol ($\uparrow, \downarrow, \uparrow, \downarrow$). The strain in the grains, measured by the neutron strain gauge, is shown as five points and dashed line. The macroscopic strain, measured with a conventional strain gauge, solid line, has an initial evolution due to the "first" visit to a stress region followed by a *strain stays in* hysteresis loop. An internal strain of about 300 microstrain in the grains is associated with a macroscopic strain of 1500 microstrain when ≈ 20 MPa uniaxial stress is applied. See Figure 10.4.

Based on a qualitative picture, with strong confirmation in a limited domain (sandstones?), there is the suggestion that mesoscopic elastic elements come to dominate the behavior of a material when they reside in the material in such a way that the strain cannot be shunted around them. These elastic elements carry the hysteresis and the extreme nonlinearity. This observation is not intended as dogma but as a prod to thinking as we proceed.

10.2
Quasistatic Stress-Strain Data; Hysteresis

Consider the stress-strain data shown in Figures 10.3, 10.5, 10.6, 10.7, and 10.8. They are data from Fontainebleau sandstone [8], Berea sandstone [9], Castlegate

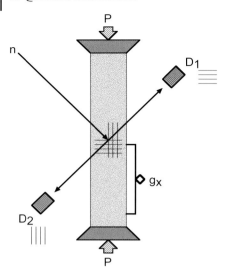

Fig. 10.4 Two-Strain Gauge Measurement. A sample, subject to an applied uniaxial stress, is outfitted with an external, macroscopic strain gauge, g_x. A neutron beam is incident on the sample so that Bragg scattering from suitably oriented planes of atoms can be monitored, detectors D_1 and D_2. The evolution of the spacing between planes of atoms (with attending evolution of the Bragg pattern) as the stress changes comprise an internal strain gauge.

sandstone [10], a plasma-sprayed thermal barrier coating [11], and a soil [12] under uniaxial stress. With the exception of the soil these stress-strain loops close on the time sale of measurement and are repeated on subsequent repetition of the stress protocol. As illustrated in Figure 10.3, a first visit to a stress range may bring about a strain evolution that is not recovered when the stress is returned to the initial value. This could be due to nontrivial strain evolution, for example, rearrangement [13–15] (the Kaiser effect is a particular example of this [16]) or to the limited range of the stress protocol, for example, possibly the strain will recover under tension. We are particularly interested in repeated hysteresis loops over a repeated stress range as it is these that exhibit the evidence of the working of mesoscopic elastic elements.

1. The typical data set involves strain of order 1 millistrain (Castlegate, 2–5 millistrain, and soil, 1–100 millistrain, are exceptions) and stress of order 10 MPa (soil under stress of order 0.2 MPa is an exception).
2. The typical data set is collected on a time scale of $> 10^{+3}$ s. For example, the pressure protocol in the two-strain-gauge experiment, Figure 10.3, involved pressure changes at 3 MPa/minute with 15 minute stops at 5 values of the applied load; this represents a time scale of order 3 hours. Unfortunately, time is not consistently reported in experiments. Roughly speaking, quasistatic measurements are on a time scale much longer than the 10^{-3} s that characterizes dynamic measurements.

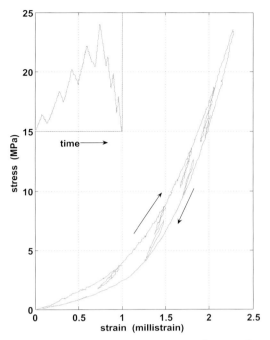

Fig. 10.5 Berea Sandstone 1. The strain (millistrain) of a room-dry Berea sandstone due to the uniaxial stress (MPa) from the protocol shown in the inset. There is a single exterior hysteresis loop with seven interior hysteresis loops. The interior loops have $d\sigma/d\varepsilon$ larger than the value of $d\sigma/d\varepsilon$ on the nearby exterior hysteresis loop. See Figure 10.10 and Figures 10.26–10.28.

3. The elastic behavior is very nonlinear. For Berea sandstone, Figure 10.5, a change in applied pressure of $\Delta P = 25$ MPa brings about a change in Young's modulus from about $E_{\text{low}} = 2$ GPa to $E_{\text{high}} = 25$ GPa or $\beta_E = (E_{\text{high}} - E_{\text{low}})/\Delta P \approx 1000$. In aluminum, a change in pressure of 15 GPa brings about a change in bulk modulus of about 60 GPa, $\beta_K \approx 4$ [17].

4. In all experiments where it can be noted the data are described by the statement *the strain stays in*. Other measurements on soils, ceramics, concrete, bone, and biomechanical systems are in accord with this.

5. In all cases these systems have endpoint memory. Most researchers call attention to the relevance of time scale to their results while not pursuing the question quantitatively. In the case of Berea sandstone a very careful test of endpoint memory, intended to distinguish it from slow relaxation, confirms this [18, 20]. See the discussion of relaxation phenomena in Section 10.3.2.

6. Moduli. There are three possible moduli that can be associated with a sample: one-way moduli, interior loop moduli, and dynamic moduli. We will use Berea sandstone as an example. See Figure 10.10.

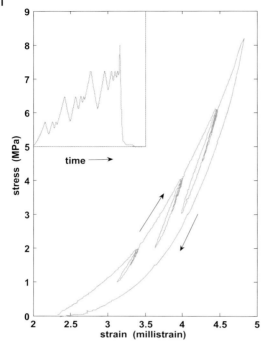

Fig. 10.6 Castlegate Sandstone. The strain (millistrain) of a room-dry Castlegate sandstone due to uniaxial stress (MPa). Note the strain axis; a strain due to a "first" visit to a stress region is not shown, cf. Figure 10.3. This sandstone has about the same strain as the Berea sandstone in Figure 10.5, although the maximum stress is only about 11 MPa. See Figures 10.26–10.28.

a. In Figure 10.5 there is a large stress-strain loop (exterior loop) with seven interior loops. At each point along the exterior loop a slope $d\sigma/d\varepsilon$ can be formed. This modulus is a *one-way* modulus in that the strain involved is that from moving the stress in one direction only. The inner loops give evidence for what happens when the stress is reversed – the strain enters the interior of the exterior loop. The one-way moduli are hysteretic with a discontinuity at stress reversal at the top/bottom of the exterior loop.

b. A two-way modulus is defined for the interior loops by drawing a line from end to end (inset) and using the resulting slope. It is apparent from Figures 10.5–10.8 that the inner loop, two-way, moduli are greater than (or possibly equal to) the one-way moduli.

c. A dynamic modulus can be defined at any point on the exterior loop from a measurement of the speed of sound and $K = \varrho c^2$, Figure 10.2 [5, 10, 12, 21]. The dynamic modulus is a two-way modulus.

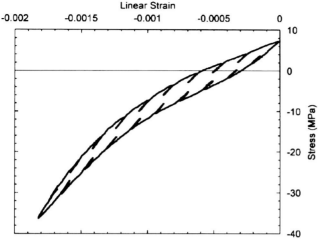

Fig. 10.7 Thermal Barrier Coating, TBC. The strain of a TBC sample, at $T = 800^\circ$ C, due to uniaxial stress (MPa). These data are reported without the rectified sign convention that is used throughout this book. These are compression tests. For a TBC, careful sample preparation is called for as the elastic system being tested starts out as a coating. There are interesting effects from a "first" visit to a stress region and stress-rate effects associated with these systems. See Figures 10.26–10.28.

d. The difference between these three moduli is the quality of the stress change, whether they are one-way or two-way, the size of the stress change (a few MPa for the two-way loops, less than 0.1 MPa for a dynamic modulus), and the time scale of the stress change (10–100 s for a two-way loop, less than a millisecond for a dynamic modulus). In a recent measurement TenCate [19] found that for some materials the quasistatic hysteresis loop was senstive to the stress rate, Figure 10.9. For Berea sandstone the loop closes as the stress rate approaches zero, whereas beyond a certain stress rate it remains approximately constant. In the case of Meule sandstone the loop size is approximately independent of stress rate.

e. While there has been a good deal of discussion in the literature about the relationship of these various moduli, it is hard to see why they are expected to be related. It is apparent that for the systems under discussion stress can bring about displacements that are not recovered on stress reversal [22]. Some of these displacements are in the one-way and two-way moduli, so that there is no appropriate comparison of these moduli with a dynamic modulus. This is too strong. Certainly the one-way moduli need have no relationship to two-way and dynamic moduli. However, if there are no issues of time scale [19], the limit of the modulus from a sequence of inner loops with smaller and smaller stress excursions should approach the dynamic modulus. The belief is that the dynamic modulus involves such small stress excursions that none of the displacements

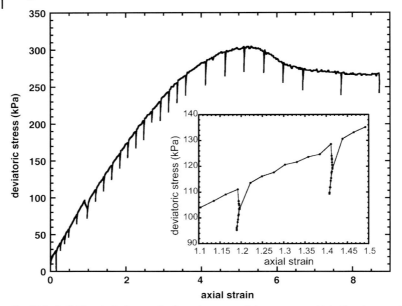

Fig. 10.8 Soil. The strain (percent) of a wet soil due to uniaxial stress (MPa). The soil sample is jacketed and a confining pressure is maintained at 68.9 kPa as an axial compressive stress is applied. The deviator stress is the difference between applied axial stress and confining pressure. Note the stress and strain scales. Interior hysteresis loops with slope very different from the slope of the "exterior" loop are studied. See inset. These interior loops show *strain stays in* behavior.

that give the hysteretic behavior, that is, displacements due to mesocopic elastic elements, are able to participate. See the discussion in Chapter 11 where this issue is brought into sharper focus.

7. The soil, Figure 10.8, represents an extreme in that there is very little strain recovery on stress reversal in the interior loops. The two-way and one-way moduli differ by an order of magnitude. Although there is no closed exterior loop in Figure 10.8 such loops are a common feature is studies of soils at lower strain levels [23, 24].

8. The qualitative ideas discussed here have application to data similar to that above on ceramics [25], concrete [26], bone [27], and biomechanical systems [28], etc.

10.3
Coupling to Auxiliary Fields

We discuss two auxiliary fields that can be applied to an elastic system, through manipulation of a known thermodynamic variable, to produce/modify the quasistatic response. These two fields are the chemical potential of the vapor of a liquid,

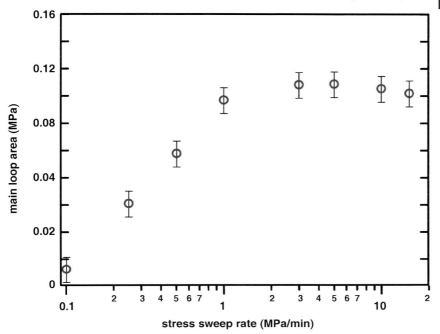

Fig. 10.9 Rate effects on hysteresis loops. The area of a stress-strain loop vs. the stress sweep rate, MPa/min. As the rate approaches zero, the area goes to zero. Beyond a sweep rate of approximately 2 to 3 MPa/min, the area of the loop remains the same [19].

which controls the saturation, and the temperature. These two fields can only be applied slowly, in contrast to the stress field, which can be applied to a system on the time scale for sound to cross it ($L/c \approx 10^{-4}$ s for L a typical system size, 10 cm, and $c \approx 10^5$ cm/s). The temperature propagates into a system difffusively, with diffusion constant $D_T \approx 0.01$ (cm)^2/s for a sandstone. A vapor pressure excess propagates into a system diffusively, with diffusion constant $D_P \approx 0.01$ (cm)^2/s for a sandstone with gas permeability $\kappa \approx 0.01$ Darcy. Thus temperature and saturation fields can be established on time scales L^2/D greater than or of order 100 s, $L = 1$ cm. Observation of the consequences of these fields usually involves time scales much greater than 1 s and endows this subject with vestiges of *slow dynamics*. Thus we will say something about slow dynamics here as well as in Chapter 11, where that subject is featured. We begin with saturation.

10.3.1
Saturation

The saturation field is a euphemism for the fluid configurations in the pore space of a sample. For a sample of volume V with pore volume V_p the porosity is $\phi = V_p/V$.

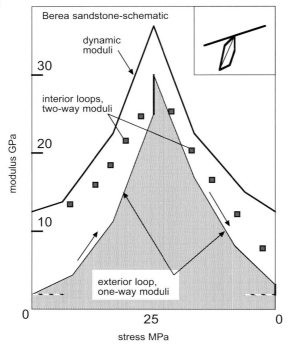

Fig. 10.10 Three Moduli. There are three possible moduli at each value of the stress. Use Figure 10.5 as an example. There are the up and down *one-way* moduli associated with the exterior hysteresis loop, *two-way* moduli associated with the interior hysteresis loops, and dynamic moduli. A *two-way* modulus (squares), from an interior hysteresis loop (inset), is necessarily greater than or equal to the nearby one-way modulus. One can conduct a sound velocity or resonant bar measurement at any stress in a stress protocol and determine a dynamic modulus. In principle, such a measurement is made at vanishing stress amplitude so that the minimum displacement accompanies the stress. One might expect that the interpolation to zero stress of a sequence of interior loops would yield a modulus in agreement with the dynamic modulus.

The saturation is the fraction of the pore volume that is filled with fluid,

$$S_W = V_F / V_p .\qquad(10.5)$$

Our concern is with the strain consequences of the system of internal stresses these fluid configurations create. The saturation is a coarse variable for this purpose. We will not give a detailed discussion of fluid configurations in pore spaces as the general idea is sketched in Chapter 5 and details are elsewhere [29–31]. Let us call attention to the major points.

1. The fluid configurations in a pore space are a hysteretic function of the chemical potential ↔ unsaturated vapor pressure ↔ relative humidity. The chemical potential, μ, plays a role for fluid configurations analogous to that of stress for hysteretic strain systems. In principle, one should create satura-

tion states, the analog of strain states, with well-defined chemical potential protocols.

2. The basic physical event that is responsible for the complexity of fluid configurations is the capillary condensation process, the precipitous filling of a pore structure at a value of the chemical potential set by local features in the pore geometry, $G_F(x)$. The emptying of a pore structure, also a precipitous event, is controlled by local features of the pore geometry, $G_E(x)$. The hysteresis in filling-emptying arises in part from $G_F(x) \neq G_E(x)$; different local features of the pore geometry influence filling-emptying.

3. On pore space filling the fluid vapor has access to the pores in the interior of the pore space and the filling process occurs more or less uniformly throughout the interior; in addition the fluid configuraions are spatially homogeneous. The pore space emptying process necessarily proceeds from the surface into the interior and involves spatially inhomogeneous fluid configurations. At a given S_W there can be two very different fluid configurations in the pore space, one on filling and one on emptying. The saturation is not a *good* variable.

These facts, very well known to practitioners of adsorption/desorption methods for characterizing pore spaces, are judged to be of little importance when discussing the influence of fluid configurations on the elastic properties of porous materials. It is not clear what evidence supports this judgement. To see what we are talking about we look at two results, Figures 10.11 and 10.12, from the seminal paper by Amberg and McIntosh on a rod of vycor glass (vycor 7930) [32]. In Figure 10.11, an adsorption isotherm, we see that the mass of water (per unit mass of vycor glass) in the pore space is a hysteretic function of the partial pressure of water vapor, P (normed by the saturated vapor pressure, P_0). Examples of complex internal hysteresis loops are found in the experiments of Hallock and coworkers [33, 34]. A Preisach space bookkeeping scheme can be used to describe the saturation as a function of a chemical potential protocol [31]. Consequently, fluid configurations necessarily obey the statement *the saturation state stays in*. Modeling of the vapor invasion process, as P/P_0 is lowered (below 0.68 in Figure 10.11) shows that in this *invasion percolation* regime the fluid configurations are very inhomogeneous [35]. Accompanying the fluid configurations in Figure 10.11 is the linear expansion (in percent) shown in Figure 10.12.

1. The linear expansion as a function of saturation is hysteretic with the basic morphology of the adsorption isotherm.

2. A strain (expansion) of order 1 millistrain is brought about by change in the saturation from a few percent to full saturation. As the Young's modulus of vycor 7930 is $Y \approx 20\,\text{GPa}$ this means that the fluid configurations exert a tensile stress of order 20 MPa!

3. At the lowest saturations, a saturation S_W causes a tensile stress given approximately by, Figure 10.12,

$$\frac{P}{Y} \approx -\frac{1}{50} S_W \,, \tag{10.6}$$

where we use pressure in place of stress to call attention to the sign. This tensile stress will appear prominently in the discussion of the influence of low saturation on the linear and nonlinear elastic properties of some mesoscopic elastic materials.

The discussion here has been particularly simplistic. There are issues of "wetting" the fluid. That is, forces of tension are brought about in part by a fluid that wets the surface of a pore space and reduces the energy per unit surface area. Some fluids wet easily (e.g. water), some only weakly (see for example, the experience of Tittman and colleagues with the influence of fluid type on attenuation [36, 37]), and some not at all (mercury). Not all pore spaces are sets of cylinders with well-defined geometry; the simplicity of the adsorption isotherm of Amberg and McIntosh may mislead [30, 38, 39]. We have made no mention of Biot theory in which the com-

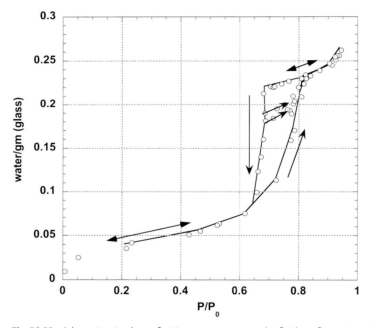

Fig. 10.11 Adsorption Isotherm for Vycor 7930. The mass of water in the pore space of Vycor 7930, $S_M = m_{H_2O}/m_{7390}$ (gm/gm), as a function of unsaturated vapor pressure (normed by the saturated vapor pressure). The chemical potential is related to the vapor pressure by $\mu \propto \log(P/P_0)$. At $P \to 0$ the sample cell is empty. The pore space is filled homogeneously on increase of P. The arrows indicate the qualitative properties of the fluid configurations. Where there are single-headed arrows the fluid configurations do not reverse on reversal of the pressure. They do in the two-headed-arrow regions. Invasion percolation occurs at the precipitous drop in S_M near $P/P_0 \approx 0.68$. The two arrows along the *invasion* drop indicate what happens to S_M when the vapor pressure is reversed along the drop. Inner loops not part of this data set are present when stresslike chemical potential protocols are used [33].

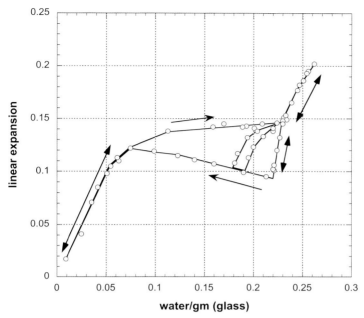

Fig. 10.12 Fluid-Induced Strain on Vycor 7390. As a Vycor 7390 sample is carried through the sequence of fluid configurations in Figure 10.11 the sample elongates as here. The hysteresis in the fluid configurations appears as a hysteresis in the strain. The stresses due to the fluid are tensile, causing the sample to expand on average as S_M increases. The one- and two-headed arrows have the same meaning as in Figure 10.11. Note $S_M \approx 0.25 \leftrightarrow S_W \approx 1$ corresponds to about a millistrain. From the known elastic constant of Vycor 7390 this corresponds to a tensile stress of order 20 MPa. See above Eq. (10.6).

pression of a pore fluid, *at fixed pore fluid configuration*, participates in the overall elastic response of a material [40]. We have also made no mention of the case where the forces exerted by the fluid configurations markedly distort the pore space so that it is not passive to pore fluid influences [41]. These topics are well developed in the literature and are mentioned to remind us of the simplicity of the leading approximation we present.

We close this section with the result in Figure 10.13 from van den Abeele and Carmeleit [42], the stress-strain relation of a Berea sandstone at five values of saturation, $S_W = 0, 0.01, 0.02, 0.05$, and 1.0. As the saturation increases, there is more strain per unit of stress, particularly at low saturation. Compare Figure 10.13 to Figure 10.12. The effective tensile stress of small saturation brings about a reduction in the strength of the linear elastic constant, the system is softer, so more strain accompanies an applied stress. This is qualitatively correct, but the results in Figure 10.13 have provided a more quantitative explanation [42]. See Section 10.4 below.

10.3.2
Temperature

The problem of the influence of temperature on the elastic properties of meso-scopic elastic materials is very complicated. Let us take the result of Ide as an introductory example [43]. In Figure 10.14 we see the sound velocity of a sandstone as a function of temperature for the case in which the temperature was carried through the protocol shown in Figure 10.15 at ambient pressure. It is believed that the sound velocity depends on the elastic state of the sample and that changes in the sound velocity follow changes in the elastic state. There are several things of note. The temperatures involved are very large, $300\,\text{K} \leq T \leq 1000\,\text{K}$. There seems to be the analog of the Kaiser effect with the temperature taking the place of stress [16]. That is, each visit to a new higher temperature produces a change in velocity that is not recovered as the temperature is lowered. Indeed the explanation Ide offered for his results accords well with that of the Kaiser effect. In Ide's case a system of elastic elements in the material have a spectrum of temperatures at which they *break*, go from stiff to soft, and do not heal. The complex system of internal strains due to thermal expansion, discussed in Chapter 5, Section 5.1, provides a qualitative picture of the source of the stresses that cause these *breaks*.

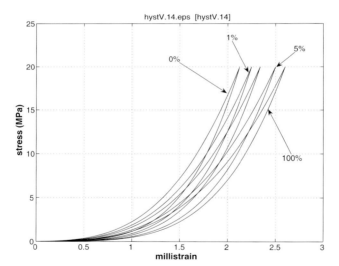

Fig. 10.13 Stress-Strain of Berea at Various S_W. The initially fully saturated pore space of a Berea sandstone is dried by evaporation. At each of several values of saturation, determined by weighting, 100%, 5%, 2%, 1%, and room dry $\approx$ 0%, the sample is taken through a uniaxial stress loop and the strain monitored. In the opposite order from that in which the data were taken: as the amount of fluid on the walls of the pore space increases, the system becomes softer. The qualitative property of hysteresis with *strain stays in* is unchanged. Most of the change due to increase in saturation occurs at the lowest saturations, say < 10%. The inverse problem associated with these data has been "solved" by Carmeliet and van den Abeele [42]. See Section 10.4.1.

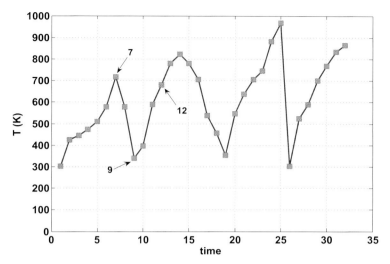

Fig. 10.14 Temperature Protocol. The temperature as a function of data point number ≈ time for the experiment of Ide on quartzitic sandstone. See Figure 10.15.

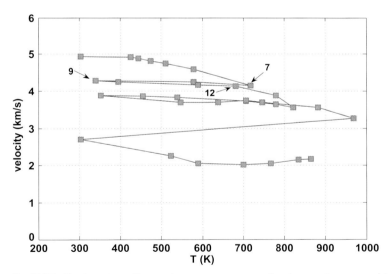

Fig. 10.15 Elastic state vs. Temperature. The velocity of sound, c, a surrogate for the elastic state, as a function of temperature for the temperature protocol shown in Figure 10.14. Each first visit to a higher temperature causes an evolution in c that is not recovered on temperature reversal. The estimate, $\gamma_T = d\ln(c)/d\ln(T) \approx -0.02$, can be made from the segments of the data in which a temperature range is retraced, for example, data points $7\ldots12$ in Figure 10.14.

From the variation of sound speed with temperature upon revisiting a temperature range, for example, time steps $7\ldots9\ldots12$ in Figure 10.15, we have the estimate $\gamma_T = (T/c)(dc/dT) \approx -0.02$.

Ide's experiment was carried out in a sample chamber that was *open to the ear*. "The specimen is placed in a cylindrical furnace... The cover of this is removed when the sample is set into vibration and the frequency of maximum acoustic response is determined by ear ..." As a consequence, as the temperature varies, the fluid configurations in the pore space vary. An initial increase in temperature would drive off volatiles, some of which would resettle as the temperature is cycled. Our concern is with the effects of temperature on the response of elastic elements. We don't want to either *break* them or entangle the effects of saturation with the effects of temperature.

We look at an experiment in the thesis of Ulrich [44]. The measurements of Ulrich were carried out on samples (again Berea sandstone was chosen) that have a well-defined temperature-saturation history. Here are some details.

1. The sample, a rectangular parallelepiped of Berea sandstone, $1.0 \times 1.2 \times 4.0\,\mathrm{cm}^3$, outfitted with a pair of transducers, is approximately free standing in a closed sample volume surrounded by a heater and temperature controller. The temperature controller responds to a thermometer in contact with the surface of the sample. A second thermometer on the wall of the sample volume tracks the primary thermometer reliably.

2. The sample is prepared for measurement by being maintained at a constant temperature of 380 K in vacuum (10^{-6} Torr) for 12 hours. The sample vol-

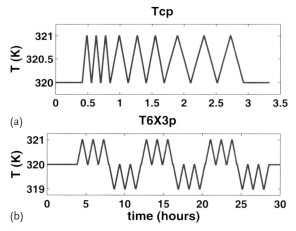

(a)

(b)

Fig. 10.16 Temperature Protocol, T-chirp. The temperature (K) as a function of time (hours) for a temperature chirp, T-chirp. There are three repeats of each of three temperature loops of amplitude +1 K and durations 500, 1000, and 1500 s. This is a positive T-chirp. A negative T-chirp starts at the same ambient temperature, 320 K, and has amplitude −1 K.

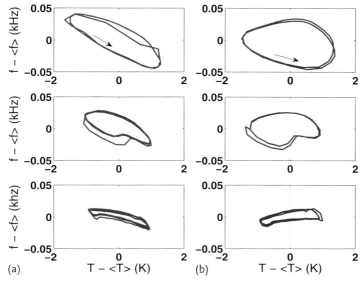

Fig. 10.17 Elastic state vs. Temperature for a T-chirp. The frequency of a resonant mode is measured every 30 s. The time average of f over each temperature loop, denoted $\langle f \rangle$, for the nine loops is determined. The values of f are reported for each loop as the shift in f from this average, $f - \langle f \rangle$. As with the velocity in Figure 10.15, this quantity stands as a sur-rogate for the elastic state. The three frames to the left are for a positive T-chirp protocol with the 500-s, 1000-s, and 1500-s loops at the top, middle, and bottom, respectively. On the right the same quantities are displayed for the negative T-chirp protocol, Figure 10.16. The estimate $\gamma_T = d \ln(f)/d \ln(T) \approx -0.0312$ can be made from the data in the lower left panel.

ume is then filled to a slight overpressure with helium gas and sealed. All temperature protocols were imposed during a 228-day period following this preparation. An example of such a protocol, in Figure 10.16, is a temperature "chirp".

3. The basic measurement that is made is the resonant frequency of a normal mode of the sample (or a suite of normal modes) as a function of temperature and time. It is believed that the frequency of a normal mode depends on the elastic state of the sample and that changes in resonance frequency follow changes in the elastic state. In a typical example, a temperature protocol lasts 24 hours and a particular normal mode, resonance frequency ≈ 77 kHz and $Q \approx 50$, is swept over at a rate of 500 Hz/s once every 30 s using strains of order 10^{-8} that make a negligible contribution to the temperature of the sample.

Temperature chirps of two signs with respect to the initial temperature, 320 K, were used. There were nine temperature loops in each chirp, three each of three durations, ≈ 500 s, ≈ 1000 s, and ≈ 1500 s. In Figure 10.17 the shift in the resonant frequency from the average resonant frequency (≈ 77 kHz) is plotted as a function of the shift in temperature from the average temperature. The three loops for each duration are plotted atop one another, the loops for each duration having a separate panel, in (a) for chirp to temperatures above 320 K and (b) for chirp to temperatures below 320 K. Several qualitative features are seen in Figure 10.17

1. The frequency (elastic state of the sample) is not able to follow the temperature. There are frequency-temperature hysteresis loops. (For $\partial T/\partial T = D_T \nabla^2 T$ the slowest mode in the time evolution is associated with time scale $\tau_1 = L^2/(\pi^2 D_T) \approx 10$ s, $L = 1$ cm. This time is a factor of 50 faster than the time scale of the fastest chirp temperature loop, ≈ 500 s.)

2. The average slope of the loops, df/dT, is negative (positive?) for temperature loops above (below) the initial temperature.

3. As the time to go through a hysteresis loop increases, the area of the loop decreases, area $\to 0$ as time $\to \infty$.

4. The area of a hysteresis loop for temperatures below the initial temperature is greater than the area of a hysteresis loop for temperatures above the initial temperature. There is an asymmetry in the response of the elastic state of the sample to the sign of a temperature change (Figure 10.18).

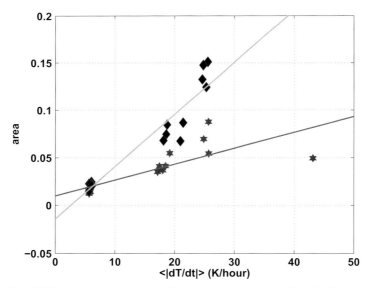

Fig. 10.18 Hysteresis Loop Area vs. Time. The area within each hysteresis loop in Figure 10.17 is plotted as a function of the magnitude of the rate of temperature change, $|dT/dt|$, for positive T-chirp protocols (stars) and for negative T-chirp protocols (diamonds). Data from T-chirp protocols with other loop time scales are included in the plot. The data are consistent with $area \to 0$ as $|dT/dt| \to 0$. An asymmetry between the evolution of the elastic state for disturbances with $\delta T = +1$ and $\delta T = -1$ is apparent.

5. From the slowest positive chirp loops we have $\gamma_T = (T/f)(df/dT) \approx -0.1$, approximately the same as the similar quantity found by Ide for a different sandstone.

The experiments of Ulrich include a number of other temperature protocols intended to explore the features remarked on here, e.g. lower panel of Figure 10.16. The essential result of an extensive set of experiments is that (1) a change in temperature sends the elastic system toward a goal that it takes a very long time to achieve (possibly logarithmically in time) and (2) there is a persistent asymmetry in the way in which the elastic system responds to temperature changes from the ambient temperature, say relatively easily for temperature increases and reluctantly for temperature decreases. See the discussion in Section 7.3.2. The asymmetry found was seen in earlier experiments by TenCate *et al.* [45], who were looking at something else, and not followed up.

There have been few careful/extensive experiments that explore the response of mesoscopic elastic systems to saturation and temperature. If this domain of exploration has something useful to say, much of it has yet to be said.

10.4
Inversion

A characteristic of much of the effort associated with understanding the source of quasistatic elastic behavior is the use of a bookeeping space, Preisach space, for tracking the history of the state of elastic elements, Chapters 4 and 6. It is in this context that one encounters the "inverse problem," an attempt to learn something about the contents of the bookkeeping space from the experimental data. This cannot be done in a model-independent way. To the degree that a minimalist model is employed, the results will be most useful. However, they may be few. One tack for getting more out of analysis is to complement a minimalist model with an elaborate stress protocol. We will discuss both of these cases. Finally, we sketch a simple analysis of the inverse problem that provides an estimate of numbers that are useful in evaluating the connection between quasistatic response and the nonlinear dynamic response. We look at the uniaxial stress experiment of Hilbert *et al.* [9], Figures 10.5, a similar experiment by Plona and Cook [10], Figure 10.6, two experiments by Boitnott [21, 46] one of which involves an elaborate stress protocols, Figures 10.19 and 10.21, the saturation experiment of Carmeliet and van den Abeele [42], Figure 10.13, and the experiment of Dillen [47], Figure 10.22.

10.4.1
Simple $\sigma - \varepsilon$ Protocol and Minimalist Model

The typical data set of interest is an exterior stress-strain loop, for example, Figures 10.5, 10.13, and 10.19. The inverse problem, i.e. finding $\varrho(X, Y)$, is solved using the model of McCall and Guyer [48]:

1. The macroscopic strain at a particular macrosopic stress is the sum of the strains of identical, two-state hysteretic elastic elements. (This result, sum of the strains, is the effective medium result for these elastic elements, Chapter 4.)
2. Each elastic element supports a stress field equal to the macroscopic stress field.
3. The elastic elements differ from one another only by the stress pair at which they change state.

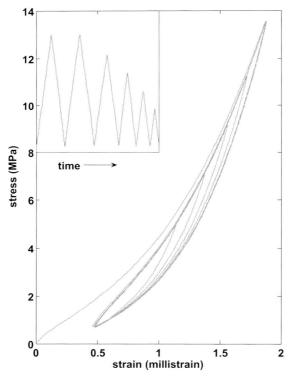

Fig. 10.19 Stress-Strain for Berea Sandstone. The strain (millistrain) as a function of stress (MPa) for a room-dry sample in uniaxial compression. The stress protocol is shown in the inset. Compare to Figure 10.5. Approximate treatment of the inverse problem for the exterior loop is associated with Figures 10.26–10.28.

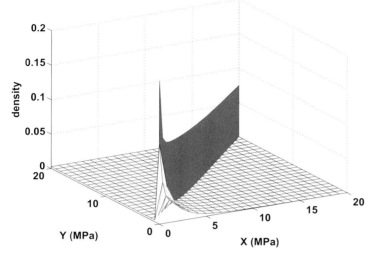

Fig. 10.20 Filling of Preisach Space. The density $\varrho(X, Y)$, where X and Y are in MPa. The diagonal term, which is singular, has been scaled to a finite value for display purposes [42].

The argument for the adequacy of the MG model was that with grain sizes of order $100\,\mu\text{m}$ a typical sample $100\,\text{cm}^3$ has of order 10^8 elastic elements within it. Can the detailed motion of any of these be important? So within the domain of this very simple model there is an inverse problem and a Preisach space density associated with a given experiment.

This model allows one to go from an experimental data set to a filling of a Preisach space with relative ease, Eqs. (6.17) and (6.18). There is a bit of a wrinkle as Eqs. (6.17) and (6.18) are integral equations of a type that is ill-posed [49]. Some of what is seen in the various treatments is a number of strategies for handling an ill-posed problem. Details are found in [21, 42, 50]. The data sets in Figures 10.5 and 10.19 have inner loops that can be used to test a solution. Ultimately the solution is presented as a density $\varrho(X, Y)$. This density typically has a diagonal term that is singular as here

$$\varrho(X, Y) = A(X)\delta(X - Y) + \alpha(X, Y). \tag{10.7}$$

An illustration of the behavior of $\varrho(X, Y)$ is shown in Figure 10.20. The $Y < X$ part of the Preisach space, where $\alpha(X, Y) \geq 0$, is in the foreground. On the other side of the diagonal, where the delta function part of the density, $A(X)\delta(X - Y)$, is erected, one sees $\alpha(Y > X) = 0$. The classical elastic constant increases with pressure so A is seen to decrease as X increases along the diagonal. The off-diagonal density is largest at the lowest pressures and decreases to be unimportant as X, Y increase to beyond about 10 MPa. In this example the hysteretic component of the strain is largest at low pressure. From the delta function part of the density one can find the linear and nonlinear classical elastic constants [21, 48]. See Section 10.4.3. The

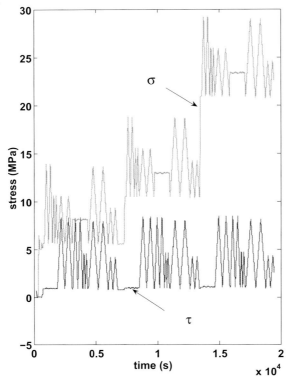

Fig. 10.21 Elaborate Stress Protocol I, Boit-nott. An extensive set of stress-strain curves, measured by Boitnott, uses the mean stress, shear stress protocol shown here. At three values of the ambient mean stress, 5 MPa, 10 MPa, and 20 MPa, there is a series of stress loops. In all cases there are two passes over each loop. The combination of manipulations of the two stresses yields mean stress, shear stress, uniaxial stress, and uniaxial strain loops. Some segments of the very complex stress-strain curve are shown in the paper by Boudjima *et al.* [51]. The horizontal axis provides evidence of the time scale over which these measurements were made, ≈ 6 hours.

results shown here are from the analysis of a Berea sandstone by Carmeliet and van den Abeele [42] at $S_W = 0$.

The data of Carmeliet and van den Abeele in Figure 10.13 [42], involving the saturation, pose a further problem. These data are part of a study of quasistatic stress-strain and dynamic modulus over the parameter space $0 \leq \sigma \leq 25$ MPa, $0.001 < S_W \leq 1$. A possible model for the effect of saturation is that it produces an effective pressure that simply shifts the origin of pressure, that is, when the sample is carried through a stress protocol an effective stress $\sigma_e = \sigma + P(S_W)$ is to be used. From the discussion above, Section 10.3.1, we expect $P(S_W)$ to be a tension. Carmeliet and van den Abeele use the minimalist model along with a practical analytic form for $\varrho(X, Y)$. They find that the notion of an effective pressure is too simplistic, although the qualitative idea of saturation ↔ tension is supported. Their analysis is backed up by adsorption isotherm and mercury intrusion data so that Carmeliet and van den Abeele are able to relate the saturation to models for the important pore space features. Independent of details it is apparent that the major influence of saturation is that as S_W increases from $S_W = 0$, tension due to thin fluid films markedly reduces the elastic constants (the dynamic modulus is reduced) and increases the amount of hysteresis in $\sigma - \varepsilon$.

10.4.2
Elaborate $\sigma - \varepsilon$ Protocol and Minimalist Model

Consider the experiments of Boitnott and Dillen. Figures 10.21 and 10.22 are of the stress protocols associated with these experiments. Boitnott measured the compressive strain and the shear strain as a Berea sandstone was carried through a stress protocol having compressive stress, shear stress, uniaxial stress, and uniaxial strain segments at three ambient pressures (values of the compressive stress). A different view of this experiment is provided in Figure 10.23, a picture of the region of compressive stress-shear stress space covered by the experiment. Up to this

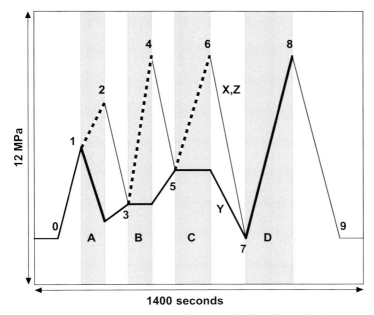

Fig. 10.22 Elaborate Stress Protocol, Dillen. Dillen employed an apparatus that allowed the X, Y, and Z pressures to be set independently. As the sample, a Colton sandstone, is taken through this pressure protocol, a variety of sound velocities were measured, for example, the velocity of a shear wave propagating in the X-direction with polarization in the Y-direction. The data, say a sound speed over a complex trajectory in (X, Y, Z) space, requires a two-component bookkeeping space.

point we have been discussing uniaxial stress experiments and implicitly imagining the minimalist model to involve elastic elements that brought about compressive strain in response to this stress. The model that connects these data to the bookkeeping space must be enlarged. Boudjema *et al.* [51] successfully did the inverse problem posed by these data by adopting a two Preisach space model, a set of elastic elements that give compressive strain in response to compressive stress and a set of elastic elements that give shear strain in response to shear stress. Each set of elastic elements has the properties of the MG elastic elements. Two Preisach spaces are called for. The stress range covered by this experiment is similar to that in earlier work. The numbers are much the same. Several important points:

1. Decomposition of a collection of elastic elements into two groups, compressive and shear, allows only an approximate solution to the inverse problem. See below.
2. The shear elastic elements appear to be more hysteretic than the compressive elastic elements.
3. In agreement with earlier results, the amount of hysteretic displacement decreases markedly with an increase in ambient stress.

To see the limitation of a simple two Preisach space description we look at a small portion of the data in the experiment by Boitnott in some detail, Figures 10.24 and 10.25. In Figure 10.24 we show the early part of the stress protocol in Figure 10.23. There is a point near ($\sigma \approx 8, \tau \approx 4.5$) that is reached by two paths, one at constant $\sigma \approx 8$ MPa and one at constant radial stress. We denote these stress values on the

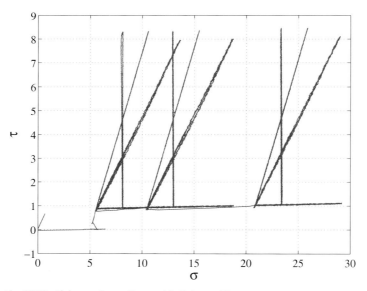

Fig. 10.23 Elaborate Stress Protocol II, Boitnott. The stress protocol in Figure 10.21 as a set of trajectories in (σ, τ) space.

upper panel of Figure 10.25 with open circles and the strain values at these stress values in the lower panel with open circles. In all cases the *strain stays in*. Note that a compressive stress loop at constant shear strain produces no shear stress. But a shear stress loop at constant compressive stress produces both shear strain and compressive strain. Shear stress and compressive stress are not always decoupled. The proper venue for tackling this problem is that provided by Helbig and Rasolofoasaon [52]. These authors develop a description of the stress-strain relationship in terms of the eigenstress vectors and the associated eigenstrain vectors. For a linear elastic system these are independent stress-strain relations. If one imposes an eigenstress, the output is the corresponding eigenstrain. This description suggests that quasistatic experiments should involve eigenstress protocols with the corresponding strain measured. In place of the two Preisach spaces used by Boudjima *et al.* one would use say six Preisach spaces, one for each of the six eigenstresses of the compliance tensor. This seems an ambitious program. It has the virtue of being precisely formulated. (As an aside, the experiment/calculation of Boitnott/Boudjima *et al.* is in principle the recipe of Helbig and Rasolofoasaon for the special case where the compliance tensor is appropriate to an isotropic system.)

The experiment of Dillen involves measurement of dynamic moduli of a Colton sandstone as the stress is carried through a complex protocol [47]. While these data are very interesting, it is not possible to unscramble them using current methods. A characteristic of the data sets of Boitnott and others is that there are multiple

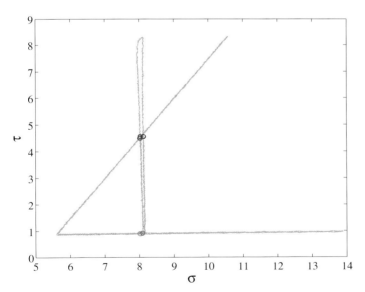

Fig. 10.24 Elaborate Stress Protocol for the First 4000 s. The first 4000 s of the stress protocol in Figure 10.21 as a set of trajectories in (σ, τ) space. The circles near $(\sigma \approx 8.0, \tau \approx 4.50)$ and $(\sigma \approx 8.0, \tau \approx 1)$ are also seen in stress time and strain time in Figure 10.25.

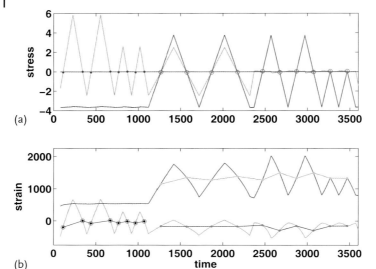

(a)

(b)

Fig. 10.25 Stress and Strain for the First 4000 s. (a) The compressive stress and the shear stress as a function of data point ≈ time (seconds) for the early time part of the protocol in Figure 10.21 (Figure 10.24). The two stresses have been shifted so that the ($\sigma \approx 8.0, \tau \approx 4.50$) point is near the origin. This point is visited at constant compressive stress and at constant radial stress. The strain at this point is shown in (b) (circles). The three segments of the stress protocol are (1) constant shear stress, (2) constant radial stress, and (3) constant compressive stress. (b) The compressive and shear strain as a function of data points. The points on the stress protocol are noted on the strain curves (circles).

visits to the regions of stress space. As first visits can lead to strain, evolution that is not recoverable, a second visit is necessary. The inverse analysis we have discussed uses data from second visits only. It would be of interest to undertake an experiment like that of Dillen using stress protocols that employ multiple visits.

But there is another tack one can take. Imagine a model for hysteretic elastic units, a van der Waals model, an asperity model, etc. With such a model one could make a theory of a stress-strain relationship. As it is a model with hysteretic elastic units, a bookkeeping space might be called for, or convenient, but it is not required [53, 54]. Perhaps Preisach space is a crutch for the unimaginative.

10.4.3
The Relationship of $\sigma - \varepsilon$ Data to Dynamics

10.4.3.1 Approximate Treatment of $\sigma - \varepsilon$ Data
Let us sketch an approximate scheme to extract the essential content in finding a solution to the inverse problem. We assume that the solution is given approximately by

$$\varrho(X, Y) = A(X)\delta(X - Y) + \overline{a} , \qquad (10.8)$$

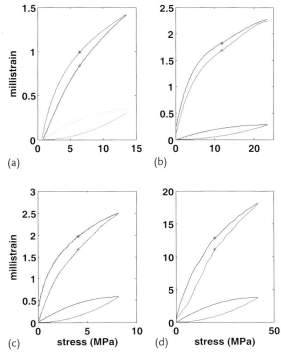

Fig. 10.26 Four Exterior Loops. The exterior stress-strain loops from the data in Figures 10.5, 10.6, 10.7, and 10.19. The stars on the loops are the stress-strain points at $P_{1/2}$, Eq. (10.11), from which the estimate of the hysteretic contribution to the strain (lower curve in each panel) is made.

that is, the diagonal density, corresponding to the classical elasticity, is possibly nonlinear but the off-diagonal density is constant. It then follows that on stress increase (decrease), Eqs. (6.13) and (6.14), we have

$$\varepsilon_\uparrow(P) = \varepsilon_0 I(P) + \varepsilon_0 \overline{a} \frac{P^2}{2} , \tag{10.9}$$

$$\varepsilon_\downarrow(P) = \varepsilon_0 I(P) + \varepsilon_0 \overline{a} \frac{2P_m P - P^2}{2} , \tag{10.10}$$

where $I(P) = \int_0^P A(X) dX$ and $0 \le P \le P_m$. Then at $P = P_{1/2} = P_m/2$ we have

$$\varepsilon_\downarrow(P_{1/2}) - \varepsilon_\uparrow(P_{1/2}) = \varepsilon_0 \overline{a} \times P_{1/2}^2 ; \tag{10.11}$$

the product $\varepsilon_0 \overline{a}$ can be estimated as

$$\varepsilon_0 \overline{a} = \frac{\varepsilon_\downarrow(P_{1/2}) - \varepsilon_\uparrow(P_{1/2})}{P_{1/2}^2} . \tag{10.12}$$

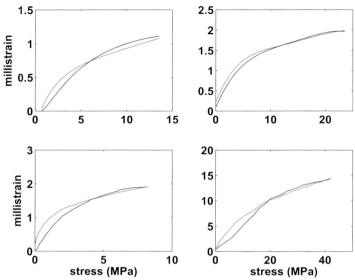

Fig. 10.27 Four Exterior Loops, Corrected. The exterior stress-strain loops from the data in Figures 10.5, 10.6, 10.7, and 10.19 with the hysteretic contribution to the strain subtracted. The arrows show the stress down curve. In all cases the constant background density approximation, Eq. (10.8), underestimates the amount of hysteretic strain at low stress. It very slightly overestimates the hysteretic strain at high stress. The average of the two curves here is used to form the approximate classical stress-strain curve, Figure 10.28.

We can test $\alpha(X, Y)$, a constant, by subtracting the estimates in Eqs. (10.10) from the exterior loop data to form $\varepsilon_\uparrow^c$ and $\varepsilon_\downarrow^c$. For $\varepsilon_\uparrow^c = \varepsilon_\downarrow^c$ the constant $\overline{\alpha}$ approximation is good. Departure from this result gives an indication of how $\alpha(X, Y)$ behaves. If the departure is not too great, the average $\overline{\varepsilon}^c = (\varepsilon_\uparrow^c + \varepsilon_\downarrow^c)/2$ is a good first estimate of the classical elasticity.

10.4.3.2 Dynamics

From the discussion in Section 6.3.1.2, Eq. (6.30), we have the expectation of a linear frequency shift in the resonance frequency of a bar based on

$$\sigma = K_0 \left[1 - \varepsilon_0 \alpha K_0^2 \varepsilon_m + O(s(\dot{\varepsilon})) \right] , \qquad (10.13)$$

where ε_m is the amplitude of the strain. That is,

$$\frac{\delta f(\varepsilon_m)}{f_0} \sim -\frac{\varepsilon_0 \alpha K_0^2}{2} \times \varepsilon_m = -\alpha_f \varepsilon_m . \qquad (10.14)$$

If $\overline{\alpha}$ is a reasonable estimate of α, then we can use quasistatic data to learn about dynamics.

10.4.3.3 **Quasistatic Dynamics**

In Figures 10.26, 10.27, and 10.28, we show the result of simple analysis of four data sets, those in Figures 10.5, 10.6, 10.7, and 10.19. In the cases where there are interior loops, these were removed from the data and the exterior loop alone examined. In Figure 10.26 we show the exterior stress-strain loops, marked with the strain pair involved in Eq. (10.11). When the hysteretic contribution to the strain is subtracted from the exterior loop, the results shown in Figure 10.27 are obtained. In Figure 10.28 we show $\bar{\varepsilon}_c$, below Eq. (10.12), and a polynomial fit to these data.

1. The four sets of curves in Figure 10.26, three sandstones and a TBC, are much the same. Perhaps the sandstone of Hilbert, a Berea, is least like the other three. This sample is least hysteretic. It is also apparent that the hysteretic contribution to the strain is not uniform over stress.

2. The difference curves for the four samples show that the constant α approximation typically finds too much hysteretic strain at large stress and too little hysteretic strain at low stress. Note the arrows on the figures that follow the $\varepsilon_\downarrow$ curves.

The numbers found from analysis of the four stress-strain curves are summarized in Table 10.1. These estimates are approximate as they use data from the entire stress-strain curve. Comparison of static to dynamic measurements is usually made at low pressure for which the numbers here will be lower limits; in the α_f column

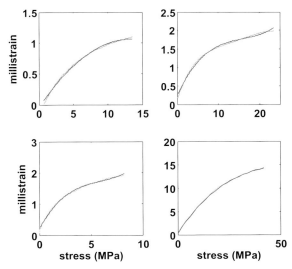

Fig. 10.28 Four Exterior Loops, Classical Stress-Strain Curve. An approximation to the classical stress-strain curve is found according to the prescription below Eq. (10.12). From a numerical fit to strain as a function of stress one can find estimates of the classical linear and nonlinear elastic constants.

258 | *References*

Table 10.1 Elastic parameters from Figures 10.26, 10.27 and 10.28.

Sample	$P_{1/2}$ MPa	$\varepsilon_0\alpha$ m-strain/(MPa)^2	K_0 GPa	$|\beta|$	α_f
Castlegate	4.1	0.017	1.7	244	
Berea (B)	6.4	0.0037	5.9	218	130 (1000)
Berea (H)	11.8	0.0010	4.2	243	
TBC	20.0	0.0042	1.5	25	

we show the result from Eq. (10.14) using $\varepsilon_0\alpha$ from (c) and the result from careful treatment of the inverse problem in parenthesis [21, 50].

References

1 Bourbie, T., Coussy, O., and Zinszner, B. (1987) *Acoustics of Porous Media*, Butterworth-Heinemann, New York.
2 Lide, D.R. (2006) *CRC-Handbook-Chemistry-Physics*, CRC, New York.
3 Mavko, G., Mukerji, T., and Dvorkin, J., (2003) *The Rock Physics Handbook: Tools for Seismic Analysis of Porous Media*, Cambridge Univ. Press, New York.
4 Landau, L.D. and Lifshitz, E.M. (1987) *Fluid Mechanics*, 2nd edn, Butterworth-Heinemann, New York.
5 Gist, G.A. (1994) Fluid effects on velocity and attenuation in sandstones. *JASA*, **96**, 1158–1173.
6 Zinszner, B. (unpublished).
7 Frischbutter, A., Nevo, D., Scheffzuk, C., Vrana, M., and Walther, K. (2000) Lattice strain measurements on sandstones under load using neutron diffraction. *J. Structural Geol.* **22**, 1587–1600.
8 Darling, T.W., TenCate, J.A., Brown, D.W., Clausen, B., and Vogel, S.C. (2004) Neutron diffraction study of the contribution of grain contacts to nonlinear stress-strain behavior. *Geophys. Res. Lett.*, **31**, L166041–4.
9 Hilbert, L.B., Hwong, T.K., Cook, N.G.W., Nihei, K.T., and Myer, L.R. (1995) *Rock Mechanics*, (eds J.J.K. Daeman and R. Schultz), Balkema, Rotterdam.

10 Plona, T.J. and Cook, J.M. (1995) *Rock Mechanics*, (eds J.J.K. Daeman and R. Schultz), Balkema, Rotterdam.
11 Rejda, E.F., Socie, D.F., and Itoh, T. (1999) Deformation behavior of plasma-sprayed thick thermal barrier coatings. *Surf. and Coat. Technology*, **113**, 218–226.
12 Lu, Z. (2005) Role of hysteresis in propagating acoustic waves in soils. *Geophys. Res. Lett.*, **32**, L14302.
13 Lemaitre, A. (2002) Rearrangements and dilatancy for sheared dense materials. *Phys. Rev. Lett.*, **89**, 195503.
14 Falk, M.L. and Langer, J.S. (1998) Dynamics of viscoplastic deformation in amorphous solids. *Phys. Rev. E*, **57**, 7192.
15 Makse, H.A., Gland, N., Johnson, D.L., and Schwartz, L.M. (1999) Why effective medium theory fails in granular materials. *Phys. Rev. Lett.*, **83**, 5070–5073.
16 Li, C. and Nordlund, E. (1993) Experimental verification of the Kaiser effect in rocks. *Rock Mechanics and Rock Engineering*, **26**, 333–351.
17 Mehl, M.J. (1993) Pressure dependence of the elastic moduli in aluminum-rich Al-Li compounds. *Phys. Rev. B*, **47**, 2493–2500.
18 Claytor, K.E., Koby, J.R., and TenCate, J.A. (2009) Limitations of Preisach Theory: Elastic aftereffect, congruence, and end point memory, Geophys. Res. Lett., **36**, L06304. doi:10.1029/2008GL036978.

19 TenCate, J.A. (unpublished).

20 Vakhnenko, O.O., Vakhnenko, V.O., and Shankland, T.J. (2005) Soft-ratchet modeling of end-point memory in the nonlinear resonant response of sedimentary rocks. *Phys. Rev B*, **71**, 174103–174117.

21 Guyer, R.A., McCall, K.R., and Boitnott, G.N. (1995) Hysteresis, discrete memory, and nonlinear wave propagation in rock: A new paradigm. *Phys. Rev. Lett.*, **74**, 3491–3494.

22 Walsh, J.B. (1965) The Effect of Cracks on the Compressibility of Rock. *J. Geophys. Res.*, **70**, 399–411.

23 Vucetic, M. (1990) Normalized behavior of clay under irregular cyclic loading. *Can. Geotech. J.*, **27**, 29–46.

24 Ishihara, K. (1996) *Soil Behavior in Earthquake Geotechnics*, Clarendon Press, Oxford.

25 Fugimagari, E., Horibe, S., and Matsuzawa, S. (2001) Analysis of stress-strain hysteresis loops in 3Y-TZP. *J. Mat. Sci. Lett.*, **20**, 2065–2066.

26 Watanabe, K., Niwa, J., Yokota, H., and Iwanami, M. (2003) Formulation of Hysteresis Loop for Concrete Subjected to Compressive Cyclic Loading. *Proceedings of JSCE*, **767**, 143–159.

27 Bonfield, W. and O'Conner, P. (1978) Anelastic deformation and the friction stress of bone. *J. Mat. Sci.*, **13**, 202–207.

28 Taber, L.A. (2004) *Nonlinear Theory of Elasticity: Applications in Biomechanics*, World Scientific Publishing, Singapore.

29 Rouquerol, J., Rouquerol, F., and Sing, K.S.W. (1999) *Adsorption-Powders-Porous-Solids Applications*, Academic Press, New York.

30 Guyer, R.A. (2006) *Science of Hysteresis III*, (eds G. Bertotti and I.D. Mayergoyz), Academic Press, New York.

31 Flynn, D., McNamara, H., O'Kane, P., and Pokrovskii, A. (2006) *Science of Hysteresis III*, (eds G. Bertotti and I.D. Mayergoyz), Academic Press, New York.

32 Amberg, C.H. and McIntosh, R. (1952) A study of absorption hysteresis by means of length changes of a rod of porous glass. *Can. J. Chem.*, **30**, 1012–1032.

33 Lilly, M.P., Finley, P.T., and Hallock, R.B. (1993) Memory, congruence, and avalanche events in hysteretic capillary condensation. *Phys. Rev. Lett.*, **71**, 4186–4189.

34 Wootters, A.H. and Hallock, R.B. (2003) Superfluid avalanches in nuclepore: constrained versus free-boundary experiments and simulations. *Phys. Rev. Lett.*, **91**, 165301.1–165301.4.

35 Wilkinson, D.J. and Willemsen, J. (1983) Invasion percolation: a new form of percolation theory. *J. Phys. A*, **16**, 3365–3376.

36 Clark, V.A. (1980) *Effects of Volatiles on Seismic Attenuation in Sedimenrary Rock* (thesis), Texas A and M.

37 Tittmann, B.R., Clark, V.A., Richardson, J.M., and Spencer, T.W. (1980) Possible mechanism for seismic attenuation in rocks containing small amounts of volatiles. *J. Geophys. Res.*, **85**, 5199–5208.

38 Dolino, G., Bellet, D., and Faivre, C. (1996) Adsorption strains in porous silicon. *Phys. Rev. B*, **54**, 17919–17929.

39 Ma, J., Qi, H., and Wong, Po-zen (1999) Experimental study of multilayer adsorption on fractal surfaces in porous media. *Phys. Rev. E*, **59**, 2049–2059.

40 Johnson, D.L. and Plona, T.J. (1982) Acoustic slow waves and the consolidation transition. *JASA*, **72**, 556–565.

41 Reichenauer, G. and Scherer, G.W. (2000) Nitrogen adsorption in compliant materials. *J. of Non-Crystal. Solids*, **277**, 162–172.

42 Carmeliet, J. and van den Abeele, K. (2004), Poromechanical approach describing the moisture influence on the non-linear quasi-static and dynamic behaviour of porous building materials. *Concrete Sci. and Engr.*, **37**, 271–280.

43 Ide, J.M. (1937) The velocity of sound in rocks and glasses as a function of temperature. *J. of Geol.*, **45**, 689–716.

44 Ulrich, T.J. (2005) (thesis), University of Nevada, Reno. Ulrich, T.J. (2005) (thesis), University of Nevada, Reno.

45 TenCate, J.A., Smith, D.E., and Guyer, R.A. (2000) Universal slow dynamics in granular solids. *Phys. Rev. Lett.*, **85**, 1020–1023.

46 Boitnott, G.N. (1997) Experimental characterization of the nonlinear rheology of rock. *Int. J. Rock Mech. and Min. Sci.*, **34**, 379–388.

47 Dillen, M.W.P. (2000) *Time-lapse Seicmic Monitoring of Subsurface Stress Dynamics*, thesis, Delft.

48 McCall, K.R. and Guyer, R.A. (1994) Equation of state and wave propagation in hysteretic nonlinear elastic materials. *J. Geophys. Res.*, **99**, 23887–23897.

49 Press, W.H., Teukolsky, S.A., Vetterling, W.T., and Flannery, B.P. (2002) *Numerical Recipes*, Cambridge Univ. Press, New York.

50 Guyer, R.A., McCall, K.R., Boitnott, G.N., Hilbert, G.N., and Plona, T.J. (1997) Quantitative implementation of Preisach-Mayergoyz space to find static and dynamic elastic moduli in rock. *J. Geophys. Res.*, **102**, 5281–5293.

51 Boudjema, M., Santos, I., McCall, K.R., Guyer, R.A., and Boitnott, G.N. (2003) Linear and nonlinear modulus surfaces in stress space, from stress-strain measurements on Berea sandstone. *Nonlinear Proc. in Geophys.*, **10**, 589–597.

52 Helbig, K. and Rasaolofosaon, P.N.J. (2000) *Anisotripy 2000*, Society of Expl. Geophys. Tulsa.

53 Aleshin, V. and van den Abeele, K. (2005) Micro-potential model for stress-strain hysteresis of micro-cracked materials. *J. of the Mech. and Phys. of Solids*, **53**, 795–824.

54 Aleshin, V. and van den Abeele, K. (2007) Microcontact-based theory for acoustics in microdamaged materials. *J. of the Mech. and Phys. of Solids*, **55**, 366–390.

11
Dynamic Measurements

This chapter is about dynamic measurements. We begin in Section 11.1 with a description of experiments on the interaction of dynamic elastic fields with quasistatic elastic fields. A quasistatic elastic field may arise from the application of external stress, Section 11.1.1, or from auxiliary fields, temperature, Section 11.1.2, and saturation, Section 11.1.3. In Section 11.2 we discuss experiments that involve the interaction of dynamic elastic fields with one another. Collinear and crossed-beam experiments are described in Sections 11.2.1.1 and 11.2.1.2. In preparation for a description of resonant bar experiments we detour in Section 11.1.2 to discuss *fast dynamics* and *slow dynamics*. The idea of a nonequilibrium steady state (NESS) is introduced. Then, in Section 11.1.3 we describe the sequence of experiments that provide our current understanding of the interaction of dynamic elastic fields. The concepts of anomalous fast dynamics and slow dynamics appear prominently. In Section 11.3 experiments on a variety of materials that exhibit anomalous fast dynamics and slow dynamics are described.

11.1
Quasistatic-Dynamic

Slowly varying changes in the elastic state of a system can be brought about by an applied pressure, a change in temperature, or a change in saturation state. Dynamic strain fields, a tone burst, the AC strain field in a resonant bar, ..., are often used as the detector of presence and consequences of applied quasistatic fields. We will describe the case in which the quasistatic field is a stress (pressure) field in some detail and make mention of the temperature and saturation cases to emphasize the coherence of the underlying physical picture.

11.1.1
Pressure-Dynamic

The typical example of the consequence of the coupling of a dynamic strain field to a quasistatic strain field is shown in Figure 11.1 [1], the compressional modulus of granite (calculated as $K = \varrho(c_L^2 - 4c_T^2/3)$ from the measurement of c_L and c_T)

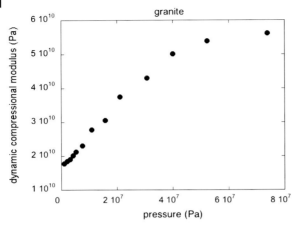

Fig. 11.1 Modulus of granite. Compressional modulus as a function of confining pressure from measurements of the sound velocities and density. Representing this modulus in the form $K = K(0)\left(1 + \beta p - \delta p^2\right)$, $p = P/K(0)$, where P is the pressure, leads to $\beta \approx 1000$ and $\delta \approx 10^5$.

as a function of confining pressure. A confining pressure of 20 MPa brings about a change in modulus by a factor of 2, cf. Figure 10.2 in Section 10.1. A phenomenological description of the result of this measurement might use

$$K(P) = \varrho\left(c_L^2 - \frac{4}{3}c_T^2\right) = K(0)(1 - \Lambda_P U_x) = K(0)\left(1 - \beta_P \frac{P}{K(0)}\right), \quad (11.1)$$

where on the right we define β_P using one of the conventional scalar representations of nonlinearity; $\beta_P \approx -1000$ from the data. A formal description of the measurements required to form K would use the apparatus developed in Chapter 3, Section 3.3, with

$$\boldsymbol{u} = \begin{pmatrix} U_x + u_x & 0 & 0 \\ 0 & U_x & 0 \\ 0 & 0 & U_x \end{pmatrix}, \quad (11.2)$$

where U_x is a uniform quasistatic strain (caused by the applied pressure) and u_x is a dynamic strain. For the case that u_x propagates in the x-direction we would have

$$\varrho\ddot{u} = \left(K + \frac{4}{3}\mu\right)\frac{\partial^2 u}{\partial x^2} + \left(5K + \frac{8}{3}\mu + 2A + 10B + 6C\right) U_x \frac{\partial^2 u}{\partial x^2} = c_L^2(U_x)\frac{\partial^2 u}{\partial x^2}. \quad (11.3)$$

A similar treatment of a shear wave traveling in the x-direction yields

$$\varrho\ddot{v} = \mu\frac{\partial^2 v}{\partial x^2} + (3K + 2\mu + A + 3B) U_x \frac{\partial^2 v}{\partial x^2} = c_T^2(U_x)\frac{\partial^2 v}{\partial x^2}. \quad (11.4)$$

Table 11.1 Linear and nonlinear elastic constants.

	K GPa	μ GPa	A GPa	B GPa	C GPa	β_K	(e_V/μ)
Berea (B)	3.7	3.7	−6233	−2425	−1045	−2250	−1433
Limestone 1083	29.6	20.6	−9730	−6434	−1868	−634	−416
Westerly granite	29.9	23.6	14071	−20227	−1150	−1325	−960
Glass beads	8.0	5.5	46	−701	−15	−177	−125
Polystyrene	3.8	1.4	−10	−8	−11	−10.3	−6.71
Agar (P1)	8.25	9.0 (kPa)	−68 (kPa)	−12	67	13.7	7.72
Water	2.28	0	0	−2.28	−4.78	5.21*	

If we take the uniform strain to arise from a uniform compressional stress, we have $U_x = -P/(3K)$, Eq. (3.9), where $\sigma_{xx} = \sigma_{yy} = \sigma_{zz} = P$. Assembling $K(P)$ and comparing to Eq. (11.1)

$$\beta_P = \frac{K + 2A/3 + 6B + 6C}{3K}. \tag{11.5}$$

There are two points to note here. (1) Measurements like that in Figure 11.1 get at the classical nonlinear elastic constants. (2) For granite, β_P is of order 10^3, so $|2A/3 + 6B + 6C| \approx 3 \cdot 10^3$ K, and it would seem that the classical nonlinear elastic constants are several orders of magnitude greater than the linear elastic constants.

Of course, the compressional modulus is not enough. There are well-understood recipes for learning the complete set of third-order elastic constants, A, B, and C, from measurement of the velocity of dynamic strain fields in the presence of quasistatic strain fields [2], Section 3.1.2 and Table 3.1. The relatively recent experiment of Winkler and Liu [3] contains a nice description of the experimental procedure, a description of the major steps in data analysis, and a comparison of many different types of materials. (There are several sets of definitions of third-order elastic constants [4–6], see [7]. Recent workers have typically used the A, B, and C of [4], as will we.) In Table 11.1 we list the value of A, B, and C for several isotropic elastic systems, Berea sandstone, limestone, Westerly granite, cemented glass beads, polystyrene [3], Agar-gelatin-based phantom [8], and water. All of the consolidated granular materials have large values of A, B, and C using polystyrene as standard. The tissuelike material, Agar-gelatin-based phantom, has a very small A coefficient, cf. water.

Let us check some numbers. If we use A, B, and C for granite from Table 11.1 in Eq. (11.5), we would make the estimate $\beta_P \approx -1325$ (reported in column 7 of the table). This is in reasonable agreement with the estimate below Eq. (11.1) from the slope in Figure 11.1.

Quasistatic-dynamic coupling is described in Chapter 3, Section 3.3 and Figure 3.1, as the coupling of a $k \to 0$, $\omega \to 0$, (quasistatic) strain field with a dynamic strain field. The description of *time-of-flight* tomography in Section 9.2.1 is of a localized quasistatic strain field coupling to a dynamic strain field. These two situations are essentially the same, differing only in the details surrounding the description of the quasistatic field.

The discussion to this point has been about third-order nonlinearity ↔ the 3-phonon process ↔ β ↔ A, B, C. Let us say a few words about the fourth-order nonlinearity, δ_P, defined by

$$K(P) = K(0)\left[1 - \beta_P \frac{P}{K(0)} + \delta_P \left(\frac{P}{K(0)}\right)^2\right] = K(0)\left[1 - \beta_P(P)\frac{P}{K(0)}\right].$$

(11.6)

A simple estimate of the order of magnitude of δ_P comes from the general appearance of the stress-strain curves. From Figure 11.1 the value of β at low pressure, $\beta \approx -1000$, goes over to $\beta \approx 0$ at a pressure of order 8 MPa. Thus

$$\delta_P = -K(0)\frac{d\beta_p}{dP} \approx -2 \times 10^6.$$

(11.7)

A second example, from the result of Gist, Chapter 10, leads to $\delta_P \approx 10^7$. The order of magnitude of these estimates is in accord with similar estimates on a wide variety of rocklike materials. Generally from quasistatic data $|\delta_P| \approx 10^6$–10^7, or very approximately $\delta_P \approx -\beta_P^2$.

11.1.2
Temperature-Dynamic

The coupling of a dynamic strain to temperature is fundamental to the experiments of Ide [9] and Ulrich [10] described in Section 10.3.2. In both experiments the frequency of a resonance peak, examined with low-amplitude drive, is used as a monitor of the elastic state. Ulrich reports the frequency directly, whereas Ide reports a velocity deduced from the frequency. Analogously to the description above, the logic is that δT produces a slowly varying strain field, U_x^T, to which the dynamic strain couples. We might write

$$\boldsymbol{u} = \begin{pmatrix} U_x^T + u_x & 0 & 0 \\ 0 & U_x^T & 0 \\ 0 & 0 & U_x^T \end{pmatrix},$$

(11.8)

where $U_x^T = \alpha\delta T$ and α, the thermal expansion, is a constant for a homogeneous isotropic material. See Section 5.1. Thus there will be equations for c_L^2 and c_T^2 similar to those written out in detail above, Eqs. (11.3) and (11.4), and the analog of Eq. (11.1):

$$K(T) = K(0)\left(1 - \Lambda_P U_x^T\right) = K(0)\left(1 - \Lambda_T \frac{\delta T}{T_0}\right),$$

(11.9)

where $\varLambda_T = \alpha T_0 \varLambda_P$. The discussion in Chapters 5 and 10 call attention to the possibly very complex nature of the internal forces that temperature can bring about in a porous material. So the result here is intended to suggest the nature of the influence of temperature on a dynamic strain field and to call attention to the similarity of that influence to the influence of pressure.

11.1.3
Saturation-Dynamic

From the discussion in Chapter 10 we know that fluid configurations deliver internal forces that result in a macroscopic strain. At low saturation we write $U_x^S = \gamma S_W$, Chapter 10, Eq. (10.6), and

$$
\boldsymbol{u} = \begin{pmatrix} U_x^S + u_x & 0 & 0 \\ 0 & U_x^S & 0 \\ 0 & 0 & U_x^S \end{pmatrix} . \tag{11.10}
$$

Then

$$
K(S_W) = K(0)\left(1 - \varLambda_P U_x^S\right), = K(0)(1 - \varLambda_S S_W), \tag{11.11}
$$

where $\varLambda_S = \gamma \varLambda_P$.

Equations (11.1), (11.9), and (11.11) describe the change in modulus due to pressure, temperature, and saturation in terms of a dimensionless measure of the applied field, P/K_0, $\delta T/T_0$, and S_W, respectively, and a dimensionless coupling constant, 1, αT_0, and γ, respectively.

1. Using $\varLambda_P = -\beta_p \approx 1000$ and $\alpha \approx 10^{-6}$ K^{-1} we have $\varLambda_T \approx 0.3$. This leads to

$$
\frac{T_0}{\delta T} \frac{\delta K(T)}{K(0)} \approx -0.3 , \tag{11.12}
$$

at $T_0 = 300$ K, a number in order-of-magnitude agreement with results from Ide and Ulrich (Chapter 10).

2. From Amberg and McIntosh [11], Chapter 10, Eq. (10.6), we have $U_x^S \approx 0.02 S_W$ and

$$
\frac{\delta K(S_W)}{K(0)} \approx -20 S_W , \quad S_W \ll 1 , \tag{11.13}
$$

using $\varLambda_P = 1000$. This estimate is made combining a number from porous glass, U_x^S, and a number from Berea sandstone, $\varLambda_P = 1000$. What is suggested is that for a wetting fluid that induces substantial tension at low saturation, there should be a large reduction in the elastic constant as the saturation increases from $S_W = 0$.

3. In Figure 11.2 we show a schematic diagram of K as a function of (S_W, P). At low pressure and low saturation the elastic system is relatively stiff. It becomes stiffer on an increase in pressure at $S_W = 0$. It becomes softer

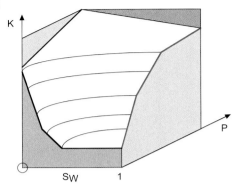

Fig. 11.2 Modulus as a function of P and S_W. On increase in pressure, the modulus increases. On increase in saturation, a wetting fluid exerts tensions (negative pressure) causing the modulus to decrease. The elastically softer system at $S_W = 1$ is changed more by pressure than the elastically stiff system at $S_W = 0$.

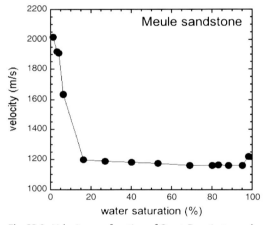

Fig. 11.3 Velocity as a function of S_W at $P = 1$ atmosphere, Meule sandstone. On increase in saturation (by water) the modulus decreases. The decrease is most rapid at the lowest saturations as the initial *layers* of wetting fluid decrease the surface energy and produce tensionlike forces.

on an increase in saturation at $P = 0$. Since it is softer at large saturation, $S_W = 1$, an increase in pressure at $S_W = 1$ brings about a relatively larger change in modulus than that at $S_W = 0$. As Figure 11.2 suggests, the proper venue for discussing the modulus of a porous elastic system is (P, S_W) space. This is rarely done. In general, there is little control of the saturation state in experiment and little recognition of the burden elastic hysteresis

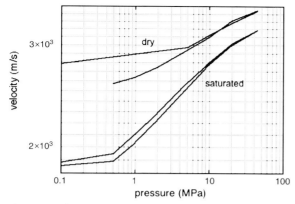

Fig. 11.4 Velocity as a function of pressure at $S_W = 0$ and at $S_W = 1$, Meule sandstone. In accord with the qualitative picture in Figure 11.2 at $S_W = 1$, the elastic constant is softer and more easily changed by pressure than at $S_W = 0$.

and fluid configuration hysteresis put on the choice and execution of experimental protocols.

4. In Figures 11.3 and 11.4 we show two results on Meule sandstone that agree with the general picture developed above [12, 13]. There are many data sets [14] in qualitative accord with the expectation exhibted in Figure 11.2.

5. As was remarked on earlier, our concern here is with changes in elastic state brought about by pressure or by the auxiliary fields, T and S_W. These changes can be learned through studies of the coupling of a dynamic strain field to the quasistatic strain field. The subject of wave propagation in porous media in which the fluid in a pore space contributes its elasticity to the elasticity of the whole, the provenance of Biot theory [15], is outside the range of this discussion.

11.2
Dynamic–Dynamic

There are two standard ways in which to have dynamic strain fields interact with one another. You can broadcast strain fields so that they encounter one another momentarily in some volume of space where there is a nonlinear coupling, or you can put strain fields in a region of space where they continually encounter one another through a nonlinear coupling. The first case is wave mixing as in Jones and Kobett [16], Section 3.4.2, and the second case is the resonant bar, Section 3.4.3. We look at wave mixing experiments first.

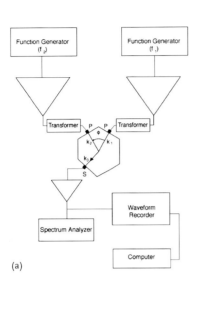

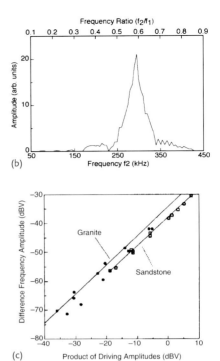

Fig. 11.5 Noncollinear wave mixing experiment. (a) Schematic diagram of apparatus. Details of the geometry of the 3-phonon process $L(\omega_1) + L(-\omega_2) = T(\omega_3)$ are in Figure 11.6. (b) The amplitude of the detected transverse wave as a function of x as in Eq. (11.15). The input and output angles are fixed, the sound velocities c_L and c_T are fixed, so the test of the $L(\omega_1) + L(-\omega_2) = T(\omega_3)$ selection rule is the frequency of the detected transverse wave, Eq. (11.15). (c) The amplitude of the detected transverse wave, A_3, as a function of the product of the amplitude of the two longitudinal waves A_1 and A_2. The result of a second experiment on a granite are also shown.

11.2.1
Dynamic–Dynamic: Wave Mixing

11.2.1.1 Noncollinear Wave Mixing

The prototypical wave mixing experiment using dynamic strain fields is that of Rollins, Taylor, and Todd [16–18] and [19]. For isotropic materials, for example, polycrystalline aluminum, these authors confirm the *3-phonon* selection rules discussed earlier, 3.4.2. Let us look at a similar experiment on a sandstone sample [20, 21]. In Figure 11.5 we show an experimental system involving a sandstone, cut with flat faces to facilitate contact with the transducers. Two longitudinal tone bursts are launched from faces 1 and 2 at relative angle $\phi = 34°$. A shear wave detector is fixed on face 3 at $\gamma = 34.5°$, Figure 11.6. The physical process that will launch a shear wave from the interaction volume toward the shear detector is case II of Taylor and Rollins [17], $L(\omega_1) \rightarrow L(\omega_2) + T(\omega_1 - \omega_2)$, rearranged to read

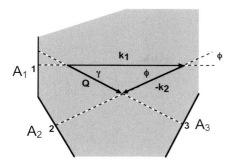

Fig. 11.6 Geometry of wave mixing experiment, $L(\omega_1) + L(-\omega_2) = T(\omega_3)$. Since
$k(-\omega_2) = -k(\omega_2)$ the two-input compressional
wave, amplitudes A_1 and A_2, yield a transverse wave, amplitude A_3, having wave vector
$Q = k_1 - k_2$, $\omega_Q = \omega_3 = \omega_1 - \omega_2$. The angles ϕ
and γ characterize the geometry of the scatter-
ing and dictate how the rock is to be cut (or if
cut they dictate the frequencies involved in the
scattering process, Eqs. (11.14) and (11.15)).

$L(\omega_1) + L(-\omega_2) \rightarrow T(\omega_1 - \omega_2)$. For the detected shear wave, (wavevector, frequency) = (Q, Ω)), momentum and energy conservation yield $Q = k_1 - k_2$ and $\Omega = \omega_1 - \omega_2$
since $k(-\omega_2) = -k(\omega_2)$. Using the acoustic approximation for all frequencies and
$r = c_T/c_L$ the conservation laws lead to

$$(1 - r^2)(1 + x^2) = 2(1 - r^2 \cos\phi)x , \tag{11.14}$$

where $x = |k_2|/|k_1| = \omega_2/\omega_1 \leq 1$. Combining the conservation of momentum and
the law of sines, Figure 11.6, we have a relationship between the geometry of the
faces and the frequencies

$$\tan\gamma = \frac{x \sin\phi}{1 - x \cos\phi} . \tag{11.15}$$

For the angles $\phi = 34°$, $\gamma = 34.5°$, and $r = 0.64$ (for Berea sandstone) the solution
to these equations is $x \approx 0.62$. Thus the experiment to confirm the description of
wave mixing is conducted with (1) broadcast at amplitude A_1 and fixed frequency
$f_1 = 500\,\text{kHz}$ from face 1, (2) broadcast at amplitude A_2 and swept frequency f_2,
$50\,\text{kHz} \leq f_2 \leq 450\,\text{kHz}$, from face 2, and (3) detection of the amplitude A_Q at $f_1 -
f_2$ on face 3. The result, Figure 11.5b, confirms the prediction of Eqs. (11.14) and
(11.15), as do further tests, for example, the amplitude dependence of the signal on
face 3, $A_Q \propto A_1 A_2$, the time of arrival of the signal on face 3, etc.

From the expression for the nonlinear elastic energy, Eq. (2.67), the known
strength of the three strain fields involved and the values of the linear and nonlinear elastic constants in Table 11.1 for a Berea sandstone we have the estimate

$$\frac{e_V}{\mu} = R\varepsilon_1\varepsilon_2\varepsilon_Q = -1433\varepsilon_1\varepsilon_2\varepsilon_Q , \tag{11.16}$$

where $\varepsilon_1 = k_1 A_1$, $\varepsilon_2 = k_2 A_2$, and $\varepsilon_Q = Q A_Q$. To give an idea of the elastic energy
involved in a typical *3-phonon* process, we report the strain energy amplitude, R,

for the $Q = k_1 - k_2$ process, calculated from the values of the linear and nonlinear elastic constants in columns 2–6 of Table 11.1, column 8. We note that Berea sandstone, limestone, Sierra White granite, and glass beads are much more nonlinear than polystyrene, Agar-gelatin-based phantom, and water. For these highly nonlinear materials the coefficient B makes a dominating contribution to β_P and the strain amplitude R.

11.2.1.2 Collinear Wave Mixing

(a) In Figure 11.7 we show the experimental system employed by Meegan *et al.* [22] to investigate collinear wave mixing, $L(\omega_1) + L(\omega_2) \rightarrow L(\omega_3)$. A Berea sandstone 2 m long and 6 cm in diameter was driven from one end with tone bursts of various (frequencies, durations), $(f_0, \Delta t)$. The time train at each of 11 detectors along the sample was recorded. Fourier analysis of the time trains provides the amplitudes $A(f_0) = A_1$, $A(2 f_0) = A_2$, $A(3 f_0) = A_3$, etc. at each detector. The description of this experiment, Section 3.4.2, Eq. (3.46), shows that a second harmonic is expected with amplitude

$$A_2 \sim \beta (k_0 A_1)^2 x \,, \tag{11.17}$$

where x is the distance from the drive, $k_0 = 2\pi f_0/c$, and $\beta/K \approx -4000$ is given by Eq. (3.29) and Table 11.1. In preparation for testing the functional form in Eq. (11.17), the displacement field at the sample surface was monitored with a fiber optic probe that looked through a hole in the piezoelectric transducer. When the source was driven at 13.7 kHz, the spectrum of the sample surface displacement, for drive amplitudes spanning almost two decades, was as shown in Figure 11.8. Most notable is the background at frequencies other than the drive frequency. As the amplitude at the drive frequency increases, this background remains constant at $< 10^{-8}$. Meegan *et al.* confirmed the dependence of A_2 on x, A_1^2, and $k_0^2 \propto f_0^2$ and made the estimate $|\beta|/K \approx 7000$ in reasonable accord with 4000. As an illustration of their results we show the amplitude A_2 as a function of distance from the source in Figure 11.9. The results of Meegan *et al.* establish the presence of a nonlinear scatterer with coupling functionally equivalent to

$$+\beta \frac{\partial}{\partial x} \left(\frac{\partial u}{\partial x} \right)^2 \,, \tag{11.18}$$

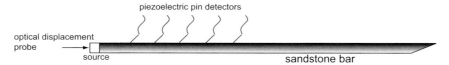

Fig. 11.7 Collinear wave mixing in Berea sandstone, sample. Pulse mode tone bursts are launched from the left using a piezoelectric source with tantalum backload. A small hole is drilled through the backload/transducer, on center, and an optical probe that can look directly at the rock surface is used to measure input wave displacement. In order to measure the development of the wave with distance, a series of pin transducers is embedded along the length of the sample with epoxy.

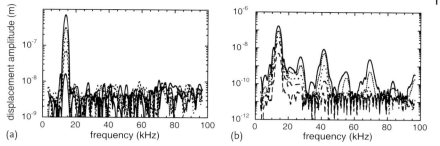

Fig. 11.8 Amplitude as a function of frequency. (a) Source spectrum for tone bursts at 13.7 kHz at progressively increasing drive amplitude (different line types) from optical probe, Figure 11.7. (b) Spectrum measured at 0.58 m from the source at progressively increasing drive amplitude.

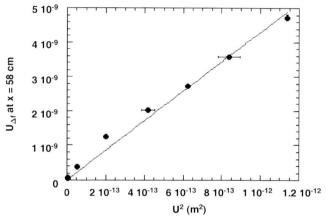

Fig. 11.9 Scaling of second harmonic. The amplitude of the second harmonic, $u_{\Delta f}$ (Figure 11.8b), measured at 0.58 m from the source as a function of the square of the amplitude of the source, U^2.

Eq. (3.28), throughout the sample. A followup experiment by TenCate *et al.* [23] found interesting and somewhat different results for what was essentially the same experiment. This calls attention to a most important point. In the experiment of TenCate *et al.*, the source spectrum was much less clean than that shown in Figure 11.8. Consequently there were higher harmonics in the sample that were not created by its nonlinearity and that could not easily be separated out from those that were.

In an experiment in which a strain field interacts with itself, $f_1 = f_2$, great care must be taken to be sure that $2f_1$, $3f_1$, ... did not come from the drive. In a wave mixing experiment in which two different frequencies are mixed, the result, $f_1 \pm f_2$, is unlikely to be a multiple of f_1 or f_2.

(b) A practical example of collinear two-wave mixing is provided by LeBas *et al.* [25] who used a Berea sandstone in a parametric array experiment [24]. By suitably phasing the two fundamentals, LeBas *et al.* were able to steer the array. The transducer set is shown in Figure 11.10 and an illustration of the evidence for mixing is shown in Figure 11.11.

(c) The recent very careful experiment of d'Angelo *et al.* [26, 27] sets the standard for collinear wave mixing experiments.

1. The apparatus consisted of a $1 \times 1 \times 2\,\mathrm{m}^3$ water-filled volume containing source and receiver transducers. Prior to placing samples in this volume it was studied experimentally and modeled with the KZK equation [24]. Understanding of signal broadcasts in this volume entailed understand-

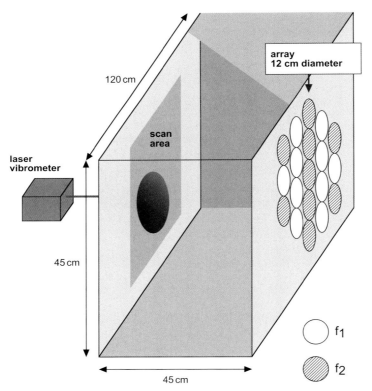

Fig. 11.10 Parametric array apparatus (not to scale). A large Berea sandstone sample ($0.45\,\mathrm{m} \times 0.45\,\mathrm{m} \times 1.2\,\mathrm{m}$) has a 19-element source array, diameter 12 cm, attached to one side. The 19 tranducers are addressed independently, 10 with frequency f_1 and 9 with frequency f_2. The signal from the array is measured over the scan area ($0.38\,\mathrm{m} \times 0.45\,\mathrm{m}$) across from the transducer array with a laser vibrometer that is stepped on a square grid $0.005\,\mathrm{m} \times 0.005\,\mathrm{m}$.

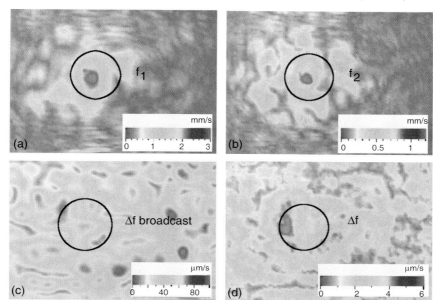

Fig. 11.11 Parametric array results. (a) Amplitude of direct arrival on the scan area of signal $f_1 = 65\,\text{kHz}$, broadcast from 10 transducers (open circles) in the array, Figure 11.10. (b) Amplitude of direct arrival on the scan area of signal $f_1 = 60\,\text{kHz}$, broadcast from 9 transducers (shaded circles) in the array, Figure 11.10. (c) Amplitude of direct arrival on the scan area of signal $\Delta f = f_2 - f_1 = 5\,\text{kHz}$, broadcast from **all** 19 transducers in the array. (d) Amplitude of arrival on the scan area of signal at $\Delta f = f_2 - f_1 = 5\,\text{kHz}$ when $f_1 = 65\,\text{kHz}$ is broadcast from 10 transducers (open circles) in the array and $f_2 = 60\,\text{kHz}$ is broadcast from 9 transducers (open circles) in the array. The black circle on each panel is the projection of the source array onto the scan area. (Please find a color version of this figure on the color plates)

ing transducer voltage-fluid pressure conversion, attenuations, and incipient nonlinearities. When all of this was satisfactory, the samples were placed in the volume.

2. Seven samples – PMMA, Portland sandstone (dry and wet), Indiana limestone (dry and wet), and Berea sandstone (dry and wet) – were examined. Prior to being the subject of a collinear wave mixing experiment, each of these samples was studied with a set of quasistatic-dynamic measurements, Section 11.1.1, that allowed determination of the third-order elastic constants, A, B, and C.

3. Each sample was placed between source and receivers and subjected to overlapping 50 µs tone bursts at frequencies of $f_1 = 1.05\,\text{MHz}$ and $f_2 = 0.95\,\text{MHz}$. The detected amplitudes at $f_1, f_2, 2f_1, 2f_2, f_1 + f_2$, and, most importantly, $f_1 - f_2 = 100\,\text{kHz}$ were studied as a function of the incident wave amplitude, distance from source to receivers, and lateral sample position in order to provide a detailed understanding of wave propagation through the

samples. Part of the refinement of this understanding employed feedback between experimental observations and their description using the KZK equation.

4. In general the results found are in reasonable accord with expectation; there were signals detected at $2f_1$, $2f_2$, $f_1 + f_2$, and $f_1 - f_2$, their amplitudes scaled with the source amplitude in essentially the proper way, and the wet samples (saturated) were more nonlinear than the dry samples. But all was not perfect. Of most importance is the observation that the nonlinear coupling constant, β, found from a wave mixing experiment was considerably smaller than the value one would predict using the measured values of the third-order elastic constants, A, B, and C. As both experiments were done on each sample, this conclusion is very strong.

5. A most intriguing possibility, suggested by d'Angelo *et al.* [27], is that the nonlinear coupling of two dynamic strains (each at about 1 MHz) calls for the elastic system (elastic elements?) to respond on a very different time scale from that involved in a quasistatic-dynamic measurement, that is, the measurement of A, B, and C. Thus, in a two-wave mixing experiment, the frequency of the strain fields and duration of their encounter may be of importance.

We turn now to a discussion of measurements involving dynamic strains that are continuously in interaction with one another, resonant bar measurements.

11.2.2
Dynamic–Dynamic, Resonant Bar, Preliminaries: Fast Dynamics and Slow Dynamics

Slow dynamics is a central part of the discussion of the behavior of a resonant bar. So we take a serious detour to give extra care to the meaning of the language that we will use in this context. To model the physics of the resonant bar, we recall the lumped element model (Chapter 6) and use a mass/nonlinear spring for which we take the equation of motion

$$\ddot{u} + \frac{1}{\tau}\dot{u} = -\frac{\Gamma_0}{m}\left[(1 + \gamma_{SD}(t))u - \delta u^3\right] + F_P(t) + F_p(t), \tag{11.19}$$

where u is the displacement field, $F_P(t)$ the pump source, F_p the probe source, Γ_0 the linear elastic constant (spring constant), δ the coefficient of quartic nonlinearity, and γ_{SD} a slowly varying change in the linear elastic constant having a variety of sources (see below). Equation (11.19) is supplemented by an equation of motion that describes the time evolution of the slow dynamics field, $\gamma_{SD}(t)$. We rewrite the nonlinear term in the form $\delta u^3 = u^3/u_0^2$ ($\delta = 1/u_0^2$), so that we have, as a simple displacement measure of the importance of the nonlinearity,

$$\ddot{u} + \frac{1}{\tau}\dot{u} = -\omega_0^2\left[(1 + \gamma_{SD}(t))u - \varepsilon\frac{u^2}{u_0^2}u\right] + F_P(t) + F_p(t), \tag{11.20}$$

where $\omega_0^2 = \Gamma_0/m$ and $\varepsilon = 1(0)$, depending on whether or not we want to consider the quartic nonlinearity.

In order to be as clear as possible, we will talk through a series of examples.

11.2.2.1 Fast Dynamics: Linear

Consider a situation in which $F_P(t) = 0$, $\varepsilon = 0$, and $\gamma_{SD}(t) = 0$. For the probe $F_p(t) = A_0 \exp(-i\omega t)$ the response is taken to be the displacement amplitude at the frequency of the probe. This response, called the *fast dynamics* response, is

$$u_p(t) = A(\omega) \exp(-i\omega t), \quad A(\omega) = \frac{1}{\omega_0^2 - \omega^2 - \frac{i\omega}{\tau}} A_0. \tag{11.21}$$

More precisely, this is the *linear fast dynamics response* since $A(\omega) \propto A_0$. The fast dynamics response is simply the displacement response at the probe frequency. It is called *fast* because the probe frequency is *fast* by some measure.

(We need to have a quantitative understanding of the time scale necessary to undertake the measurement that establishes a resonance frequency. Take $\omega_0 = 2\pi f_0$, $f_0 = 1000$ Hz, and $\tau = 0.016$ s. Then $Q_0 = \omega_0\tau \approx 100$, and we could establish steady state at a drive frequency of ω_0 in, say, $3Q_0$ periods or 0.3 s. If we were to use 100 frequency values to sweep over a resonance, it would take about 30 s to establish a resonance frequency. We use this time, call it the sweep time τ_s, as the time scale necessary to establish a resonance frequency. See the discussion in Chapter 8.)

Near resonance, $\omega = \omega_0$, $|A(\omega_0)| = A_r = \tau A_0/\omega_0$. The importance of the neglected nonlinear term near resonance, where it is largest, depends on A_r/u_0. Suppose $(A_r/u_0)^2$ is large enough to be of interest. We have case 2.

11.2.2.2 Fast Dynamics: Nonlinear

Consider a situation in which $F_P(t) = 0$, $\varepsilon = 1$, and $\gamma_{SD}(t) = 0$. The probe is $F_p(t) = A_0 \exp(-i\omega t)$. To see the influence of the nonlinear term, we use the simple approximation

$$\overline{uuu} = 3\overline{uu}\,u, \quad \frac{u^2}{u_0^2}u \rightarrow 3\frac{\overline{uu}}{u_0^2}u, \tag{11.22}$$

where $\overline{uu}$ is the average of u^2 over one period of the probe. [In the phonon language of Chapter 3 we take from the 4-phonon process the three terms in which three phonons $(\pm\omega)$ coalesce to produce a fourth phonon at ω, for example, $\omega + \omega - \omega \rightarrow \omega$.] The equation of motion for the displacement is

$$\ddot{u} + \frac{1}{\tau}\dot{u} = -\omega_0^2 \left[1 - \frac{u^2}{u_0^2}\right] u + F_p(t),$$

$$= -\omega_0^2 \left[1 - 3\frac{\overline{uu}}{u_0^2}\right] u + F_p(t). \tag{11.23}$$

The displacement response at the frequency of the probe, the fast dynamics response, is

$$u_p(t) = A(\omega) \exp(-i\omega t), \tag{11.24}$$

$$A(\omega) = \cfrac{1}{\omega_0^2 \left[1 - 3\dfrac{\overline{u_p^2}}{u_0^2}\right] - \omega^2 - \dfrac{i\omega}{\tau}} A_0 \,,$$

(11.25)

$$\overline{u_p^2} = \frac{|A|^2}{2} \,.$$

(11.26)

For $\overline{u_p^2} \ll u_0^2$ the resonance frequency is at $\omega \approx \omega_0$ so that, evaluating $\overline{u_p^2}$ at ω_0, we have resonance frequency

$$\omega^2 = \omega(A_0)^2 \approx \omega_0^2 \left[1 - \frac{3}{2}\left(\frac{\tau A_0}{\omega_0 u_0}\right)^2\right] \equiv \frac{\Gamma(A_0)}{m} \,.$$

(11.27)

This is an example of *fast nonlinear dynamics* in which the response at the probe frequency is influenced by the amplitude of the fast displacement that follows at the probe frequency, $A \propto A_0 + O(A_0^3) + \dots$

[As the resonance is swept over on time scale τ_s the resonance frequency shifts, following $\omega(A)^2 = \omega_0^2(1 - 3\overline{uu}/u_0^2)$, on the time scale associated with u_p reaching steady state, approximately τ. The frequency shift is largest at resonance, where $\overline{uu}$ is largest, and small well below (above) resonance, where $\overline{uu}$ is small, that is, the resonance curve is a *peak bending* resonance curve.]

11.2.2.3 Slow Dynamics; External Source

Consider a situation in which $F_P(t) = 0$, $\varepsilon = 0$, and $\gamma_{SD}(t) \neq 0$. The elastic constant Γ_{SD} varies slowly in time in response to the time evolution of an auxiliary field like temperature or saturation. Then the equation of motion for the displacement is

$$\ddot{u} + \frac{1}{\tau}\dot{u} = -\omega_0^2 \left[1 + \gamma_{SD}(t)\right] u + F_p(t) \,,$$

(11.28)

where the slow dynamics spring constant varies negligibly over the time, approximately τ, required to establish a steady-state displacement response to the probe, $F_p(t) = A_0 \exp(-i\omega t)$. Then the displacement response at the frequency of the probe, *the fast dynamic response*, is

$$u_p(t) = A(\omega) \exp(-i\omega t) \,, \quad A(\omega) = \cfrac{1}{\omega_0^2[1 + \gamma_{SD}(t)] - \omega^2 - \dfrac{i\omega}{\tau}} A_0 \,.$$

(11.29)

[If, as a resonance is swept over, taking time τ_s, the slow dynamics spring constant changes negligibly, then the time evolution of the frequency shift, $\omega_0^2\gamma_{SD}(t)$, governed entirely by the dynamics of the auxiliary field, can be used to monitor the time evolution of that field [10, 28].]

11.2.2.4 Slow Dynamics; Internal Source

Consider a situation in which the probe source is zero, $F_p(t) = 0$, $\varepsilon = 0$, $F_P(t) \neq 0$, and $\gamma_{SD}(t) \neq 0$. The pump source, $F_P(t) = A_0 \exp(-i\omega t)$, causes a change in the elastic state of the system. The evidence of this is that after the pump is turned off, well after time on the order of $3Q_0$ periods, a probe source will see evidence of the pump having been on. Now pump and probe are the same thing, just sources with a time dependence of the form $\exp(-i\omega t)$. But the difference is in the strain amplitude that the source causes. There is a strain amplitude threshold, ε_T. When the source produces strain amplitudes below the threshold, it leaves the elastic state of the system unchanged, and the source is a valid probe. When the source produces strain amplitudes above the threshold, it changes the elastic state of the system and the source cannot be regarded as a probe. We call it a pump. A probe necessarily leaves the elastic state of the system unchanged. When two sources are used at the same time, the larger is usually called the pump.

Let us refine this description. We adopt a particular model for $\gamma_{SD}(t)$ and sharpen the language of the discussion. [The displacement in the lumped element model of Eq. (11.20) is a surrogate for the strain, $\varepsilon_T \leftrightarrow A_T$.]

1. Suppose the system is in thermal equilibrium for $t < 0$. Turn $F_P(t) = A_0 \exp(-i\omega t)$ on at $t = 0$ and monitor the displacement at ω until a steady state is reached. Then, for (ω, A_0) such that the displacement amplitude exceeds A_T, we have

$$u_P(t) = A(\omega)\exp(-i\omega t), \quad A(\omega) = \frac{1}{\omega_0^2[1 - X_{SS}(|A|)] - \omega^2 - \frac{i\omega}{\tau}} A_0 .$$

(11.30)

where $X_{SS}(|A|)$ is the steady-state value of $\gamma_{SD}(t)$ when the amplitude of the displacement response is A. The pump source (ω, A_0) maintains the system in a *nonequilibrium steady state* (NESS) that is set up on a time scale $\tau_{SD} \gg \tau$. (A possible model for this might be the *ratchet* model of Chapter 7, although the discussion here does not depend on any particular model.) For simplicity of discussion we adopt one feature of the ratchet model; in the ratchet model for $\omega\tau_{SD} \gg 1$ the steady-state value of $X_{SS}(|A|)$ is independent of ω. We take this to be the case so that the NESS is characterized simply by the displacement amplitude caused by the pump, for example, $X_{SS}(|A|)$ depends only on $|A|$.

2. When the system is being maintained in a NESS, an elastic state different from the equilibrium elastic state, the nature of this state can be determined by two schemes that leave the NESS undisturbed.

a. Fixed pump and weak probe.

   i. With $F_P(t) = 0$ sweep over a resonance with $F_p(t) = a_0 \exp(-i\omega t)$, $a_0 \ll A_T$.

   ii. Establish the system in a NESS with $F_P(t)$ for which $\gamma_{SD} = X_{SS}(|A|)$.

   iii. Maintain the system in the NESS and employ $F_p(t) = a_0 \exp(-i\omega t)$ to sweep over the resonance a second time. The equation of motion for the displacement response to $F_p(t)$ is

$$\ddot{u} + \frac{1}{\tau}\dot{u} = -\omega_0^2(1 - X_{SS}(|A|))u + F_p(t)$$

(11.31)

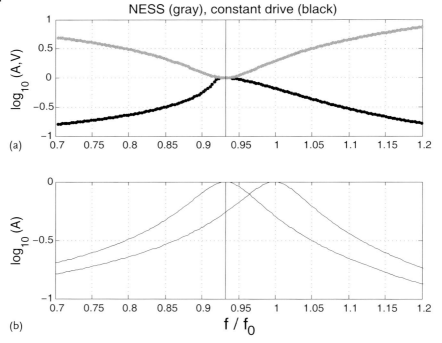

Fig. 11.12 Resonant bar; amplitude as a function of frequency. (a) High-strain amplitudes (back curve). Amplitude as a function of frequency for constant large drive voltage (log scale, the amplitude is scaled by the amplitude at resonance, $|A_r|$). The resonance, at $f_r = 1$ at low drive amplitude, is shifted to $f_r = 0.935$ at high drive amplitude. The resonance curve has a *bent tuning fork* shape (gray curve). Voltage as a function of frequency for the constant strain state (NESS) $\varepsilon_r \leftrightarrow |A|_r$ (log scale, the voltage values are scaled by the voltage at resonance). The resonance curve is symmetric around $f^2 = f_r^2(|A_r|)[1 + 1/(2Q_r^2(|A_r|))]$, Eq. (11.33). (b) Low-strain amplitudes. When the system is in the NESS at $|A_r|$, a frequency sweep at low drive voltage (probe) will have resonance frequency shifted from $f_r = 1$ (pump off) to $f_r = 0.935$.

and

$$u_p(t) = a(\omega)\exp(-i\omega t), \quad a(\omega) = \frac{1}{\omega_0^2(1 - X_{SS}(|A|)) - \omega^2 - \frac{i\omega}{\tau}}a_0.$$

$$(11.32)$$

The elastic state of the system in the NESS is determined from the resonance frequency, the frequency at which the response at constant probe amplitude (voltage) is a maximum. The difference in this resonance frequency for pump-off and pump-on yields $X_{SS}(|A|)$.

b. Constant strain pumping. The pump source is swept over resonance at constant $|A|$ or strain. At each frequency ω the amplitude A in Eq. (11.30) is kept fixed by adjusting the strength of the pump, A_0 (typically a voltage

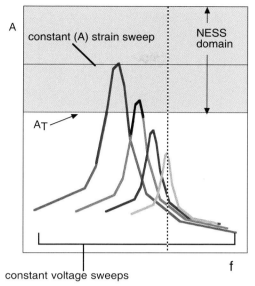

Fig. 11.13 Nonequilibrium steady state. Nonequilibrium steady states can be established with drive amplitudes that produce strains above $A_T \leftrightarrow \varepsilon_T$. For a frequency sweep at constant strain, drive voltage varies with frequency (gray curve in Figure 11.12a). For a frequency sweep at constant voltage the strain amplitude varies with frequency along a resonance curve. For relatively high constant voltage the system enters the NESS domain temporarily. A modification of the resonance curve results that depends on the amount of time spent in the NESS domain. A variety of experimental protocols, drive voltage, and frequency as a function of time have been developed to unravel the consequences of encounters with the NESS domain. See Figures 11.14 and 11.25.

in experiments). The elastic state of the system in the NESS is determined from the resonance frequency, the frequency at which the pump voltage is a minimum, Figure 11.13.

$$A_0 = \left[\omega_0^2 (1 - X_{SS}(|A|)) - \omega^2 - \frac{i\omega}{\tau} \right] A . \qquad (11.33)$$

While an experiment of this type can probe the nature of the NESS, it cannot establish that the elastic state examined at constant $|A|$ is a NESS. To learn that you have to see the elastic state move in time.

3. Observations.
   a. To establish the large displacements for which $A > A_T$ one typically operates $F_P(t)$ at a resonance. As a matter of principle this is not required.
   b. The resonance frequency of a resonance is generally taken as the measure of the elastic state. The change in elastic state associated with the NESS will be detectable by a probe sweep at any resonance, not just the resonance used to pump as in (a).
   c. It is possible to carry out a constant A study of a resonance for $A < A_T$.

d. When the elastic system has traditional nonlinearity as well as a frequency shift due to establishing a NESS, one might have

$$u_P(t) = A(\omega) \exp(-i\omega t),$$ (11.34)

$$A(\omega) = \frac{1}{\omega_0^2 \left[1 - X_{SS}(|A|) - \frac{3}{2}\frac{|A|^2}{u_0^2}\right] - \omega^2 - \frac{i\omega}{\tau}} A_0$$ (11.35)

in place of Eq. (11.30). A weak probe of this system as in 2(a) would be unchanged, yielding the result in Eq. (11.32) as $|A|^2$ is negligible for a weak probe.

4. Sweeping over a resonance at constant a_0, for which $a(\omega) < A_T$, $\forall \omega$, leaves the system unchanged. Sweeping over a resonance at constant a_0, for which $a(\omega) > A_T$ for some (all?) ω, has the system changing as the sweep is conducted (entering and then leaving the NESS domain) and produces results that require care in their interpretation, Figure 11.13.

5. The slow dynamics of the NESS are typically studied by establishing the NESS, turning the pump off abruptly, and following a resonance with a weak probe over time.

11.2.3
Dynamic–Dynamic: Resonant Bar, Data

We look at a sequence of experiments, notable because of the attention given to issues of strain amplitude, frequency protocols, and time scale, that helped to flesh out the current picture.

11.2.3.1 Linear
In Figure 11.38, which has eight resonance curves, the curve for graphite is typical of what is seen for a linear material. A series of amplitude-frequency curves that appear the same for a sequence of drive voltages. The other seven panels in the figure, to be discussed below, show qualitative evidence of nonlinearity. When at a sequence of drive voltages the resonance curves look as they do for graphite, one knows that the strain fields within the bar, largest at the bar center, do not activate a noticeable amount of nonlinearity. Then, the resonance frequency is an effective measure of the elastic constants of the bar. Resonant ultrasound spectroscopy (RUS) uses precise measurement of a set of linear resonance frequencies to infer accurate values of the linear elastic constants of a material [29].

11.2.3.2 First Examination of Slow Dynamics (Figure 11.14)
The seminal work on slow dynamics is that of TenCate and Shankland [30]. In the course of a set of resonant bar measurements on a cylindrical sample of Berea sandstone, conducted with frequency sweeps at constant drive voltage, these authors noticed slow dynamics and stopped to pay attention to it. The sample, 50 mm in diameter and 0.3 m long, had a low-amplitude resonance at $f_r \approx 3920\,\text{Hz}$ (strain

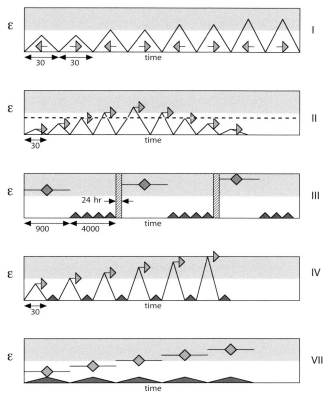

Fig. 11.14 Experimental protocols; slow dynamics. In each of five panels an experimental protocol designed to be aware of the NESS domain and the slow dynamics associated with entering this domain is shown schematically. These are the experimental protocols used in the discussion in Sections 11.2.3.2–11.2.3.6. The figures are of the amplitude (strain) as a function of time. At constant voltage a frequency sweep produces an amplitude change indicated by the tooth of a saw. The direction of a frequency sweep is indicated by an arrow. A small black saw tooth represents a probe frequency sweep. All times not otherwise noted are in seconds and are approximate, intended to give a rough idea of the time scale.
I. The experimental protocol of TenCate and Shankland, Figure 11.15.
II. The experimental protocol employed by Guyer, TenCate, and Johnson to produce data for constant strain analysis, Figure 11.17.

III. The experimental protocol employed by TenCate, Smith, and Guyer in the study of the recovery from a NESS. The NESS, established at essentially constant pump voltage and pump frequency, is watched for approximately an hour after the pump is turned off with small-amplitude probe sweeps, Figure 11.19.
IV. The experimental protocol of Pasqualini *et al.* for investigating the elastic state of a resonant bar at the lowest strain amplitudes. Returning to a low-strain probe sweep after each constant voltage sweep makes it possible to monitor for the influence of entry into the NESS domain. An approximate location of the boundary of the NESS domain can be established, Figure 11.20.
V. The experimental protocol employed by Haller and Hedberg. The frequency is swept at constant strain by adjusting the drive voltage. The constant strain values begin below the NESS domain, Figure 11.24.

at resonance of order 10^{-8}). As the resonance was swept over [(down followed by up) 4000 to 3800 Hz to 4000 Hz in 2-Hz steps (200 frequency points) at 300 ms per step] with increasing drive voltage, the resonance frequency was seen to shift to lower values, for example, $f_r \approx 3850$ Hz when the strain at resonance was of order 10^{-5}. Most importantly, it was noted that when the drive voltage produced a strain amplitude greater than about 10^{-7} during a frequency sweep, the subsequent strain behavior was modified as the frequency sweep continued. A frequency sweep of 4000 to 3800 Hz was different from a frequency sweep of 3800 to 4000 Hz, Figure 11.15. The modification was seen to be dependent on the amount of time the sample spent at strains greater than about 10^{-7}. The frequency sweep rate, $1/\tau_s$, was encountering a physical rate, associated with the elastic system comparable to itself. An example of an experiment that confirmed the intuition that developed from this set of experiments is shown in Figure 11.16.

1. At constant drive voltage the frequency was swept (on the time scale above) from 4000 Hz (3800 Hz) over the resonance peak to a point about 20 Hz below (above) the peak.
2. The drive voltage was set to zero, where it remained for 30 s.
3. The original drive voltage was reestablished, and the frequency sweep was completed.

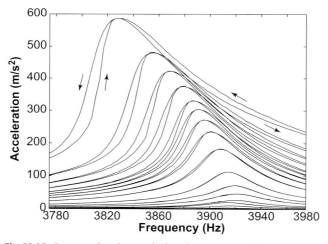

Fig. 11.15 Resonant bar data; with slow dynamics. See protocol I in Figure 11.14. At fixed voltage V_1 the frequency is first swept down over the resonance and then swept back up over the resonance. The voltage is increased to V_2 and the down/up frequency sweeps repeated, etc. At a voltage producing strains of order 5×10^{-7} it is very apparent that the up and down sweeps give different resonance curves. The total time for an up/down sweep pair is about 60 s. So times on this scale are involved in the elastic response to AC strains of order 5×10^{-7}. The fast dynamical response, $f \approx 3900$ Hz, sees the slow dynamics of the elastic state.

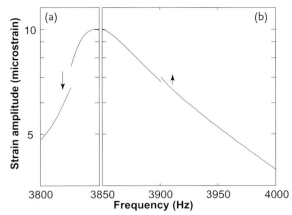

Fig. 11.16 Test of slow dynamics phenomenology. (a) At constant drive voltage, starting at 4000 Hz the frequency is swept over the resonance down to 3830 Hz, where the drive voltage is turned to zero for 30 s. On passage through the NESS domain near the resonance peak, the elastic constant is softened. During the 30 s the drive is off the elastic constant begins to return to its equilibrium value so that on reestablishing the drive voltage and resuming the frequency sweep the system appears to be on a resonance curve of higher resonance frequency, hence a jump down. (b) At constant drive voltage, starting at 3800 Hz the frequency is swept over the resonance up to 3900 Hz, where the drive voltage is turned to zero for 30 s. On passage through the NESS domain near the resonance peak, the elastic constant is softened. During the 30 s the drive is off the elastic constant begins to return to its equilibrium value so that on reestablishing the drive voltage and resuming the frequency sweep the system appears to be on a resonance curve of higher resonance frequency, hence a jump up.

When the drive voltage was reestablished, the displacement amplitude was found to be shifted downward (upward) by an amount that was taken to be due to relaxation of the elastic state created by the large amplitudes that preceded step 2. The sign of the shift in displacement amplitude is determined by the sign of the frequency shift caused by the large displacements. This experiment has a simple explanation using the model from above. However, because of the way in which it was conducted, sweeping frequency at constant voltage, the system was passed through the NESS domain rather than being established in one NESS.

From this experiment two important conclusions follow.

1. In the conduct of a resonance measurement by conventional means, sweep frequency at constant voltage, it is important to be aware of the possibility of entering a strain domain in which slow time evolution of the elastic state ensues.

2. An elastic system can be placed in an elastic state that is characterized by a pump strain amplitude. To examine the system in this state, it would be desirable to look with methods that do not disturb it. R. O'Connell [31] suggested conducting resonant bar measurements at constant strain (Section 11.2.3.6). An alternative, in the spirit of O'Connells's suggestion but

not equivalent to it, is constant strain analysis, described in Chapter 8 and below.

11.2.3.3 Constant Strain Analysis (Figure 11.14)

The result reported by Guyer, TenCate, and Johnson [32] is of the analysis of an experiment conducted with conscious awareness of the need to monitor for slow dynamics. The experiment was on a Berea sandstone (nominally of the same size as that in the TenCate–Shankland experiment) that was maintained in a controlled environment (under vacuum of approximately 2 mTorr and temperature controlled to 0.1 K). The experiment was conducted at constant voltage with frequency sweeps of approximately 30 s. The drive voltage was taken through a sequence of 23 values, $V_1 \rightarrow V_{11}, V_{12}, V_{11}, \rightarrow V_1$, in approximately $23 \times 30 \approx 700$ s. Each of 11 resonance curves was visited twice in this sequence, and a data set (set of resonance curves) was judged to be free of slow dynamics if for each constant voltage curve the resonance frequency was the same on both visits to the voltage. An example of the data so acquired is shown in Figure 11.17. If free of slow dynamics, the measured strain (displacement) values can be taken to be equilibrium values and the data can be examined on a trajectory of constant strain (displacement). This has been done for a family of strain values that cross Figure 11.17 horizontally, $\varepsilon_1, \ldots, \varepsilon_N$. For each resulting constant strain resonance curve both the frequency at resonance and the value of Q^{-1} can be determined. Details of the analysis employed are described in Chapter 8. To the degree that a constant strain resonance curve ε_n is described adequately by two parameters, $(f_r(\varepsilon_n), Q^{-1}(\varepsilon_n))$, one can say that strain is a good variable for describing the behavior of the elastic state and in particular the shift in $(f_r(\varepsilon_n), Q^{-1}(\varepsilon_n))$ away from (f_r, Q^{-1}) at the lowest strains. The results of this type of analysis for the data in Figure 11.17 is shown in Figure 11.18.

There are several points to make.

1. When one chooses to examine resonance data along a trajectory defined by a physical variable (strain, strain rate, . . .), one immediate outcome is the determination of whether or not the variable chosen is physically relevant to the observed behavior. In the example above, all points on a constant strain curve were described by the same resonance frequency and Q^{-1}. Had this not been the case, other possibilities could have been tested, see Chapter 8. The RTMF [33] scheme, also discussed in chapter 8, offers a similar opportunity to examine the nature of the underlying physical process. See also Section 11.2.3.6.

2. If a suitable field for describing the elastic behavior is found, it is straightforward to determine both f and Q^{-1}. Conventional definitions of Q, related to the width of a resonance curve, need to be helped along when a resonance curve looks like the large strain curves in Figure 11.15.

3. When there is the possibility that some elastic elements are present, having hysteretic response to dynamic strains, for example, Hertz–Mindlin contacts, there is the expectation of a connection between frequency shift and

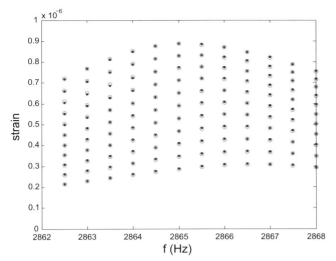

Fig. 11.17 Resonant bar data: for constant strain analysis. The 22 resonance curves used in the constant strain analysis experimental data appear as 11 curves. Open circles are the resonance curves for the up voltage sequence, $V_1, \ldots, V_{11}$, and the dots are the resonance curves for the down voltage sequence, $V_{11}, \ldots, V_1$, Figure 11.14 II. The curve at maximum drive voltage, visited only once, is not shown. The dots, generally within the circles, are evidence for the absence of the influence of slow dynamics on the data. As the frequency sweep of each resonance curve took approximately 30 s, the conclusion is slightly weaker. There is no evidence in the data for effects more long lived than 30 s having an influence on the data.

nonlinear attenuation. So neither of these variables should be regarded as a second-class.

4. The results shown in Figure 11.18 show the frequency shift to be a linear function of the strain $f(\varepsilon) = 2881.0 - 2.3\varepsilon$ and Q^{-1} to be a linear function of the strain $Q^{-1}(\varepsilon) = Q^{-1}(0) + 0.00030\varepsilon$ ($Q(0) = 350$), over the limited strain range of the data, $0.3\,\mu$ strain $< \varepsilon < 0.9\,\mu$ strain. Thus we have

$$-\frac{f(0)}{f(0) - f(\varepsilon)} \left(\frac{1}{Q(\varepsilon)} - \frac{1}{Q(0)} \right) \approx 0.40. \tag{11.36}$$

The scaling of the frequency shift and Q^{-1} with ε accords with a simple theoretical expectation. The same simple theory would have the ratio in Eq. (11.36) almost equal to 1. The numerical value, 0.40, is similar to that found for many other purported hysteretic elastic systems, Section 11.2.3.6 and Table 11.2.

5. The results above are conditional on the evidence from Figure 11.17 that there is no slow dynamics in the data set, that is, on time scales on the order of 30 s, the time between frequency sweeps at adjacent voltage values. Either in the experiment, at constant voltage, the sample never crossed the

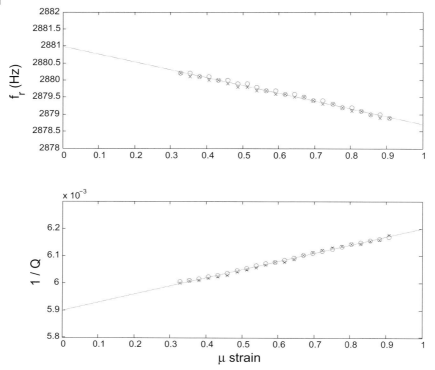

Fig. 11.18 Resonant frequency and Q^{-1} as a function of strain. The resonance frequency and Q^{-1} that characterize a set of constant strain curves constructed from the data in Figure 11.17 are shown as a function of strain. Over a limited strain range f_r and Q_r^{-1} are well described by polynomials in strain of order 1. The success of the constant strain analysis on these data established the strain as a good variable for describing the nonlinear elasticity involved.

threshold into the NESS domain or on the relevant time scale of 30 s the evidence of this entrance was no longer present in the system.

11.2.3.4 Slow Dynamics and log(t) (Figure 11.14)

If the creation and decay of a NESS were characterized by a single (or a few) time scale(s), you could get beyond them, shorter than the shortest, longer than the longest. But early on it was recognized that over a substantial strain range the NESS created by a pump would decay away slowly, essentially as the logarithm of time. This means that there is a broad spectrum of time scales in the dynamics of the NESS and that experimental time scales are more or less in the middle of these. This was established in the experiment by TenCate, Smith, and Guyer [37] in which a Berea sandstone that was maintained in a controlled environment (under vacuum of ≈ 2 mTorr and temperature controlled to 0.1 K) was subjected to the following experimental protocol.

Table 11.2 Ratio of $1/Q$ shift to frequency shift. Column 2 from anomalous fast dynamics measurements and column 3 from slow dynamics measurements.

	AFD	SD
Berea	0.40	–
Granite	0.20	–
Sandstone	0.22	–
Glass beads	0.6–3	–
Marble	0.42	0.43
Gray iron	0.35	0.25
Alumina	0.66	0.26
Quartzite	0.28	0.36
Pyrex	1.1	0.30
Sintered metal	0.31	0.25
Perovskite	0.66	0.23

1. The sample was maintained at thermal equilibrium at temperature T for at least 24 h. Toward the end of this time a resonance peak was probed with frequency sweeps at low voltage (maximum strain field less than 10^{-8}) to establish the elastic state of the sample.
2. A large voltage drive at fixed frequency (near resonance) was turned on for 15 min. This pump produced a NESS with strains of order 10^{-6}.
3. At the end of 15 min the pump was turned off and the transducer employed to probe a resonance of the sample at low voltage (maximum strain field less than 10^{-8}) to establish the elastic state of the sample.
4. Probe sweeps over the resonance continued for times out to about 4000 s (an hour), at which point the resonance frequency differed undetectably from that established in step 1.

The typical results are shown in Figure 11.19. The elastic state created/maintained by the pump, NESS, is seen to return to the thermal equilibrium state after the pump is turned off essentially as $\log(t)$ over the approximate time scale 10 s $<$ t $<$ 1000 s. In the experiment, a number of additional important features were established [37].

1. A NESS was observed in a variety of samples under a variety of conditions: intact and damaged concrete, a Berea sandstone under room-dry conditions and under vacuum, several sandstones, etc. In all cases the NESS decayed on removal of the pump approximately at a rate of $\log(t)$.
2. In the one case studied extensively, Berea sandstone, the amplitude of the decay of the NESS scaled with temperature being larger at high tempera-

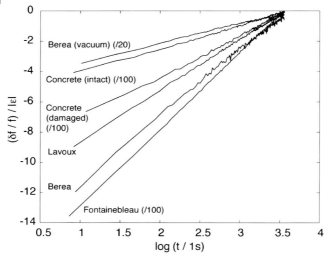

Fig. 11.19 Recovery from a NESS. The frequency shift is plotted as a function of time measured from the moment of turning off the drive voltage that established the NESS, Figure 11.14 III. The frequency shift is measured relative to the resonance frequency in equilibrium, f_0, and scaled by the strain in the NESS, typically $> 10^{-6}$. The time scale is logarithmic with $t_0 = 1$ s; the closest time to the time of turn-off is about 10 s [44].

ture. That is, for $[f_r(t) - f_r(t_0)] \approx B(T) \log(t/t_0)$, $t \gg t_0$, $B(T) \sim T^\nu$, and $\nu > 0$.

3. The amplitude of the decay of the NESS was seen to depend on the strain in the NESS seeming to vanish at a strain of order a few times 10^{-7}. That is, for $[f_r(t) - f_r(t_0)] \approx B(\varepsilon_P) \log(t/t_0)$, $B(\varepsilon_P) \sim (\varepsilon_P - \varepsilon_0)$ for $\varepsilon_P > \varepsilon_0$ and $B(\varepsilon_P) = 0$ for $\varepsilon_P < \varepsilon_0$.

4. This experiment, carried out in accord with the sense, developed in case 4, of what was going on, Section 11.2.2, leads to four important conclusions.

   a. There is a strain threshold for nonequilibrium steady states.

   b. Temperature plays an important role in the time evolution of the NESS. This suggests the presence of creep, a thermally activated process that drives a system toward equilibrium. While the evidence is not overwhelming, it is provocative.

   c. The $\log(t)$ time dependence is usually associated with the operation of phenomena that involve a broad range of time scales. When this is the case, one comes to realize that careful attention to time is necessary. So it seems unavoidable that this attention must be a part of any serious experimental exploration.

   d. Considerations such as these may be relevant to many materials.

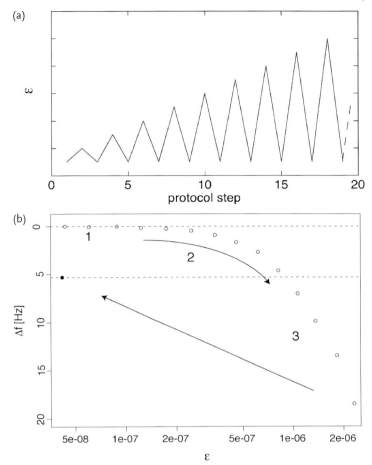

Fig. 11.20 Drive voltage protocol for low strain. To examine of the influence of the NESS domain on an investigation of low-strain resonance data, the experimental protocol in Figure 11.14 IV was employed. The open circles are the frequency shift found as a function of strain at resonance (constant voltage frequency sweeps). The resonance frequency at the lowest strain is the test resonance frequency. When the probe frequency sweep following a pump frequency sweep is unchanged from its initial value, the pump frequency sweep has not carried the system into the NESS domain, and there is no slow dynamics contamination of the data. When this is not the case, the probe frequency sweep has a frequency shift, for example, a 5-Hz shift in the resonance frequency of the probe sweep is caused by entering the NESS domain up to strain 10^{-6} (Fontainebleau sandstone).

11.2.3.5 Low-Strain Behavior (Figure 11.14)

Following the observation by Sutin [38], that the picture associated with the experiment of Guyer, TenCate, and Johnson [32] broke down at low strain, Pasqualini *et al.* [39] undertook a very careful experiment to establish the elastic state of a vari-

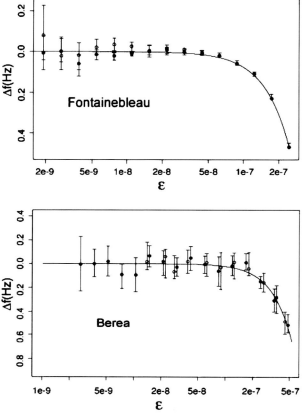

Fig. 11.21 Resonant frequency as a function of strain; low strain. Frequency shift as a function of strain at resonance for $\varepsilon < \varepsilon_T$: (top) Fontainebleau sandstone and (bottom) Berea sandstone. The two sets of data points are from two runs conducted using the protocol described in Figure 11.20a. The threshhold of the NESS domain is slightly different for a Fontainebleau sandstone ($\varepsilon_T \approx 2 \times 10^{-7}$) and Berea sandstone ($\varepsilon_T \approx 5 \times 10^{-7}$).

ety of samples at the lowest measureable strains. Further, aware of slow dynamics, in this experiment great effort was placed on finding the strain domain (at low strains) where there was no evidence of slow dynamics. The experiment was carried out using conventional constant voltage frequency sweeps (here called pump sweeps). After each frequency sweep at constant voltage the system was probed with a frequency sweep at very low drive voltage (here called a probe sweep). See panel IV of Figure 11.14. The behavior of the resonance curve (resonance frequency) found with the probe sweep was used to monitor for the influence of slow dynamics produced by the pump sweep. If after a pump sweep the probe sweep returned to its initial behavior (that found before any pump sweeps were used), the pump sweep was taken to not have carried the system into the NESS domain. Several points are worthy of note.

1. Over the strain domain, which was clearly free of slow dynamics, the behavior of the resonance frequency is sensibly an analytic function of the strain. That is, the elastic state, defined as the modulus, is an analytic function of the strain, in accord with conventional nonlinear elasticity.

2. In the low-strain regime the behavior of the resonance curves are taken to be described by the lumped element model

$$\ddot{u} = -\Omega^2 u - \gamma u^3 - 2\mu\dot{u} + F\sin\omega t \,, \tag{11.37}$$

and the parameter γ was found by fitting the data. For Berea (Fontainebleau) sandstone $\gamma = -8 \times 10^{19}$ m^{-1} s^{-1} (-5×10^{19} m^{-1} s^{-1}). The connection between this γ and the δ parameter usually used for the 4-phonon process, Eq. (3.54) (Chapter 3), is found from

$$\ddot{u} = -c^2 \frac{\partial}{\partial x} \left(\varepsilon + \delta_\gamma \varepsilon^3 \right) + F\sin\omega t \,, \tag{11.38}$$

$$\ddot{u} = -c^2 \frac{1}{L} \left(\frac{u}{L} + \delta_\gamma \left[\frac{u}{L} \right]^3 \right) + F\sin\omega t \,. \tag{11.39}$$

Thus $\delta_\gamma \approx L^4\gamma/c^2$, and using $c \approx 1.5 \cdot 10^3$ m/s and $L \approx 0.35$ m we have $\delta_\gamma \approx -10^{11}$. This value is of the same sign and about three to four orders of magnitude greater than the δ found from examining quasistatic data, Section 11.1.1. Remember that δ_γ involves dynamic strains less than 10^{-7}, whereas δ involves strains typically greater than 10^{-5}.

3. From the sensitivity to "contamination" from entering the NESS domain a fuzzy boundary on this domain is set at $\varepsilon > 2(5) \times 10^{-7}$ for Fontainebleau (Berea). The boundary is fuzzy because of the finite time scale, approximately 30 s, associated with establishing the elastic state of the system with measurements that change the elastic state.

4. A major point established by this experiment was that the approach to high strain from below is critical. Many early experiments tended to infer low-strain behavior from extrapolation to the low-strain of behavior seen at high strain. As it is hard to make measurements at low strain (see the experimental discussion in this paper), who could blame you? But the slow dynamics of high strain, the NESS domain, can contaminate the results and make extrapolation from high to low very difficult.

So one conclusion might be that all nonclassical elastic behavior results from failure to properly account for slow dynamics. All results are described by classical nonlinear elasticity + carelessness. Is there nonlinear elastic behavior, independent of slow dynamics, that is not classical? This question might be answered by experimental methods that set the elastic state of the system and interrogate it by means that do not disturb it.

11.2.3.6 **Constant Strain Measurement** (Figure 11.14)
A major advance over most previous experiments is due to Haller and Hedberg [40], here and in Section 11.2.3.7. Haller and Hedberg conducted frequency sweeps at

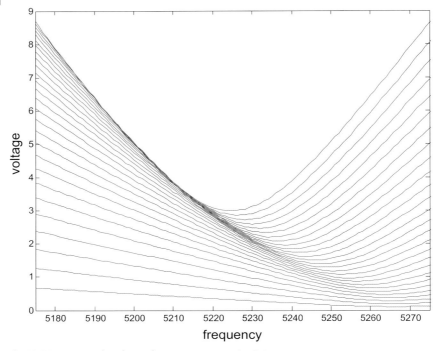

Fig. 11.22 Resonant bar data: taken at constant strain. Drive voltage as a function of frequency for frequency sweeps at constant strain. The resonance frequency, at the voltage minimum, shifts from $f_r \approx 5267$ Hz ($\varepsilon = 3 \times 10^{-8}$) to $f_r \approx 5225$ Hz ($\varepsilon = 8 \times 10^{-7}$).

constant strain. The sample was a granite cylinder, 50 mm in diameter × 42 cm, that was maintained at constant temperature and saturation. As remarked on above, this style of measurement entails carrying the system through a sequence of frequency values and with each step in frequency making the change in drive voltage necessary to maintain the strain, Eq. (11.33). An example of the results is shown in Figure 11.22, drive voltage as a function of frequency for $3 \times 10^{-8} < \varepsilon_P < 8 \times 10^{-7}$. Unlike conventional resonance curves, the drive voltage as a function of frequency is upside down. Off resonance it requires a relatively large drive voltage to maintain a strain, whereas on resonance it requires a relatively small drive voltage.

1. It is possible that the strain is a good variable for describing all of the elastic states examined in this experiment, those below the threshhold to the NESS domain and those above the threshold. For example, the elastic state maintained at pump strain $\varepsilon_P = 0.76\,\mu$ strain, is described by the simple function of ω^2 implied by Eq. (11.32). In Figure 11.23 the drive voltage squared is plotted as a function of the frequency (measured from the resonance frequency). The smooth curve through the data points (open circles) is given

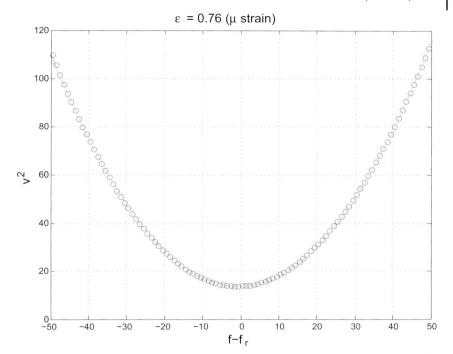

Fig. 11.23 Strain as a good variable. As the frequency is swept over a resonance at constant strain the voltage required to maintain the strain varies as in Eq. (11.40). Here V^2 is plotted as a function of $f - f_r$ (open circles) for the constant strain curve at $\varepsilon = 0.76\,\mu$ strain. The gray curve is the fit to these data with a polynomial quadratic in f^2, Eq. (11.40). Each constant strain curve can be fit to a polynomial of the form in Eq. (11.40) and the resonance frequency and Q^{-1} appropriate to the curve found, Eq. (11.40). The results of this kind of analysis of the data in Figure 11.22 are in Figure 11.24.

by

$$V^2 = C \left[1 + a(\varepsilon_P)\omega^2 + b(\varepsilon_P)\omega^4\right], \tag{11.40}$$

where the constant C is the calibration constant, which relates drive voltage to displacement output, and the two constants $a(\varepsilon_P)$ and $b(\varepsilon_P)$, one pair for each constant strain curve, are related to the resonance frequency, $f_r(\varepsilon_P)$, and attenuation, $Q^{-1}(\varepsilon_P)$, which describe that curve. Thus, since a pair of numbers $(f_r(\varepsilon_P), Q^{-1}(\varepsilon_P))$ describes each resonance curve, ε_P is a good variable and the data acquired in the experiment are acquired on trajectories of constant value of this variable.

2. f and Q.

a. The resonance frequency, the frequency at which the drive voltage is a minimum, is plotted as a function of ε_P, $3 \times 10^{-8} < \varepsilon_P < 8 \times 10^{-7}$, in

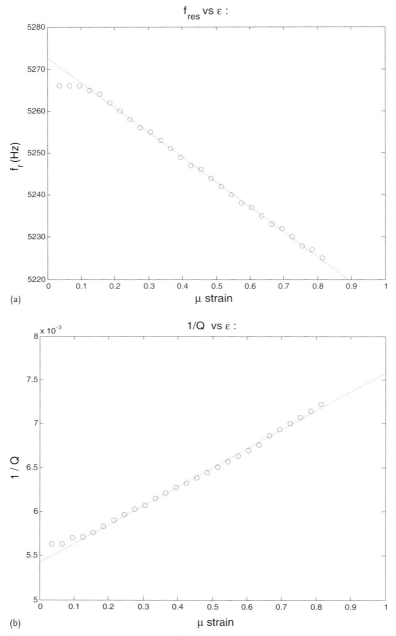

Fig. 11.24 Resonant frequency and Q^{-1} as a function of strain; constant strain. (a) The resonance frequency as a function of strain. (b) The value of Q^{-1} as a function of strain. See Figures 11.22 and 11.23.

Figure 11.24. Above $\varepsilon_P \approx 2 \times 10^{-7}$ the resonance frequency decreases linearly with ε_P, $f_r(\varepsilon_P) = 5273 - 58.9\varepsilon_P$, cf. Figure 11.18.

b. The value of Q^{-1} associated with each constant strain curve is plotted as a function of ε_P, $3 \times 10^{-8} < \varepsilon_P < 8 \times 10^{-7}$, in Figure 11.24. Above $\varepsilon_P \approx 2 \times 10^{-7} Q^{-1}$ increases linearly with ε_P, $Q^{-1}(\varepsilon_P) = Q^{-1}(0) + 0.0022\varepsilon_P$, $Q^{-1}(0) = 185$, cf. Figure 11.18.

c. From the behavior of the resonance frequency and Q^{-1} we have

$$-\frac{f(0)}{f_r(\varepsilon_P) - f(0)} \left(\frac{1}{Q(\varepsilon_P)} - \frac{1}{Q(0)} \right) \approx 0.20 . \tag{11.41}$$

This number is to be compared to that found above for Berea sandstone, 0.40, and the expectation from the Preisach space model of about 1, Table 11.2.

3. There is no slow dynamics in this experiment to distinguish the behavior of low values of ε_P from high values of ε_P. At constant strain you can't learn about slow dynamics and the location of the NESS domain.

11.2.3.7 Slow Dynamics and log(t) Again (Figure 11.25)

The second significant advance over previous work is associated with learning about slow dynamics. To do this Haller and Hedberg [41] again placed the system in a constant strain state. Then, one frequency at a time, they probed the system for the low-strain response. The protocol is shown in Figure 11.25.

1. The system is brought to constant strain with a pump source of amplitude A_1 for $\tau_i = 30$ min. It is to be probed with drive amplitude a_0, which produces minimal strain.

a. Then the pump is turned off.

  i. The probe is turned on with drive amplitude a_0 at frequency f_1 and the displacement is measured. This takes minimum time $\tau_p^{(1)} = 200$ ms (to *ring down* the pump and *ring up* the probe).

  ii. Immediately upon measuring the displacement at (a_0, f_1) the probe is turned off and the pump is returned to drive amplitude A_1, where it remains for time $\tau_r = 5$ s.

  iii. Then the pump is turned off.

  iv. The probe is turned on with drive amplitude a_0 at frequency f_2 and the displacement is measured.

  v. Immediately upon measuring the displacement at (a_0, f_2) the probe is turned off and the pump is returned to drive amplitude A_1, where it remains for time $\tau_r = 5$ s.

  vi. etc.

  vii. This sequence of steps, $(\tau_p^{(1)}, \tau_r)(\tau_p^{(1)}, \tau_r) \dots$, is carried out for a set of frequencies that sweep over the resonance at probe amplitude a_0.

b. Then, still at pump amplitude A_1, the probe frequency sequence is repeated with the displacement measured at time $\tau_p^{(2)}$ after the pump is turned

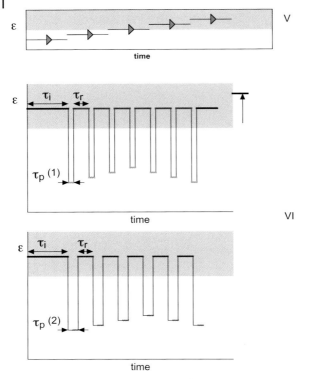

Fig. 11.25 Experimental protocol to probe slow dynamics. In order to minimally disturb slow dynamics, associated with probing a NESS, a probe frequency sweep is conducted one frequency at a time. The NESS, strain ε_1, is established for time $\tau_i \approx 30$ min with pump voltage V_1. (i) V_1 is turned to the probe voltage, V_p, at frequency f_1, and the displacement amplitude measured at $\tau_p^{(1)} \approx 200$ ms. (ii) The pump voltage is reestablished for time $\tau_r = 5$ s. (iii) V_1 is turned to the probe voltage, V_p, at frequency f_2, and the displacement amplitude measured at $\tau_p^{(1)} \approx 200$ ms. (iv) etc. In this way the NESS associated with V_1 is probed at time $\tau_p^{(1)}$ after the pump is turned off. The procedure is repeated for the NESS associated with V_1 at time $\tau_p^{(2)}$. Then, the entire procedure is repeated for NESSs of strain $\varepsilon_2, \varepsilon_3, \ldots$

off. This sequence of steps, $(\tau_p^{(2)}, \tau_r)(\tau_p^{(2)}, \tau_r)\ldots$, is carried out for a set of frequencies that sweep over the resonance at probe amplitude a_0. The value of $\tau_p^{(*)}$ is shifted to a larger value and the entire procedure repeated, $\tau_p^{(*)} = 0.2, 0.5, 1.0, 2.0, 5.0$ s. The elastic state present at time $\tau_p^{(*)}$, after the pump is turned off, is taken to be determined by the resonance frequency found from the probe frequency sweep at $\tau_p^{(*)}$.

2. This entire procedure is carried out for a series of pump source amplitudes $A_1, A_2, \ldots$ that maintain steady states with pump strain $6.7 \times 10^{-10} < \varepsilon_P < 4.3 \times 10^{-7}$.

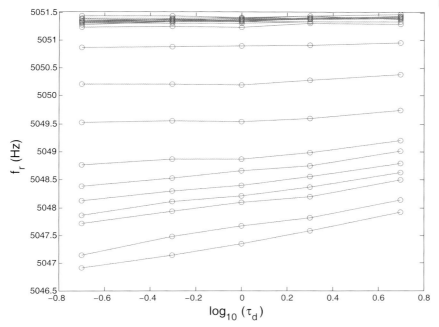

Fig. 11.26 Resonant frequency as a function of time as a NESS decays. The resonant frequency as a function of $\tau_p^{(*)}$ for NESS having strains $6.7 \times 10^{-10} < \varepsilon < 4.3 \times 10^{-7}$. The time scale is logarithmic. Strain evolves from the top of the figure 6.7×10^{-10} (f_r approximately independently of $\tau_p^{(*)}$) to the bottom 4.3×10^{-7}, ($f_r \propto \log_{10}(\tau_p^{(*)})$).

The results of this experiment are very provocative.

1. In Figure 11.26 the resonance frequency is plotted as a function of $\log_{10}(\tau_p)$ for values of the pump strain $6.7 \times 10^{-10} < \varepsilon_P < 4.3 \times 10^{-7}$. Large pump strain values, at the bottom of the plot, suggest $f_r \propto \log_{10}(\tau_p)$, whereas small pump values, at the top of the plot, suggest f_r independent of τ_p and ε_P. The data in the strain range $\varepsilon_P < 1 \times 10^{-7}$ (upper 13 curves in Figure 11.26) were fit to

$$f_r(\varepsilon_P, \tau_p) = A(\varepsilon_P) + B(\varepsilon_P)\tau_p, \tag{11.42}$$

and those at strain $\varepsilon_P > 1 \times 10^{-7}$ (lower 10 curves in Figure 11.26) were fit to

$$f_r(\varepsilon_P, \tau_p) = A(\varepsilon_P) + B(\varepsilon_P)\log_{10}(\tau_p). \tag{11.43}$$

2. In Figure 11.27 the coefficient B is plotted as a function of ε_P (closed circles are B from Eq. (11.42) and the open circles are B from Eq. (11.43)). From this figure it is apparent that there is a sharp onset to slow dynamics behavior near $\varepsilon_P \equiv \varepsilon_T \approx 0.3$ (consistent with an observation made earlier on Berea

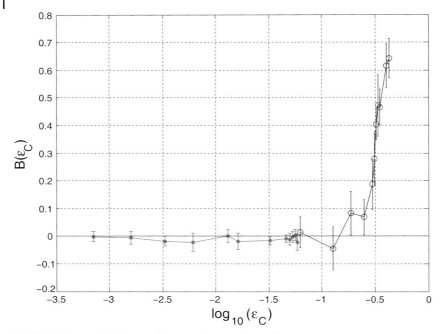

Fig. 11.27 Decay of NESSs as a function of strain. The coefficient B as a function of the strain associated with a NESS. The grey points are B from the linear fit, Eq. (11.42), and the open circles are from the $\log(t)$ fit, Eq. (11.43).

sandstone [37], in an experiment in which τ_p stood much further away from $\tau_p = 0$). Quite possibly the onset is of the form

$$B(\varepsilon_P) \sim |\varepsilon_P - \varepsilon_T|^\nu, \quad \nu < 1. \tag{11.44}$$

Whether or not this is the case awaits further detailed experimental investigation.

3. In Figure 11.28 the coefficient $A = f_r(\varepsilon_P, 0)$ is plotted as a function of ε_P (closed circles). Also on the figure as open circles is the value of $f_r(\varepsilon_P, 0.2)$ from Eq. (11.43) since $\tau_p \rightarrow 0$ doesn't make sense for this equation. The black line is a fit of the polynomial $a + b\varepsilon_P + c\varepsilon_P^2$ to the low-strain data. The open circles are above this line as they are expected to be. The approximately quadratic frequency shift seen above $\varepsilon_P = 3 \times 10^{-8}$ is of the same order of magnitude as the similar frequency shift observed in Fontainebleau sandstone and Berea sandstone. There is no striking difference in the elastic state of the system as the pump carries it into the NESS domain.

11.2.3.8 Pump/Probe Study of the NESS (Figure 11.14)

Nazarov *et al.* [42] have studied the NESS in steady state. The experiment, conducted on a sandstone from an oil and gas production site, used an apparatus that

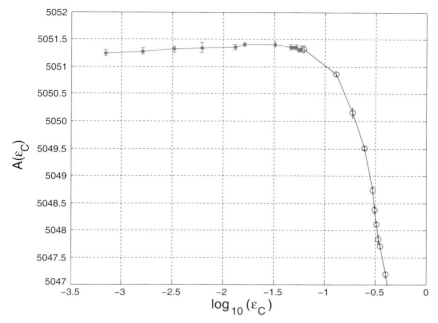

Fig. 11.28 Resonant frequency as a function of pump amplitude. Resonance frequency as a function of the strain associated with a NESS. The closed points are A from the linear fit,
Eq. (11.42), and the open circles are the value of f_r at $\tau_p^{(1)}$, the
closest measured time point to $t = 0$, Eq. (11.43).

allowed simultaneous study of two modes of a resonant bar. One of these modes
was driven at high amplitude and served as the pump. This mode, maintained at
constant strain, established the NESS. Then, the frequency was swept over the second mode resonance at constant low voltage, the probe. To show that the NESS
is created by a large-amplitude AC strain more or less independent of frequency,
Nazarov *et al.* used the fundamental resonance as the pump and the fourth mode
as the probe and then reversed the role of pump and probe (pump at fourth mode
and probe at fundamental resonance). The basic result is in Figure 11.29; (upper
panel) the resonance curves of the fourth mode probe as a function of the strain of
the pump at the fundamental mode, (lower panel) the resonance curves of the fundamental mode probe as a function of the strain of the pump at the fourth mode.
In both cases the resonance frequency of the probe mode shifts to a lower frequency as the strain of the NESS is increased. The qualitative and quantitative behavior
of the probe resonance curves are of interest.

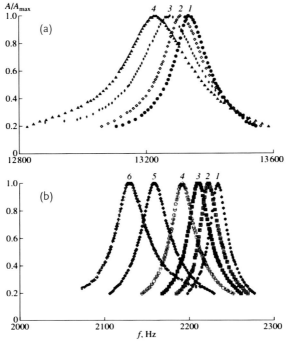

Fig. 11.29 Resonance of first and fourth modes. (a) Amplitude of the probe, fourth mode, as a function of frequency for four values of the pump strain (the pump is driving the first mode, $0 < \varepsilon_P \leq 2.5 \times 10^{-6}$, and $\varepsilon_p \leq 10^{-8}$). (b) Amplitude of the probe, first mode, as a function of frequency for six values of the pump strain (the pump is driving the fourth mode $0 < \varepsilon_P \leq 2.7 \times 10^{-6}$, and $\varepsilon_p \leq 10^{-8}$). In both plots the amplitude is scaled by the amplitude at the maximum. Data like these yield the results shown in Figure 11.30.

1. Qualitative:
   a. The resonance frequency of the probe resonances (both fundamental and fourth mode) decrease approximately linearly with the strain of the pumped mode. (NESS).
   b. The shift in $1/Q$ of the probe mode is to larger values (smaller Q) approximately linearly in the strain of the pumped mode (for strains beyond 1μ strain). (Nazarov *et al.* are able to fit the shift in $1/Q$ to a power slightly greater than 1 over the larger strain range of the complete data set.)
2. Quantitative:
   a. For the frequency shift of the probe mode

$$\frac{f_r(0) - f_r(\varepsilon_P)}{f_r(0)} \sim 0.015\varepsilon_P . \tag{11.45}$$

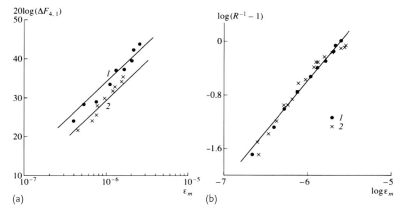

Fig. 11.30 Frequency shift and $1/Q$, data in Figure 11.29. (a) The frequency shift of the probe as a function of pump strain found from data like those in Figure 11.29, 1(2) for probe at the fourth (first) mode. (b) The value of $1/Q$ for the probe found from the change in amplitude at resonance due to pump strain, that is, $Q(0)/Q(\varepsilon_P) = A_r(0)/A_r(\varepsilon_P)$, with the probe voltage the same for all values of the pump strain. The frequency shift scales approximately linearly with ε_P, while the shift in $1/Q$ scales with ε_P to a power slightly larger than 1.

b. For the shift of $1/Q$ of the probe mode ($Q(0) \approx 150$)

$$\frac{1}{Q(\varepsilon_P)} - \frac{1}{Q(0)} \sim 0.0033\varepsilon_P , \quad \varepsilon_P > 1\mu . \tag{11.46}$$

3. For the relationship of frequency shift to $1/Q$ shift, Eq. (11.41),

$$-\frac{f(0)}{f_r(\varepsilon_P) - f(0)} \left(\frac{1}{Q(\varepsilon_P)} - \frac{1}{Q(0)} \right) \approx 0.22 . \tag{11.47}$$

Compare this number to that in Eq. (11.41), Table 11.2.

11.2.3.9 Case Study: Designer Elastic Media

In a series of papers, Jia and coworkers [34, 43] have studied the linear and nonlinear behavior of a sequence of samples formed as monodisperse glass beads under varying applied pressure. As such, the elastic properties of these samples are controlled by the elasticity of the system of Hertz–Mindlin contacts within them.

Hertz–Mindlin Elasticity in Principle   The essentials of the involvement of the contacts between glass beads in the dynamics is suggested by writing a wave equation that uses the contact stress-strain relations as the driving forces. For example, compressive forces are brought to bear due to compressive strain ε, Chapter 4,

$$\sigma = \sigma(h_0 + \delta h) = \sigma_0 \left[1 + \frac{1}{2\varepsilon_0}\varepsilon + \frac{3}{8\varepsilon_0^2}\varepsilon^2 + \ldots \right] , \tag{11.48}$$

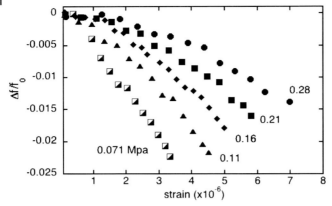

Fig. 11.31 Frequency shift vs. dynamic strain, L. The shift in the resonance frequency of the longitudinal resonance, as a function of the dynamic strain, for samples formed with applied forces $5\,\text{N} < F < 1500\,\text{N}$ $(0.007\,\text{MPa} < P < 2\,\text{MPa})$ [35].

where h_0 is the displacement brought about by the prestress, $\sigma_0, \varepsilon_0 \sim (\sigma_0/K_0)^{2/3}, \delta h$ is the dynamic displacement, and $\varepsilon = \delta h/R$, where R is the nominal bead radius. Shear forces are brought to bear in a resonant bar due to shear strain, $\varepsilon = \delta s/R$,

$$\tau = \tau_0 \left[\varepsilon - \frac{K_0}{\mu\sigma_0}\varepsilon_{\max}\varepsilon - S(\dot{\varepsilon})\frac{K_0}{\mu\sigma_0}\left(\varepsilon_{\max}^2 - \varepsilon^2\right) \right], \tag{11.49}$$

where τ is the shear stress and $X = \tau_{\max}/\mu\sigma_0$. See Eq. (6.59) in Chapter 6.

Observations

1. From Eq. (11.48) we expect to find the linear elastic constant scaling as $\sigma_0^{1/3}$ and the speed of sound scaling as $\sigma_0^{1/6}$. This is borne out in experiment [43].
2. The nonlinear term in Eq. (11.48) yield $\beta \sim \varepsilon_0^{-1}$ or $\beta \sim \sigma_0^{-2/3}$. This is borne out by quasistatic measurements and in wave mixing experiments [43].
3. From Eq. (11.48) we expect a nonlinear contribution to the behavior of the resonant bar like that described in Chapter 6 and above. Indeed this behavior is seen both when the bar is put into compressional resonance and when the bar is put into shear resonance [34]. The results for the behavior of the frequency shift and the shift in $1/Q$ are shown in Figures 11.31–11.34.

1. Over the external stress range of approximately three orders of magnitude, $10^{-3}\,\text{MPa} \le \sigma_0 \le 2\,\text{MPa}$, both the frequency shift and shift in $1/Q$ scale approximately as ε_d, where ε_d is the dynamic strain [34].
2. The ratio of the frequencys shift to the shift in $1/Q$, cf. Eq. (11.47), lies in the range 0.5–3. See Table 11.2.
3. From Eq. (11.49) we expect the amplitude of the anomalous fast dynamics to scale as $1/\sigma_0$. Indeed it does [34].

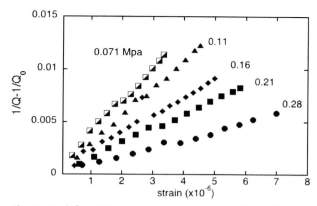

Fig. 11.32 Shift in $1/Q$ vs. dynamic strain, L. The shift in $1/Q$ of the longitudinal resonance, as a function of the dynamic strain, for samples formed with applied forces $5\,\text{N} < F < 1500\,\text{N}$ $(0.007\,\text{MPa} < P < 2\,\text{MPa})$ [35].

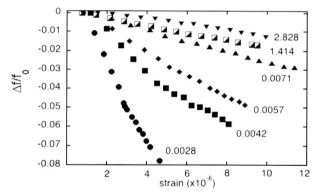

Fig. 11.33 Frequency shift vs. dynamic strain, T. The shift in the resonance frequency of the transverse resonance, as a function of the dynamic strain, for samples formed with applied forces $50\,\text{N} < F < 2000\,\text{N}$ $(0.07\,\text{MPa} < P < 2.7\,\text{MPa})$. [36]

We have examined a sequence of eight experiments/results that exhibit the important behaviors that characterize materials that have their elasticity influenced by nonlinear mesoscopic elastic features. There are two essential characteristics: (1) **anomalous fast dynamics** captured by the scaling of the frequency shift and the shift in $1/Q$ (Table 11.2) and **slow dynamics** associated with recovery from a NESS that scales as $\log(t)$. The physical systems examined were few in number, typically a bonded granular material possessing an accessible pore space, a Berea sandstone. Let us look at behavior seen in a variety of systems.

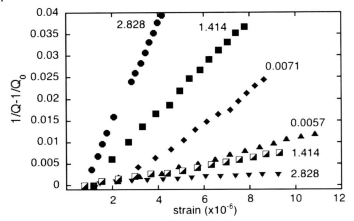

Fig. 11.34 Shift in $1/Q$ vs. dynamic strain, T. The shift in $1/Q$ of the transverse resonance, as a function of the dynamic strain, for samples formed with applied forces $50\,\mathrm{N} < F < 2000\,\mathrm{N}$ ($0.07\,\mathrm{MPa} < P < 2.7\,\mathrm{MPa}$). [36]

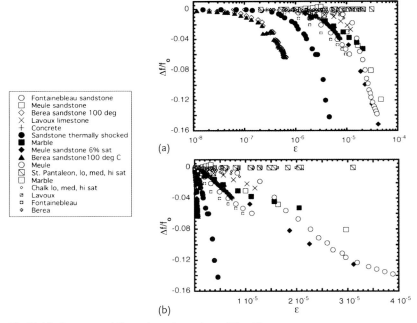

○ Fontainebleau sandstone
□ Meule sandstone
◇ Berea sandstone 100 deg
✕ Lavoux limestone
+ Concrete
● Sandstone thermally shocked
■ Marble
◆ Meule sandstone 6% sat
▲ Berea sandstone100 deg C
○ Meule
◙ St. Pantaleon, lo, med, hi sat
□ Marble
◦ Chalk lo, med, hi sat
▨ Lavoux
▫ Fontainebleau
◈ Berea

Fig. 11.35 Frequency shift vs. dynamic strain, rocklike. The shift in the resonance frequency, as a function of the dynamic strain, for a variety of rocklike materials: (a) dynamic strain scale is $\log(\varepsilon_d)$ and (b) dynamic strain scale is ε_d. On the linear dynamic strain scale the shift is approximately linear at high strain, $\varepsilon_d \approx 10^{-5}$.

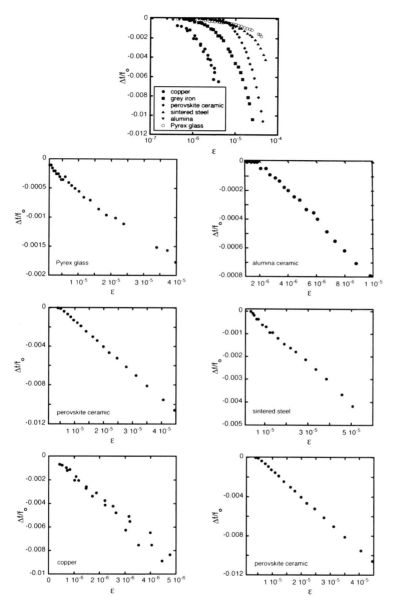

Fig. 11.36 Frequency shift vs. dynamic strain, not rocklike. The shift in the resonance frequency, as a function of the dynamic strain, for a variety of non-rock-like materials. (top) The dynamic strain scale is $\log(\varepsilon_d)$. (bottom 6 panels) The dynamic strain scale is linear in ε_d. The essential feature in all cases is that the frequency shift decreases approximately linearly with ε_d. Compare to (b) in Figure 11.35.

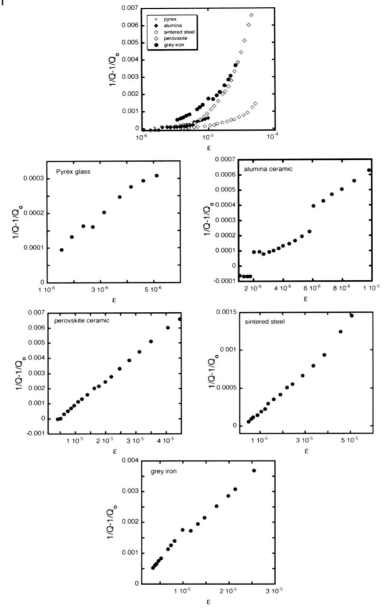

Fig. 11.37 Shift of $1/Q$ vs. dynamic strain, not rocklike. The shift of $1/Q$, as a function of the dynamic strain, for a variety of non-rock-like materials. (top) The dynamic strain scale is $\log(\varepsilon_d)$. (bottom 6 panels) The dynamic strain scale is linear in ε_d. The essential feature in all cases is that the shift in $1/Q$ increases approximately linearly with ε_d. (Note: pyrex glass intact.)

NONLINEAR RESONANCE MEASUREMENTS

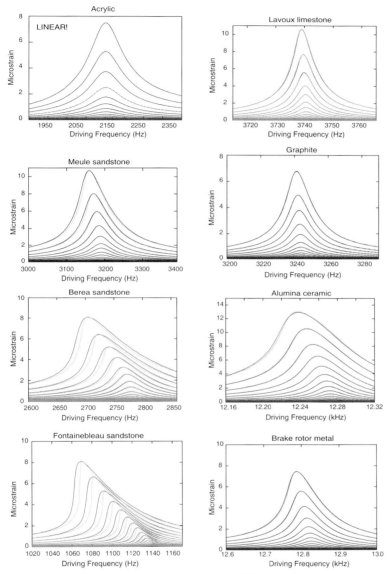

Fig. 11.38 Resonance curves. The panels are the amplitude as a function of frequency, each curve in a panel corresponding to a fixed constant drive voltage, for eight materials. In all cases the resonance curves were swept over twice, from above to below and below to above. The curves for graphite overlie one another and show there is no slow dynami-cal behavior on the time scale of a resonance sweep, 30 s. The curves for graphite show no evidence of peak bending, so there is no nonlinearity explored at the strain amplitudes involved. In the other seven curves there is evidence of peak bending and/or slow dynamics. Fontainebleau sandstone shows both behaviors very clearly.

11.3
Examples of Systems

11.3.1
Anomalous Fast Dynamics

In four experiments described above [32, 34, 40, 42] the frequency shift and shift in $1/Q$ showed anomalous behavior. What we mean by this is that the frequency shift and shift in $1/Q$ were approximately linear in the strain over a reasonable strain range, Figures 11.18, 11.24, 11.28, and 11.30. In addition, the ratio of the frequency shift and shift in $1/Q$ was a constant of order 1, Eqs. (11.36), (11.41), and (11.47) and Table 11.2.

In Figure 11.35 we show the frequency shift and the shift in $1/Q$ for a variety of systems that are rocks or rocklike, for example, ceramic, concrete, synthetic slate, etc. [44]. The four panels are linear and log plots for these quantities. In Figures 11.36 and 11.37 we show the behavior of the frequency shift, linear and log, for a variety of systems which are very different from rock. The data in these figures are in qualitative accord with the anomalous fast dynamics results from above. Notice that the strain range explored in these data are in some cases well beyond those in the studies above. In many cases the data were acquired in an era before the importance of slow dynamics was recognized. For that reason we take these data to be consistent with our expectations but not definitive.

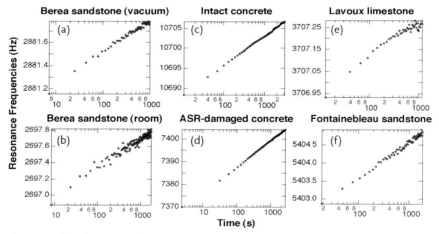

Fig. 11.39 Slow dynamics. Following establishment of a NESS in a sample, the pump that maintains the NESS is turned to zero and the resonance frequency, found with a low-amplitude probe, is monitored over time. In the six panels the resonance frequency found by the probe is plotted as a function of time, see Figure 11.40. The time scale is $\log(t)$ so that it is apparent that the resonance frequency increases (recovers) as $\log(t)$.

11.3.2
Slow Dynamics

11.3.2.1 Slow Dynamics: Review
In Figures 11.15, 11.19, and 11.26 there are various qualitative and quantitative evidences of slow dynamics, resonance curves that depend on sweep rate, log(t) recovery from a NESS. In Figure 11.38 we show sweep-rate-dependent resonance curves for eight different materials [45]. Except for graphite, a linear material, all of the sets of resonance curves show evidence of slow dynamics and peak bending. The poster child here is Fontainebleau sandstone. In Figure 11.39 we show log(t) recovery from a NESS created in a variety of porous materials [46]. Slow dynamics in non-rock-like materials is described below.

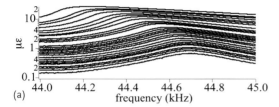

(a)

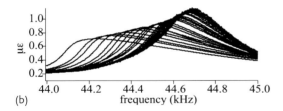

(b)

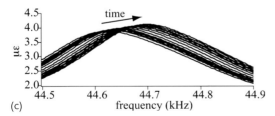

(c)

Fig. 11.40 Anomalous fast dynamics and slow dynamics. (a) The amplitude as a function of frequency for constant voltage drives that produce strain and frequency shift proportional to strain. (b) The same curves as in (a) with each scaled by the drive voltage. For the largest voltage used in (a) a NESS state is created by driving at a frequency near the resonance maximum. Once the NESS is established the pump is turned to zero. (c) The sequence of resonance curves that is seen by a low-amplitude probe as time passes appears much like those in (b). See Figure 11.14.

11.3.2.2 Anomalous Fast Dynamics and Slow Dynamics

Johnson and Sutin[44] have studied both anomalous fast dynamics and slow dynamic on seven nonrock materials, gray iron, . . . , alumina. The anomalous fast dynamics experiments were carried out as described in Section 11.2, Figure 11.14, protocol I. The slow dynamic response employed protocol III in Figure 11.14. In Figure 11.40 three sets of resonance curves are shown for experiments on quartzite. Panels (a) and (b) show the resonance curves found in the anomalous fast dynamics study, (a) the raw data and (b) the raw data scaled by the drive voltage. In (b) a frequency shift and a shift in $1/Q$ are evident. The slow dynamics response of the quartzite is shown in Figure 11.40, a sequence of probe resonance curves found as time evolves from the moment of turn-off of the pump that established a NESS. Panels (b) and (c) are qualitatively similar. The size of the frequency shift and the amplitude change in the slow dynamics recovery (c) is approximately an order of magnitude less than the similar shifts in anomalous fast dynamics (b). In Figure 11.41 the shift in frequency and the shift in $1/Q$ are shown for data, similar to that in panels (b) and (c) of Figure 11.40, for a variety of samples.

In the case of anomalous fast dynamics, we have the result in Eqs. (11.36), (11.41), and (11.47) that the ratio of the frequency shift (approximately linear in the drive strain) to the shift in $1/Q$ (approximately linear in the drive strain) is a number of order 1. In the slow dynamics response the analog of the drive strain is the time measured from the turn-off of the pump. It is apparent from Figure 11.41, panels (a) and (b), that the frequency shift is approximately linear in $\log(t)$ and the shift in $1/Q$ is approximately linear in $\log(t)$. Thus forming the ratio of the frequency shift to the shift in $1/Q$ we find the numbers in column 3 of Table 11.2.

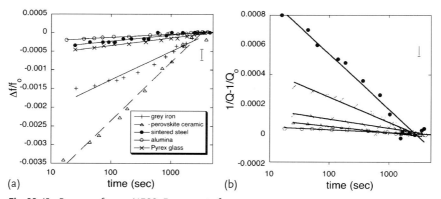

Fig. 11.41 Recovery from a NESS. From a set of resonance curves like those in Figure 11.40(c) a resonance frequency and a value of Q can be found as a function of time. These quantities are seen to evolve linearly with time just as the corresponding quantities evolve linearly with dynamic strain in Figures 11.36 and 11.37.

References

1 Zinszner, B., Rasolofosaon, P. and Johnson, P.A. (unpublished).

2 Hughes, D.S. and Kelley, J.L. (1953) Second-order elastic deformation of solids. *Phys. Rev.*, **92**, 1145–1149.

3 Winkler, K.W. and Liu, X. (1996) Measurements of third-order elastic constants in rocks. *J. Acoust. Soc. Am.*, **100**, 1392–1398.

4 Landau, L.D. and Lifshitz, E.M. (1987) *Fluid Mechanics*, 2nd edn, Butterworth-Heinemann, New York.

5 Murnaghan, F.D. (1951) *Finite Deformation of an Elastic Solid*, Chapman and Hall, New York.

6 Thurston, R.N. and Brugger, K. (1964) Third-order elastic constants and the velocity of small amplitude elastic waves in homogeneously stressed media. *Phys. Rev. A*, **133**, 1604–1610.

7 Kostek, S., Sinha, B.K., and Norris, A.N. (1993) Third-order elastic constants for an inviscid fluid. *J. Acoust. Soc. Am.*, **94**, 3014–3017.

8 Catheline, S., Gennisson, J.-L., and Fink, M. (2003) Measurement of elastic nonlinearity of soft solid with transient elastography. *J. Acoust. Soc. Am.*, **114**, 3087 3091.

9 Ide, J.M. (1937) The velocity of sound in rocks and glasses as a function of temperature. *J. Geol.*, **45**, 689–716.

10 Ulrich, T.J. (2005) (thesis), University of Nevada, Reno.

11 Amberg, C.H. and McIntosh, R. (1952) A study of absorption hysteresis by means of length changes of a rod of porous glass, *Can. J. Chem.*, **30**, 1012–1032.

12 Johnson, P.A., Zinszner, B., Rasolofosaon, P., Cohen-Tenoudji, F., Van Den Abeele, K. (2004) Dynamic measurements of the nonlinear elastic parameter A in rock under varying conditions, *J. Geophys. Res.*, **109**, 10129–10139.

13 Zinszner, B., Johnson, P., and Rasolofasoan, P.N.J. (1997) Influence of change in physical state on elastic nonlinear response in rock: Significance of effective pressure and water saturation, *J. Geophys. Res.*, **102**, 8105–8120.

14 Bourbie, T., Coussy, O., and Zinszner, B. (1987) *Acoustics of Porous Media*, Gulf Publ., Houston.

15 Johnson, D.L., Plona, T.J., Scala, C., Pasierb, F., and Kojima, H. (1982) Tortuosity and acoustic slow waves, *Phys. Rev. Lett.*, **49**, 1840–1844.

16 Jones, G.L. and Kobett, D.R. (1963) Interaction of elastic waves in an isotropic solid, *J. Acoust. Soc. Am.*, **35**, 5–10.

17 Taylor, L.H. and Rollins, F.R. (1964) Ultrasonic study of three-phonon interactions, I. Theory, *Phys. Rev.*, **136**, 591–596.

18 Rollins, F.R., Taylor, L.H., and Todd, P.H. (1964) Ultrasonic study of three-phonon interactions, II. Experimental Results, *Phys. Rev.*, **136**, 597–601.

19 Dunham, R.W. and Huntington, H.B. (1970) Ultrasonic beam mixing as a measure of the nonlinear parameters of fused silica and single-crystal NaCl, *Phys. Rev. B*, **2**, 1098–1107.

20 Johnson, P.A., Shankland, T.J., O'Connell, R.J., and Albright, J.N. (1987) Nonlinear generation of elastic waves in crystalline rock, *J. Geophys. Res.*, **92**, 3597–3602.

21 Johnson, P.A. and Shankland, T.J. (1989) Nonlinear generation of elastic waves in granite and sandstone: continuous wave and traveltime observations, *J. Geophys. Res.*, **94**, 17729–17734.

22 Meegan, G.D., Johnson, P.A., Guyer, R.A., and McCall, K.R. (1993) Observations of nonlinear elastic wave behavior in sandstone, *J. Acoust. Soc. Am.*, **94**, 3387–3391.

23 TenCate, J.A., Van den Abeele, K., Shankland, T.J., and Johnson, P.A. (1996) Laboratory study of linear and nonlinear elastic pulse propagation in sandstone, *J. Acoust. Soc. Am.*, **100**, 1383–1391.

24 LeBas, P.-Y., Guyer, R., Johnson, P., and Ten-Cate, J. (2009) The parametric array in rock: definitive experiments, *J. Acoust. Soc. Am.*, in review.

25 Hamilton, M.F. and Blackstock, D.T. (1998) *Nonlinear Acoustics*, Academic Press, New York.

26 D'Angelo, R.M., Winkler, K.W., Plona, T.J., Landsberger, B.J., and Johnson, D.L. (2004) Test of hyperelasticity in highly nonlinear solids: sedimentary rocks, *Phys. Rev. Lett.*, **93**, 214301-1–214301-4.

27 D'Angelo, R.M., Winkler, K.W., and Johnson, D.L. (2008) Three wave mixing test of hyperelasticity in highly nonlinear solids: sedimentary rocks, *J. Acoust. Soc. Am.*, **123**, 622–632.

28 Clark, V.A. (1980) (thesis) Effect of volatiles on seismic attenuation and velocity in sedimentary rocks, Texas AM University.

29 Migliori, A. and Sarrao, J.L. (1997) *Resonant Ultrasound Spectroscopy*, John Wiley & Sons, Inc., New York.

30 TenCate, J.A. and Shankland, T.J. (1996) Slow dynamics in the nonlinear elastic response of Berea sandstone. *Geophys. Res. Lett.*, **23**, 3019–3022.

31 Observation made at Nonlinear Mesoscopic Elasticity Workshop II, (1997).

32 Guyer, R.A., TenCate, J.A., and Johnson, P.A. (1999) Hysteresis and the dynamic elasticity of consolidated granular materials. *Phys. Rev. Lett.*, **82**, 3280–3283.

33 Smith, E. and TenCate, J.A. (2000) Sensitive determination of nonlinear properties of Berea sandstone at low strains. *Geophys. Res. Lett.*, **27**, 1985–1988.

34 Laurent, J. (2007) (thesis) Université Paris-Est Marne-La-Vallee.

35 Johnson, P. and Jia, X. (2006) Nonlinear dynamics, granular media and dynamic earthquake triggering. *Nature*, **473**, 871–874

36 Laurent, J. and Jia, X., in preparation.

37 TenCate, J.A., Smith, D.E., and Guyer, R.A. (2000) Universal slow dynamics

in granular solids. *Phys. Rev. Lett.*, **85**, 1020–1023.

38 Sutin, S. (unpublished).

39 Pasqualini, D., Heitmann, K., TenCate, J.A., Habib, S., Higdon, D., and Johnson, P.A. (2007) Nonequilibrium and nonlinear dynamics in Berea and Fontainebleau sandstones: Low-strain regime. *J. Geophys. Res.*, **112**, B01204.1–B01204.16.

40 Haller, K.C.E. and Hedberg, C.M. (2008) Constant strain frequency sweep measurements on granite rock. *Phys. Rev. Lett.*, **100**, 068501.1–068501.4.

41 Haller, K.C.E. (2008) (thesis) Blekinge Institute of Technology 371 79 Karlskrona, Sweden. Haller, K.C.E. (2008) (thesis) Blekinge Institute of Technology 371 79 Karlskrona, Sweden.

42 Nazarov, V.E., Radostin, A.V., and Soustova, I.A. (2002) Effect of an intense sound wave on the acoustic properties of a sandstone bar resonator. Experiment. *Acoust. Phys.*, **48**, 76–80.

43 Brunet, T. (2005) (thesis) Université Paris-Est Marne-La-Vallee.

44 Johnson, P.A. and Sutin, A. (2005) Slow dynamics and anomalous nonlinear fast dynamics in diverse solids. *J. Acoust. Soc. Am.*, **117**, 124–130.

45 Tencate, J.A. (unpublished).

46 TenCate, J.A., Smith, E., Byers, L., and Shankland, T.J. (2000) *Nonlinear Acoustics at the Turn of the Millennium, the 15th International Symposium on Nonlinear Acoustics*, Gottingen, Germany, (eds Lauterborn, W. and Kurz, T.), AIP Conf. Proc. **524**, 303.

12
Field Observations

In this short chapter we describe field measurements that are akin to the laboratory measurements above. In Section 12.1 we describe field measurements in which the earth is the subject of investigation using deliberate man-made, active probes, whereas in Section 12.2 we discuss passive probes of the earth, earthquakes. It is not the earthquake that is passive but rather the observer of it.

12.1
Active Probes

12.1.1
Wave Mixing in the Earth

A wave mixing experiment was conducted in 1991 by an oilfield service company in collaboration with Los Alamos National Laboratory and Lawrence Berkeley Laboratory. In the experiment, thumper trucks were placed in a hexagonal pattern

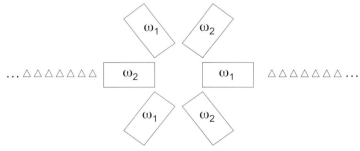

Fig. 12.1 Experimental configuration for a field experiment employing frequency mixing using 20-t vibrator sources (Vibroseis) oriented as asterisk-shaped arrays, each emitting either angular frequency ω_1 or ω_2 as noted. The detector array, oriented along a line emanating outward from the source array (triangles), comprised 160 receivers with a maximum offset of 536 m from the source array. The vibrator sources are coupled to the soil by approximately 2×2 m^2 plates located near the center of each source (figures courtesy of Tom Daley).

Nonlinear Mesoscopic Elasticity: The Complex Behaviour of Granular Media including Rocks and Soil. Robert A. Guyer and Paul A. Johnson
Copyright © 2009 WILEY-VCH Verlag GmbH & Co. KGaA, Weinheim
ISBN: 978-3-527-40703-3

with the front of the trucks pointing to the hexagon center (Figure 12.1). The vibrators moved vertically and were phase locked. A 20-s step-sweep was employed with three vibrators sweeping 50–90 Hz while the other three vibrators were sweeping 90–50 Hz. This procedure was intended to guard against difference frequencies arising from the mechanical or electrical behavior of the driving system. Thus the recorded signals at the difference frequencies, 40–0–40 Hz, are taken to arise from nonlinear mixing in the earth. The data, received on a linear string of geophones straddled by the thumper trucks (Figure 12.1), were processed in a standard reflection seismology manner, that is, averaged over detector subgroup and multiple repetitions of the step-sweep. The resulting data were then correlated with a synthetic difference frequency signal constructed from the step-sweeps of the vibrators. The amplitude at the difference frequency is shown in Figure 12.2a as a function of position and time. Both direct compressional and surface waves as well as compressional wave reflections are present (and noted in the figure). Figure 12.2b shows the Fourier transform for one of the traces. While one might have expected increased amplitude at the difference frequency with distance from the source, there is no evidence for this. Thus it is inferred that most of the frequency mixing took place in the near-source region.

In a study employing one vibrator driven at a single frequency, Beresnev and Nikolaev [1] found an increase in harmonic content, relative to the fundamental driving frequency, as a function of distance from the vibrator. This increase suggests additive nonlinear mixing in the earth as the fundamental propagates away from the source. In a similar study conducted on surface waves in southern Cali-

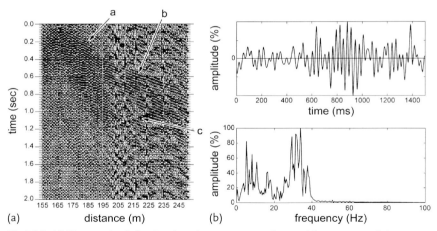

Fig. 12.2 (a) Time vs. depth for signals at the difference frequency. Noted are **a** compressional waves, **b** surface waves and Rayleigh waves, and **c** backscattered surface waves probably off subsurface heterogeneity or topography. (b) Time series signal (top) and spectrum (bottom) for one trace of data are shown at left. The primary energy is from 0 to 40 Hz [3] (data from experiments conducted by Los Alamos and data analysis courtesy of Tom Daley and Tom McEvilly, unpublished).

fornia by Lawrence *et al.* [2], a dense accelerometer array was deployed within meters of a shaker source. The wave field was dominated by Rayleigh surface waves and ground motions were strong enough to produce observable nonlinear changes in wave velocity. It was found that as the force load of the shaker increased, the Rayleigh wave-phase velocity decreased by as much as 30% at the highest frequencies used (up to 30 Hz). Phase velocity dispersion curves were inverted for S-wave velocity as a function of depth using an isotropic elastic model to estimate the depth dependence of changes to the velocity structure. The greatest change in velocity occurred within the 4 m nearest the surface.

12.1.2
The Earth as Resonant Bar

Experiments described in reference [4] employing a large, active shaker source show strong elastic nonlinear response in near-source sediments. The site of the experiment, located near Austin, Texas, involved 11 m of young, unconsolidated point bar sediments above bedrock (point bars are deposits formed along the inside of a river bend). A reconnaissance geophysical survey identified significant elastic impedance contrasts (interfaces between layers) at depths of 2, 4.5, 7 and 11 m (bedrock) [5]. The 7 m depth interface was the water table. The fundamental resonance frequencies corresponding to sediment columns bounded by these interfaces were estimated from numerical modeling.

   Figure 12.3 shows a schematic of the in situ resonance experiment. The shaker exerts a maximum vertical force output of approximately 267 kN (about 27 metric tons). The source couples to the sediment via a $2 \times 2 \, m^2$ plate and is capable of shaking over a broad frequency band (15 to 180 Hz) with approximately constant force. Three-component accelerometers designed for accelerations up to *2g* were deployed at the surface immediately adjacent to the source. The measured vertical accelerations are used for the results presented here.

   In these experiments, just as in laboratory experiments, one step-sweeps frequency across a band that covers the fundamental resonance modes of the sediment layers. At each frequency step, steady-state conditions are obtained and the acceleration amplitude is measured. Normed resonance curves at progressively increasing drive amplitude, measured on the accelerometer located 3 m from the source, are shown in Figure 12.4. These curves are essentially the sediment transfer functions as they result from dividing the measured acceleration in the soil by that recorded at the source.

   The most striking aspects of the sediment transfer functions shown in Figure 12.4 are the modal peaks, caused by the layer structure, and the overall shift of features in the transfer function to lower frequency as the source amplitude is increased. An open-ended system of sediment layers is a very complicated resonance system. Nonetheless, resonance peaks appear approximately where a simple calculation shows them to be. A granular material, a possible model for the layers, has nonlinear response that decreases strongly with increasing effective static stress (confining stress minus pore pressure). This dependence may explain in

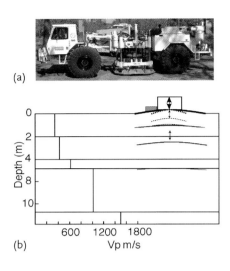

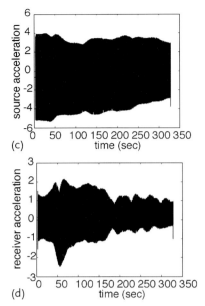

Fig. 12.3 Experimental configuration for in situ nonlinear soil studies. (a) Vibrator source. The large grey plate between the wheels is the coupling plate. The results shown in Figure 12.4 were for the source operating in com- pression. (b) Compressional wave velocity structure at site. (c) Measured acceleration on the source baseplate. (d) Measured accel- eration just adjacent to the baseplate, from a seismometer located in the soil [4].

part the smaller frequency shift seen at low frequency, that is, in the modes that reach deeper into the layers. Additionally, the deeper layers, being farther from the source, experience smaller dynamic loads. It is impossible to extract the individual lower-layer behaviors without detailed modeling that accounts for the complex distribution of the strain field in the presence of many layers. Further complexity comes from the admixture of standing and propagating waves that are in the mea- sured signals. Consequently the analogy to a resonant bar should be viewed with caution.

The largest frequency shift in the data is seen at frequencies corresponding ap- proximately to resonance of the 2 m layer (Figure 12.4b). The decrease in the am- plitude of the transfer function with increased forcing is evidence of nonlinear dis- sipation. The decrease in resonance frequency corresponding to the 2 m interface is about 20% and the amplitude decrease is about 25%. The nonlinear parameters describing the resonance frequency shift and the increased dissipation with drive level are $\alpha_f = 24\,200$ and $\alpha_Q = 4100$, respectively, Chapter 11, Section 11.3. For comparison, laboratory data on granular media taken under similar loading condi- tions yield $\alpha_f = 7200$ and $\alpha_Q = 2900$ [6].

Figure 12.4c shows the *in situ* transfer functions found in a slow dynamics ex- periment focused on the 2 m interface. Slow dynamics were induced by a large- amplitude frequency sweep. Immediately following this, a sequence of probe step-

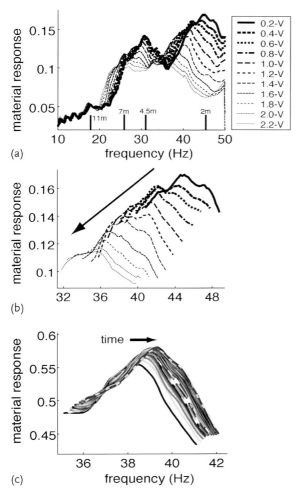

Fig. 12.4 Induced nonlinear response in layered media. (a) Resonance modal structure as a function of the normalized acceleration "material response" (receiver acceleration divided by the source acceleration measured on the source baseplate) for progressively increasing source forcing. The expected layer resonance frequencies are shown as vertical lines on the x-axis. (b) Zoom of the shallowest layer response (the 2 m layer). Note the significant change in resonance frequency as well as the decrease in amplitude. (c) The slow dynamical recovery after strong forcing. The strain range of the experiment was $8 \times 10^{-5} - 3 \times 10^{-4}$ (courtesy of F. Pearce).

sweeps all having the lowest source amplitude were applied to observe the recovery, just as in laboratory studies. There is an immediate elastic recovery from 34 to 37 Hz (not shown) followed by a slow recovery during which the resonance frequency returns to about 90% of its original value in 1 hour. Full recovery is estimated to take at least a full day.

12.2
Passive Probes

Near-surface amplification of seismic waves from earthquakes is a well-established phenomenon. Examples of this behavior include all of the most damaging earthquakes of the last two decades, including the 1985 Michoacan (Mexico), the 1989 Loma Prieta (California), the 1994 Northridge (California), and the 1995 Kobe (Japan) earthquakes. Typical detected wave strains in these earthquakes are 10^{-4} in the near-source region to 10^{-9} at teleseismic distances with frequencies ranging from 0.1 to 10^2 Hz. Strain amplitude near a free surface is amplified in the presence of low-velocity layers (at fixed energy $\propto K\varepsilon^2$ decrease in K means increase in ε) or by resonances [7, 8]. Increased strain amplitude leads to the increased importance of nonlinearity. In what follows we describe the first study published illustrating nonlinear behavior in response to a large earthquake. *The Northridge Earthquake Study.* Field *et al.* (1997) [9] described results showing that alluvium in the Los Angeles Basin reacted nonlinearly in response to the 1994 Northridge earthquake. In order to determine the material response, one must find a method to isolate the alluvium response from the source and wave propagation effects during an earthquake. The method applied in this study was the spectral ratio method, widely applied in seismology to eliminate wave dissipation effects.

The essence of treating the seismic data is as follows. Following data collection at a large number of seismometer recording sites that include hard rock and soft sediment sites, the data were corrected for dissipation effects by deconvolving the Green's function. This was accomplished by assuming a reasonable inverse dissipation ($Q(f)$) and multiplying the signals by this value. The earthquake source response was then estimated from the rock sites, assumed to have no site response. That is, it is assumed that the rock sites show the source response spectrum after they have been corrected for linear dissipation. The response at the alluvium sites was then estimated as the ratio between average alluvium and average rock sites. For small earthquakes, the ratios should be larger than unity if the response of the unconsolidated material is linear, due to the amplification effects in sediments. For large earthquakes, hysteretic effects would be present in the alluvium. If hysteretic damping is present in the alluvium, the spectral ratio should be less than unity.

Figure 12.5 is a map of southern California, showing the mainshock epicentral location and the location of the seismometers on rock and alluvium sites. In the Northridge earthquake study presented here, data were compiled from locations where both mainshock and aftershock recordings were obtained. Based on surface geology, 15 of these sites were categorized as alluvium and four as rock.

Figure 12.6 illustrates the general finding from the Northridge study. The weak motion (solid lines) and strong motion (dashed lines) site-response estimates averaged over the 15 alluvium sites and 4 rock sites are illustrated in the figure. The weak-motion response implies an amplification factor of approximately 3.1 at 1 Hz, decreasing to factors of approximately 2.5 and 1.4 at 3 and 10 Hz, respectively. The strong-motion amplification factors are systematically less, being about 1.9 at 1 Hz, about 1.3 at 3 Hz, and 0.8 (deamplification) at 10 Hz. This clear decrease in ampli-

Northridge Earthquake

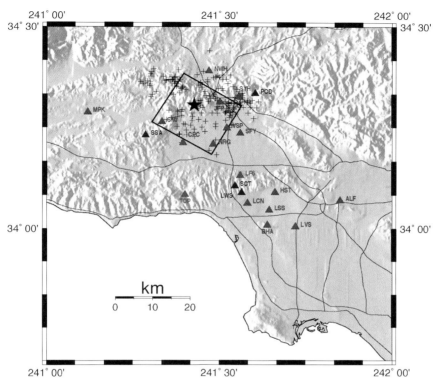

Fig. 12.5 Relief map of the study region in southern California, showing the location of 1994 Northridge earthquake (star). The alluvium recording sites are shown as red triangles and the hard rock sites as blue triangles. Aftershock epicenters are shown with black crosses and the mainshock rupture distribution is outlined by the box. The fault plane dips to the southwest, with the top edge at a depth of 5 km and the bottom edge at a depth of 20.4 km. The location of maximum earthquake slip is marked with the black star (from [9]). (Please find a color version of this figure on the color plates)

fication implies nonlinear dissipation. In fact, applying the standard t-distribution test it was found that the difference is significant at the 95% confidence level between 0.8 and 5.7 Hz, implying that there is significant nonlinear response over this frequency range (between 1.2 and 4.3 Hz, the difference is significant at the 99% level). At at least one seismic recording station, a significant resonance frequency shift was observed, Figure 12.7. This frequency shift, a factor of 3, corresponds to a shift in the elastic constant by a factor of about 10. The observation led to a large number of studies exploring the elastic nonlinear behavior of in situ materials during large earthquakes.

There are a number of other studies that show layer nonlinearity during large earthquakes. One very nice example is that of Pavlenko and Irikura [10], in which

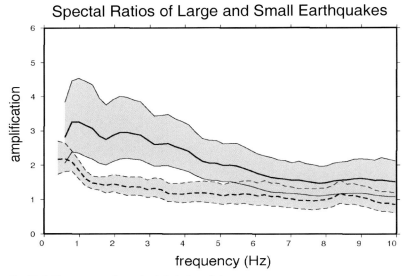

Spectal Ratios of Large and Small Earthquakes

Fig. 12.6 The mean and standard-deviation-of-the-mean confidence limits for the 15 alluvium site-amplification estimates. The solid lines represent the weak-motion results for the aftershocks, and the dashed lines represent the strong-motion results for the mainshock. The lower-frequency cutoffs reflect the lowest resolvable frequencies given seismometer and noise limitations (from [9]).

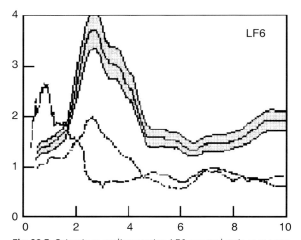

Fig. 12.7 Seismic recording station LF6 spectral-ratio response to the 1994 Northridge earthquake. The solid lines represent the weak-motion results for the aftershocks, and the dashed lines represent the strong-motion result for the mainshock. The sharp peak for the aftershocks at about 2.8 Hz shifts to about 0.9 Hz for the mainshock (from [9]).

hysteretic behavior as a function of depth is extracted from the measured acceleration signals in boreholes during the 1994 Kobe (Japan) earthquake. The great value of this study is it shows how variable the response is with depth based on the layered structure.

http://en.wikipedia.org/wiki/Seismic_source

References

1 Beresnev, I. and Nikolaev, A. (1988) Experimental investigations of nonlinear seismic effects. *Phys. Earth Planet. Interiors*, **50**, 83–87.

2 Lawrence, Z., Bodin, P., Langston, C., Pearce, F., Gomberg, J., Johnson, P., Meng, F.-Y., and Brackman, T. (2008) Induced dynamic nonlinear ground response at Garner Valley, California. *Bull. Seism. Soc. Am.*, **98**, 1412–1428.

3 Daley, T.M., Peterson, J.E., Jr, and McEvilly, T.V. (1992) *A Search for evidence of nonlinear elasticity in the Earth*, Lawrence Berkeley Laboratory Report, LBL-33313, UC-403.

4 Johnson, P.A., Bodin, P., Gomberg, J., Pearce, F., Lawrence, Z., and Menq, F.-Y. (2009) Including in situ, nonlinear soil response applying an active source. *J. Geophys. Res.*, **114**, B05304.

5 Kurtulus, A., Lee, J.J., and Stokoe, K.H. (2005) Summary report: site characterization of capital aggregates test site, University of Texas, Austin Internal Report, 1–47. Kurtulus, A., Lee, J.J., and Stokoe, K.H. (2005) Summary report: site

characterization of capital aggregates test site, University of Texas, Austin Internal Report, 1–47.

6 Johnson, P.A. and Jia, X. (2005) Nonlinear dynamics, granular media and dynamic earthquake triggering. *Nature*, **437**, 871–874.

7 Shearer, P. and Orcutt, J. (1987) Surface and near-surface effects on seismic waves-theory and borehole seismometer results. *Bull. Seism. Soc. Am.*, **77**, 1168–1196.

8 Archuleta, R., Seale, S.H., Sangas, P., Baker, L., and Swain, S. (1992) Garner Valley downhole array of accelerometers: instrumentation and preliminary data analysis. *Bull. Seism. Soc. Am.*, **82**, 1592–1621.

9 Field, E., Johnson, P.A., Beresnev, I., and Zeng, Y. (1997) Nonlinear ground-motion amplification by sediments during the 1994 Northridge earthquake. *Nature*, **390**, 599–602.

10 Pavlenko, O.A. (2003), Elastic nonlinearity of sedimentary soils, *Doklady Earth Sci.*, **389**, 294–298.

13
Nonlinear Elasticity and Nondestructive Evaluation and Imaging

Based on methods described in the previous chapters, we now turn to applications to nondestructive evaluation. In Section 13.1 we provide a general overview and perspective. Following this, Section 13.2 provides a historical overview of the development of the domain in metals, long before rocks and other materials were discovered to exhibit the same behaviors. In Section 13.3 a simple crack model based on work by a number of individuals, but primarily Igor Solodov, is described. Following this, we describe methods that are part of what we call nonlinear elastic wave spectroscopy (NEWS), Section 13.4. Contained in this group are modulation methods (Section 13.4.1), resonance methods (Section 13.4.5), signal ringdown methods (Section 13.4.6), and methods based on slow dynamics (Section 13.4.7). We then describe a number of measurements of progressive damage employing a number of NEWS methods in Section 13.5. Methods of localization and imaging follow in Section 13.6 based on harmonics (Section 13.6.1) and wave modulation (Section 13.6.2) and time reversal (Section 13.6.3). Before summarizing, two other methods of isolating the nonlinear response are described as well (Sections 13.7.1 and 13.7.2, respectively).

13.1
Overview

Over the past decade or so in particular there have been many developments in the area of nondestructive evaluation. The area is relatively well developed; however, the quantitative relationship between elastic nonlinear response and mechanical (or other) damage is still an open area that requires significant attention. That said, discerning that damage is present and localizing it are well advanced. In what follows, we first provide a historical overview of nondestructive evaluation based on nonlinear elastic means. We then describe the relatively large number of diagnostic and imaging methods by example, linking the method to rigorous development in previous chapters. Following this we describe imaging methods.

Nonlinear Mesoscopic Elasticity: The Complex Behaviour of Granular Media including Rocks and Soil. Robert A. Guyer and Paul A. Johnson
Copyright © 2009 WILEY-VCH Verlag GmbH & Co. KGaA, Weinheim
ISBN: 978-3-527-40703-3

13.2
Historical Context

Nonlinear nondestructive evaluation (NDE) is based on exploiting additional frequencies produced by material nonlinear response in the form of resonance frequency shift, harmonics, sum and difference frequency, and frequency halving. Nonlinear dissipation is used as a diagnostic as well, as are the material slow dynamics.

Nonlinear NDE has its origins in studies of nonlinear dissipation applied to determining the contribution of dislocations to (nonlinear) dissipation in materials exhibiting high dislocation density. The studies described here were the first to recognize that there were other sources of nonlinear response that tended to be larger than the intrinsic anharmonicity and were due to a different physical origin (the presence of dislocations). The first work we are aware of may be that of Read (1941), who studied the internal friction of single crystals of copper and zinc [1]. A suite of papers were published by Zener in the mid to late 1940s in which he defined anelasticity to be the property of solids where stress and strain are not elastic (what we term here elastic-nonlinear) including one paper relating fracture to stress relaxation in metals [2]. Observations of strain-amplitude-dependent attenuation date back to the work of, among others, Read and Tyndall (1946) [3] (who also observed hysteresis in the attenuation vs. strain amplitude response of a zinc crystal doped with lead and tin that had been submerged in HCL), Nowick (1950) [4], Koehler (1952) [5], and Weertman and Salkowitz (1955) [6], who worked in lead and copper alloys. For example, Figure 13.1 shows an example of strain-amplitude-dependent attenuation in lead alloy as a function of temperature, from Weertman and Salkowitz. A well-known paper by Granato and Lücke (1956) [19] described experimental studies and laid the groundwork for a theoretical description of nonlinear dissipation due to the presence of high dislocation density, as well as logarithmic recovery after dynamic excitation. In that work, the authors observed strong amplitude-dependent attenuation in aluminum before annealing, which was subsequently eliminated by the annealing process.

Chambers and Smoluchowski [8] observed variations of internal friction with excitation time and logarithmic relaxation (what is now termed slow dynamics) in zone-refined, well-annealed aluminum single crystal, as predicted by the Granato–Lücke model (zone-refining, also called zone melting, is a technique for the purification of a crystalline material in which a molten region travels through the material to be refined, picks up impurities at its advancing edge, and then allows the purified part to recrystallize at its opposite edge). Their work shows hysteresis in the decrement as a function of strain amplitude and a saturation effect during long periods of excitation (much like that observed in rock). That is, for a given excitation strain amplitude the dissipation was seen to increase gradually toward a saturation level. Upon removal of the exciting wave, the decrement decayed logarithmically in time back to the rest state. The authors also observed a strain-amplitude-dependent resonance frequency that was excitation-duration dependent. It is interesting to note that in these materials, these and other authors speak of a "breakaway ampli-

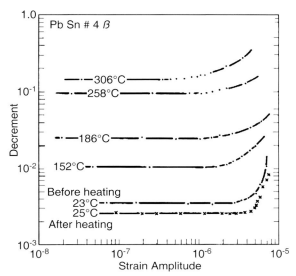

Fig. 13.1 Example of strain-amplitude-dependent attenuation in lead alloy (Pb Sn) obtained in resonance experiments from Weertman and Salkovitz [6], at a number of temperatures. Decrement is defined as equal to the energy lost per cycle divided by twice the stored energy of the system.

tude" where hysteretic effects commence. The breakaway amplitude is analogous to the transition from classical to nonclassical behaviors in rock as shown previously (e.g., in Fontainebleau and Berea sandstones). Further, they observed a slow-dynamic-like recovery process. Figure 13.2 shows some of their observations.

Strumane, De Batist, and Amelinckx (1963) observed a strain-amplitude-dependent resonance frequency shift and attenuation in NaCl crystals due to the presence of dislocations [9].

Early work exploring harmonic generation in materials with dislocations was described by Gedroitz and Krasilnokov (1963) [10], who detected a strong increase in the second harmonic generation with wave amplitude in aluminum single crystal. The mechanism of nonlinearity increase was assumed to be due to the generation of dislocations. Similar results on dislocation-induced nonlinearity were published simultaneously by Hikata, Chick, and Elbaum [11] the same year. Follow-on studies were conducted by Hikata and Elbaum in the mid-1960s [12, 13], both theoretical and experimental in nature, relating dislocation contributions to second and third harmonic generation. Additional work in this area was conducted by others, including Scorey [14] and Yermilan and colleagues [15]. A book by De Batist on the topic of internal friction in crystalline solids presents a good review of work up to the early 1970s [16]. Studies conducted at Rockwell International in the 1970s by Buck, Richardson, and colleagues are some of the earliest that we are aware of in regards to wave harmonics as applied to nonlinear NDE related to cracks and dis-

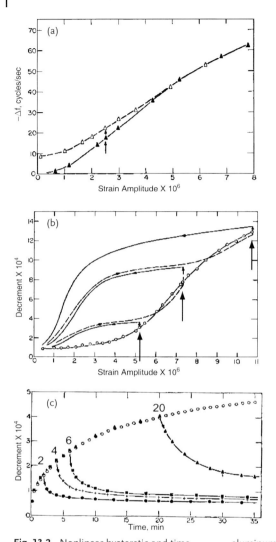

Fig. 13.2 Nonlinear hysteretic and time-dependent effects observed in aluminum and magnesium single crystal from [8]. (a) Change in resonance frequency vs. strain amplitude at 45 °C in aluminum single crystal [$-\Delta f = f - f_0$, where f_0 is the low-amplitude (linear elastic) resonance frequency]. The solid triangles are for data taken before long-duration excitation, and the open triangles were taken after wave excitation for 20 min at a strain amplitude of 2.5 × 10 – –6. (b) Hysteresis in strain amplitude vs. decrement in aluminum single crystal. The dashed and solid lines represent increasing and decreasing amplitudes, respectively. The sample was excited for 20 min at three different fixed amplitudes, as shown by the arrows. (c) Decay of the decrement as a function of time, at 60 °C. Excitation times in minutes as noted, and excitation was at fixed strain amplitude. Decrement is defined as equal to the energy lost per cycle divided by twice the stored energy of the system (modified from [8]).

bonds. In particular, the work at Rockwell was aimed at characterizing dislocations and progressive damage (e.g., [17]) and unbonded interfaces [18]. We note that the dislocation string model developed by Granato and Lücke [19] can be thought of in terms of the P-M space model. Wave modulation based on nonlinear interaction of low- and high-frequency waves that yields the combination frequency dates back to at least the mid-1960s. For instance, Zarembo and Krasil'nikov with coauthors observed frequency mixing of acoustic waves in an Al-rod resonator that was due to the presence of dislocations [20].

13.3
Simple Conceptual Model of a Crack in an Otherwise Elastically Linear Solid

In what follows, we present a simple conceptual model of waves interacting with a crack in a solid. This is followed by a description of nonlinear NDE methods, with examples.

In its simplest form, the general methodology, termed here nonlinear elastic wave spectroscopy (NEWS) [21], is based on the concept of two springs oriented perpendicular to a crack, one soft but nonlinear, with spring constants $k_1 = k_0(1 + \delta\varepsilon^2 + \ldots)$ (mechanical damage) and one stiff k_2 (surrounding material), assuming mechanical damage is localized. Here δ is the third-order, normalized nonlinear coefficient. Following the ideas of Solodov [22] and others, for illustration we consider longitudinal waves in one dimension, oriented perpendicular to the crack. At low amplitudes, the system acts linearly, and a harmonic wave $u_1 \cos(\omega_1 t - \phi)$ encountering the soft spring will maintain its shape, changing only in amplitude due to energy conservation. Figure 13.3 shows the system with the addition of Poisson-induced shearing described below.

If we transform from force–length, $F = kx$, to stress (σ)–strain (ε), $\sigma = K\varepsilon$, we can construct the nonlinear Hooke's law in one dimension, where $K_1 = K_0(1+\beta\varepsilon^2 + \ldots)$ and K_1 is the modulus. As the amplitude of a harmonic wave $\varepsilon_1 \cos(\omega_1 t - \phi)$ increases, the soft modulus will produce harmonics due to the higher-order terms in the above modulus expansion, for example, $\varepsilon_1 \cos(\omega_1 t - \phi) * \varepsilon_1 \cos(\omega_1 t - \phi)$ leads to terms corresponding to DC($\omega - \omega$) and 2ω. As the amplitude increases further, the crack, or asperities in the crack, begins to contact during the compressional phase. If the amplitude of a wave exceeds the static stress of an originally closed crack or interface, it produces cyclic, but asymmetrical, contact. The response of a prestressed contact driven by a harmonic wave strain $\varepsilon(t) = \varepsilon_0 \cos(\omega t)$ is similar to a mechanical diode and results in a pulselike modulation of its stiffness $K(t)$ (see also [23]). The waveform distortion differs from the classic sawtoothlike profile in materials whose nonlinearity is due to anharmonicity. Hooke's law becomes

$$\sigma(t) = \Delta K(t)\varepsilon = 2\Delta K[\tau/T]\left(\varepsilon_0 \cos(\omega_0 t) - \varepsilon^0\right) \sum_{n=1}^{\infty} \text{sinc}\left[n\tau/T\right]\cos(n\omega_0 t),$$

$$(13.1)$$

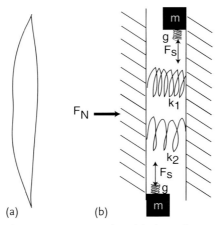

Fig. 13.3 (a) Conceptual model of a crack in a solid. (b) The corresponding mechanical system, where F_N is an oscillatory force, k_1, k_2 are the spring constants, m is the mass of a corresponding slider that mimics the frictional behavior at the crack tips, F_s are the forces due to Poisson coupling, and g are the spring constants associated with friction. Note lowercase k and g represent spring constants, and uppercase values are moduli.

where ΔK is the modulus change due to the crack, T is a half-period, and $\tau = T/4$. Here we ignore $K_1 = K_0(1 + \beta\varepsilon^2 + \ldots)$ for the sake of illustration. The concept is illustrated in Figure 13.4.

Because $K(t)$ is a rectified pulse-shaped periodic function of the driving frequency $\omega/2\pi$, the resulting spectrum of the stress induced at the crack $\sigma = K(t)\varepsilon(t)$ contains odd and even harmonics $n\omega$ whose amplitudes are modulated by a sinc-function envelope whose shape is determined by the pulse width (half of the driving frequency). If dynamic strains are large and the damage is under a small prestress, ultraharmonic generation by clapping of the damaged region can be observed [24]. At even larger driving strains, chaotic behavior can be and has been observed (I. Solodov, pers. comm.).

Due to the Poisson effect as well as the potentially complex geometry of a damage feature, the dynamics of the damaged area produces simultaneous shear traction resulting in frictional effects presumed to be dominant at asperities and at crack tips (Figure 13.3). The shear modulation in this case is due to stick-slip. Contact stiffness due to static friction in the stick phase decreases abruptly as the contact surfaces slide during oscillatory forcing. The transition from stick, to slip, to slip takes place whenever the driving force is zero, meaning twice the period of wave excitation. Therefore, the shear stress induced in the damaged area $\tau = G(t)\varepsilon(t)$ has a contact modulation $G(t)$ that is a 2ω, where ω is the forcing frequency, and the modulus ΔG is

$$\Delta G(t) = 4\Delta G(T'/T) \sum_{n=1}^{\infty} \text{sinc}\left[2nT'/T\right] \cos 2n\omega_0 t \tag{13.2}$$

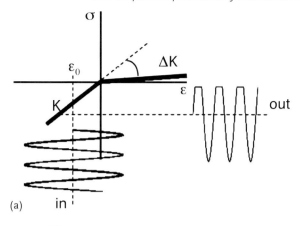

(a)

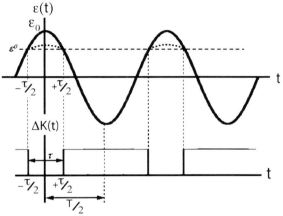

(b)   **longitudinal**

Fig. 13.4 Hooke's law in one dimension for a one-dimensional strain wave oriented perpendicular to the crack. (a) As the amplitude increases and the crack begins to close, the modulus K ($K_1 = K_0(1 + \beta\varepsilon^2 + \ldots)$) becomes bimodal, changing from ΔK to K during each cycle. (a) the input wave, with average strain ε_0, is shunted (rectified) each cycle during compression due to the closure of the crack, as seen in the output strain wave. (b) The shunted strain wave and the corresponding modulus as a function of wave period (time). The crossover strain for the bimodal modulus occurs at ε^0 (modified from Solodov ([24] and pers. comm.)).

and Hooke's law is

$$\tau(t) = 2\Delta G \varepsilon_0 (T'/T) \sum_{n=1}^{\infty} \text{sinc}\left[2nT'/T\right] \left[(\cos(2n+1)\omega_0 t + \cos(2n-1)\omega_0 t)\right],$$

(13.3)

and where T is a half-period and T' is the period at which the rectification takes place.

Thus, at large forcing when stick-slip takes place, the harmonics of 2ω form ($n2\omega$, where n is an integer), meaning that even the harmonics of the fundamental frequency are produced. These may multiply with the fundamental producing odd harmonics (e.g., $u_1 \cos[\omega_1 t + \phi]u_{2\omega n} \cos[2\omega n t + \phi] = u_1 \cos[\omega_1 t(2n - 1) + \phi] + u_n \cos[\omega t(2n - 1) + \phi]$. The above assumes a stepwise transition from stick to slip and vice versa; however, one may imagine many different forms of the stick-slip behavior that would shape the $n2\omega * \omega$ spectrum.

Moreover, if two harmonic waves $\varepsilon_1 \cos(\omega_1 t - kx)$ and $\varepsilon_2 \cos(\omega_2 t - kx)$ intersect the soft modulus, the damaged region, the low frequency modulates the high frequency, and therefore the waves multiply with each other $\varepsilon_1 \cos(\omega_1 t - kx)\varepsilon_2 \cos(\omega_2 t - kx)$, producing frequencies at the sum and difference frequencies proportional to the amplitudes of the two waves, for example, $\varepsilon_1\varepsilon_1(\omega_1 \pm \omega_2)$ (sidebands). Sidebands appear for each frequency component produced when the low frequency encounters the crack. For instance, if the low frequency is a square wave due to the hard spring/soft spring paradigm, the resulting frequency components shaped by a sinc function in the frequency domain would exhibit sidebands.

One can also imagine that both the crack opening-closing and the stick-slip behavior could be hysteretic. This would lead to an equation like Eqs. (13.1) and (13.3), which depend on the sign of the strain, $\dot{\varepsilon}, \dot{\gamma}$, respectively.

All of the above assumes the frequency is low enough for the material to respond; above a certain frequency the material would presumably not be able to respond quickly enough, and the material elastic nonlinearity may be locked. One could imagine that the cutoff frequency could be different for the compressional and sliding effects described above.

In actual observations the above, relatively simple-minded conceptual model works to varying degrees; however, independent of the physics of the dynamic wave/crack interaction, the nonlinear response can be applied to damage diagnostics. Harmonics are employed, as well as frequency mixing where sum and difference frequencies are used to discern whether or not damage is present, as well as for imaging the damage assuming it is localized. In the following sections, we explore various applications of NEWS to NDE.

13.4
Nonlinear Elastic Wave Spectroscopy in Nondestructive Evaluation (NEWS)

13.4.1
Nonlinear Wave Modulation Spectroscopy (NWMS)

This method, described in previous chapters, employs one or more waves at frequencies that are applied simultaneously to a sample, and the sum and difference frequency waves are studied to discern whether or not a localized or volumetric elastic nonlinearity is present. Here the pump waves are the fundamental frequencies and the probe waves are the harmonics and or sidebands. The word "sideband" is also applied to describe the sum and difference frequencies (the word has its

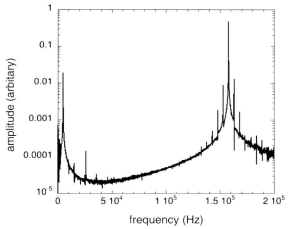

Fig. 13.5 Spectrum for a cracked steel sample. Continuous wave drive of frequencies $f_1 = 5.020$ kHz and $f_2 = 158$ kHz were used, respectively. The odd harmonics at $3f_1, 5f_1, 7f_1, 9f_1$ are visible, as are first-, second-, and third-order sidebands, at $f1 \pm f2, f2 \pm 2f1, f2 \pm 3f2$ as well as additional sidebands.

origins in radio science, where it appeared as early as 1910–1915 to describe the sum and difference frequency transmitted about a carrier frequency). In a system with a single frequency, the sum frequency is the second harmonic. In Figure 13.5 we show an example of NWMS. Modulation and harmonic results are shown for a steel-bearing cap containing a visible, surface-breaking crack. The sources and detector were all piezoceramics, bonded with epoxy to the sample. Both first- and second-order sidebands are seen, as well as a number of odd harmonics. Some sort of shaping of the harmonics is taking place and it may be sinclike in nature. The crack in this sample is able to freely flex, although it is quite stiff. Apparently, for this experimental configuration shear slip may be dominating based on the above paradigm. Like all nonlinear NDE methods, NWMS in its simplest form tells one the sample is nonlinear, but not where the nonlinearity originates.

In the next figures we show results for samples of plexiglas and for an automotive connecting rod, modified from [25]. In each case, observations for both intact and damaged samples are shown. In the experiments, two continuous-wave, separate frequencies were input into the sample simultaneously using piezoelectric transducers. The sources and detectors were bonded to the sample using epoxy. The waves were detected by an accelerometer at a separate location on the sample. The output waveform was recorded and Fourier analyzed. To illustrate the amplitude dependencies of the sidebands to the drive amplitudes, one frequency was held at a constant amplitude and the other was progressively increased.

In the plexiglas experiment a sample of dimension $110 \times 110 \times 6$ mm^3 was used, as seen in Figure 13.6. A controlled crack was induced by confining the sample

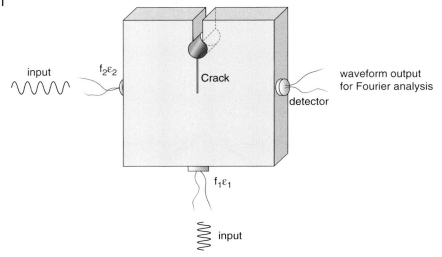

input $f_2\varepsilon_2$ Crack waveform output for Fourier analysis detector $f_1\varepsilon_1$ input

Fig. 13.6 Experimental configuration for NWMS measurements in plexiglas for measurements shown in Figure 13.7 (modified from Van Den Abeele *et al.* 2000 [25].

center and applying tension to the region of the hole (hole diameter 13 mm). Identical experiments were conducted before and after cracking. The resultant crack length was 50 mm (area $\simeq$ 250 mm^2). The two frequencies applied were $f_1 = 7$ and $f_2 = 70$ kHz, respectively. Progressively increasing drive amplitudes for A_{f_1} were applied, while A_{f_2} was held at a fixed drive level. In the figure, the sidebands and the second and third harmonic dependencies are shown as well. There is some amount of harmonic and sideband energy in the intact sample. This is due primarily to nonlinearities in the associated electronics, and some portion is due to the inherant atomic nonlinearity of the material. In contrast, the damaged sample shows large harmonics and sidebands. Interestingly, the odd and even harmonics are represented in this sample, in contrast to the cracked steel sample shown in Figure 13.5, where only the odd harmonics were observed, and the fifth harmonic was larger than the third. It is difficult to know if the difference is more than anecdotal. That is, is the observed difference due merely to how the sample is excited, to differences in the material, or to differences in the nature of the crack? The issue of crack nonlinear wave interaction is still relatively unexplored. Some related experiments will be described later where time reversal was applied to probe the crack.

In order to quantify the scaling relationships between the drive frequencies and the harmonic/modulation signals, their dependencies are plotted in Figure 13.7c–d. It is clear from the intact-plexiglas results that the system distortion plus intrinsic nonlinearity are on the order of –40 dB. The scaling in the cracked sample extracted from the measurements for the second harmonic is $A_{2f_1} \propto A_{f_1}^2$, as it should be. Note that there are few 3_{f_1} data points so the scaling is unclear,

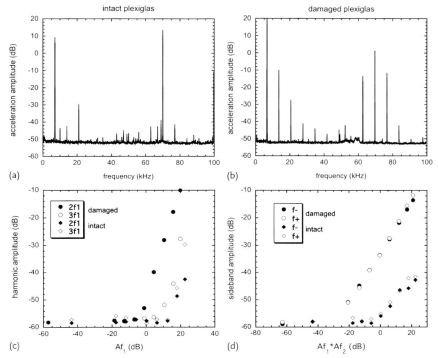

Fig. 13.7 Modulation and harmonic results in intact and cracked plexiglas. (a,b) Frequency spectra obtained in intact and cracked samples, at maximum drive levels. (c) Second and third harmonic amplitude dependencies as a function of the fundamental frequency amplitude A_{f1} in the intact and damaged samples. (d) Sideband amplitudes (in dB) $A_{f\pm}$ in both samples as a function of the product of the driving frequencies $A_{f1} * A_{f2}$ (modified from Van Den Abeele *et al.* 2000 [25]).

but it appears to be less than 3 (it should be 2 as predicted by the P-M space theory). The predicted scaling for the sidebands is a more robust result, giving $f_- = 0.90$; $f_+ = 0.95$ for the first- and second-order sidebands, respectively, close to 1, as they should be (recall $A_{2f,3f} \propto A_{f1}A_{f2}$).

In the connecting rod experiments, one intact and one damaged sample were used for comparison. An example is shown in Figure 13.8. The samples are composed of a sintered steel, as shown in the sequence of photomicrographs at different scales in Figure 13.10. The results of measurements are shown in Figure 13.9. The scaling relations extracted from the measurements in the cracked connecting rod, for the harmonics and sidebands, are $2f = 1.4$; $3f = 2$; $f_- = 1.1$; $f_+ = 1.2$, respectively. Note the measurement of the third harmonics is poor and that of the second harmonic only fair. Further, there may be an approximate sinclike shaping to the harmonics not present in the damaged plexiglas. The measured dependencies show imperfect results in relation to the theory. This is due to imperfect data and possibly a need for a modified theory. The results are interesting, because the

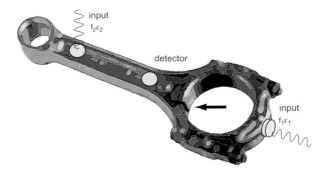

Fig. 13.8 Approximate experimental configuration for NWMS measurements in metal connecting rod for measurements shown in Figure 13.9. Arrow shows approximate crack location in damaged sample. Dimensions of crack were not measured.

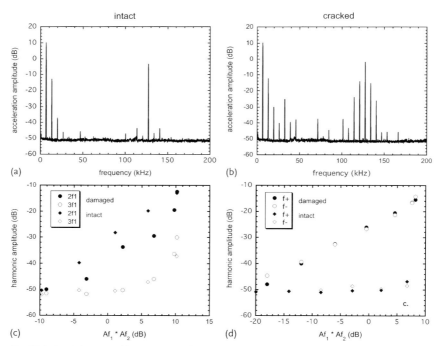

Fig. 13.9 Modulation and harmonic results in intact and cracked steel connecting rod from an automobile. (a) Modulation in intact sample. (b) Modulation and harmonics in cracked sample. (c) Sideband dependency with the product of the two input waves. (d) Sideband dependencies in damaged sample. (e) Second and third harmonic amplitudes in intact sample. (f) Second and third harmonic amplitudes in cracked sample (modified from [25]).

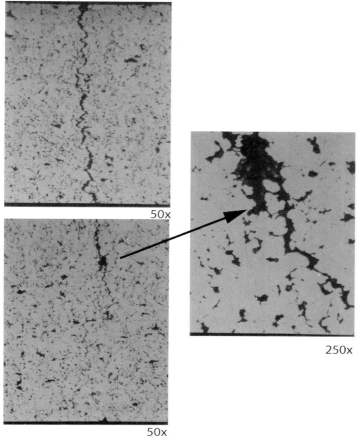

50x

250x

50x

Fig. 13.10 Photomicrographs in reflected light of connecting rod crack taken after experiment described above. The extent of the the crack is shown at left in two photos at the same magnification (50 ×). At right the crack tip region is shown in expanded view of the crack tip. The photos were taken from a sample after cutting and polishing. These objects are usually made of sintered and forged powder metallurgy, but the actual composition and provenance are proprietary. Note how little actual contact there is along the crack in this two-dimensional slice.

material is a sintered part and volumetrically nonlinear in its intact state much like a rock sample, as can be seen in the "intact" results (Figure 13.9a,c). The harmonic dependencies are very similar, as are the sidebands. When a crack is added, the material becomes much more nonlinear, as can be observed in the spectrum. Figure 13.10 shows what the crack (responsible for the above behaviors in the connecting rod) looks like a sequence of photomicrographs.

13.4.2
Harmonics

In the following example we show the results of employing harmonics as a mechanical damage diagnostic, taken from [26]. In the experiment, the sample was progressively fatigued. Before fatiguing and at predetermined intervals during the fatiguing process, the second and third harmonic amplitudes were measured as a function of drive amplitude (in a later section devoted to application of nonlinear response to progressive damage we return to this sample). The samples were thin, rectangular-shaped strips of artificial slate used in roofing construction, of dimension $200 \times 20 \times 4\,\text{mm}^3$. The material is composed primarily of Portland cement. Mineral additives and synthetic organic fibers were added for strength enhancement.

The strips were excited at their lowest-order bending resonance mode by a low-frequency, low-distortion speaker (Figure 13.11a). The speaker was positioned at 2 cm from the middle, parallel to the strip surface. The speaker was driven by a function generator through a high-power amplifier. The coupling medium between specimen and speaker was air (noncontact excitation). An accelerometer attached to one end of the strip measured the sample's out-of-plane response. The signal from the accelerometer was preamplified and recorded.

The results are shown in Figure 13.11b. As mechanical damage increases, the material harmonic response traverses the space diagonally upward to the left, indicating a progressively larger nonlinear response as manifest in both the second and third harmonics. Note that the material, like many porous Earth materials, exhibits a nonlinear response before the first damage sequence.

The addition of confining pressure to a sample can eliminate wave distortion due to the crack or delamination (in fact all elastic nonlinearity) because the feature is locked by the pressure, as was shown previously in rock samples. Fluid added to a crack or cracklike feature can dampen or eliminate the second harmonic (and again, other dynamic nonlinear response). For example, in early work by Buck, Morris, and Richardson [18] in which stacked aluminum cylinders were held under uniaxial confining pressure, and bulk compressional and shear waves were applied perpendicular to the interfaces, harmonic amplitudes were measured as a function of applied stress, both increasing and decreasing. Tests were conducted with room-dry interfaces as well as with glycerin contained in the interfaces. Figure 13.12 shows observations. There is a clear effect of pressure that diminishes the harmonic amplitudes, and the addition of the fluid eliminates the second harmonic altogether.

13.4.3
Robust NWMS

Simple pass/fail tests to discern if mechanical damage is present, by applying modulation, are extremely effective in general; however, in various materials where the nonlinear response may be small, where Q is large, or where a driving transducer is placed at or near a vibrational node of one of the applied waves, the response may be difficult to observe. Another chronic problem is discerning whether or not the observed nonlinearities originate in the sample or associated electronics. The most effective and robust diagnostic technique relies on applying a pure tone probe in combination with mechanical impact that was described as early as 1994 [27]. An example is described next.

One experimental configuration is described in Figure 13.13. The signal generator creates a continuous wave *cw* output at one or more pure frequencies. In the example described here the source was a piezoceramic bonded to the sample using epoxy. While the continuous wave source is operating on the test object, a device designed to provide a broad-frequency band time-impulse strikes the object. It is attached to a baseplate with a hinge designed so that the mechanical strike was repeatable. The larger amplitude tones excited by the mechanical impact are the lowest modes logically, and they mix with the *cw* signal. Bonded to the hammer is a piezoelectric transducer that is used as a trigger source into the oscilloscope

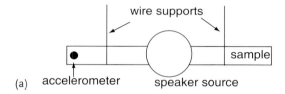

(a)

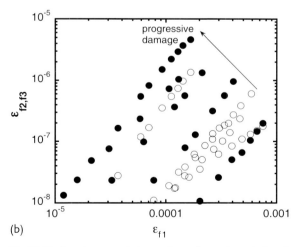

(b)

Fig. 13.11 Experiment and observations of second harmonics. (a) Experimental setup for the measurment. (b) Second (open circles) and third harmonics (closed circles) measured on a sample of synthetic slate that was progressively damaged. Note that the scaling relations between the second and third harmonics are approximately two through all damage steps, as they should be. Note that additional results will be described in a later section that addresses progressive mechanical damage in detail (modified from [26]).

or other digitizing device. When the impact source strikes the test object, the signal detector feeds the output through a high pass filter to the oscilloscope or other digitizer. The high pass filter is used to eliminate the low-frequency components due to the impact source becuase they are large in amplitude. Once the impact device strikes the object, a high-pass-filtered time signal is captured on the digitizing device. In the simplest case, a Fourier transform of the time signal is calculated, and modulation sidebands are inspected at frequencies around the pure tone(s). Figure 13.14 shows an example of this method for an experiment conducted in the sample shown in Figure 13.9.

13.4.4
NWMS Summation

If the pump *cw* tone or impact source or detector is located at a node of one or more of the primary vibrational modes, or if the pure tone frequency or the sideband(s) are located on a node, there will be diminished sideband mixing. However, if one changes the pure tone frequency(ies) stepwise and repeats the experiment, eventually one will move away from the node and toward an antinode and cap-

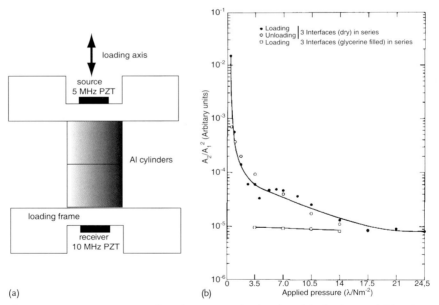

(a)   (b)

Fig. 13.12 Harmonic generation from three interfaces stacked in series. (a) Experimental configuration. Two cylinders are stacked and loaded uniaxially. They are driven by a 5-MHz piezoelectric (source). (b) Experimental result. The amplitude of the second harmonic is normalized to that of the fundamental (10 and 5 MHz, respectively). Dry interfaces show significant harmonic generation, especially close to zero pressure. Filling the interfaces with glycerine essentially eliminates the harmonics (from [18]).

ture modulation if it exists. This procedure is repeated stepwise for a number of different pure tone frequency pump waves. A Fourier transform is taken of each time signal and, at the end of the step-sweep, a summation of the signal Fourier spectra (power or magnitude) is then calculated by shifting all spectra such that the pure tone *cw* is at zero frequency. The step-frequency sweep is used to make the measurement more robust by making use of successive pure tone frequencies and their sidebands. The summation is used to improve the signal-to-noise ratio and to inspect the results of many experiments simultaneously. Because the averaging is done in the frequency domain, the phase of each successive time signal collected is unimportant.

Specifically, there are i time signals, composed of a pure tone *cw* plus the vibration due to the impact source, $t_i(t), i = n \dots m$. Here n, m are the first and last pure tone frequencies, respectively, and the frequency step interval of the pure tone signal is $\Delta f = [(m - n)/i$. There is a corresponding spectrum obtained from the Fourier transform of $t_i(t)$, termed $F_i(f)$. The spectrum $F_i(f)$ is frequency downshifted to zero frequency, then summed over the number of frequency steps i, $S_i(f) = \sum_n^m (F_i(f) - i)$. The result is a summation of spectra that is more robust in showing nonlinear mixing of the vibration spectrum and the *cw* signal than a single impact plus pure tone experiment, as illustrated in the following.

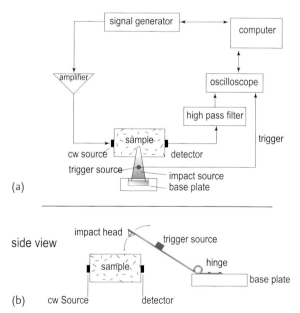

Fig. 13.13 Experimental configuration for NWMS impact plus pure *cw* tone method. (a) The setup and the bottom show the calibrated hammer. (b) Side view of hammer-sample configuration (from A. Sutin and P. Johnson, unpublished).

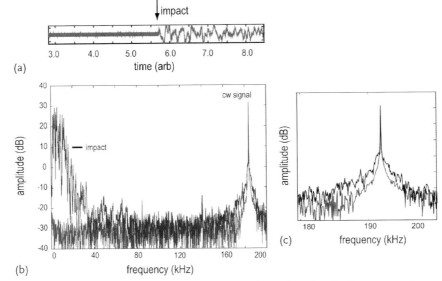

Fig. 13.14 NWMS impact plus pure *cw* tone example for sample shown in Figure 13.9. (a) A portion of the time series, showing the *cw* tone, time of impact, and resulting vibration. (b) Power spectrum of time signal, showing an undamaged sample (grey) and a damaged sample (black). No filter was applied in order to show the vibration spectrum in the damaged part, as well as the modulation about the pure tone. Note the vibrational modes range from *dc* to 40 kHz. (c) Expanded view around the *cw* tone, showing the increase in overall sideband energy due to the crack in the sample (from A. Sutin and P. Johnson, unpublished).

For an example of the method, we describe results from two bearing cap specimens, one undamaged and one cracked. An example is shown in Figure 13.15. In the experiment, the transducer emitting the pure *cw* signal was bonded with epoxy to the sample. Figure 13.16 shows examples of spectra obtained during an experiment where the *cw* frequency was stepped over a band from $n = 95$ kHz to $m = 114.5$ kHz at intervals of $\Delta f = 0.5$ kHz for a total of $i = 40$ steps. Two examples of individual frequency steps are shown in Figures 13.16a,b (105, 105.5 kHz, respectively). Note the difference in sideband and *cw* frequency amplitudes. Figure 13.16c shows the summation average $S_i(f)$ for the intact sample, and Figure 13.16d shows $S_i(f)$ for the cracked sample. It is interesting to note that the effects on nonlinear mixing can be so pronounced over such relatively small frequency changes and is a useful lesson in understanding how variable the nonlinear mixing can be due to material geometry and resulting modal structure.

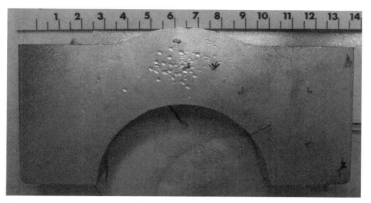

Fig. 13.15 Bearing cap sample used in the NWMS summation observations whose results are shown in Figure 13.16. Scale is in cm. Sample is excited simultaneously by a pure tone *cw* source and impact from the calibrated hammer shown in Figure 13.13. In one sample, a crack was located in the arch (photo courtesy of T.J. Ulrich).

13.4.5
Nonlinear Resonance Ultrasound Spectroscopy (NRUS)

NRUS has been described multiple times in previous chapters, in particular, in the chapter on fast and slow dynamics. In short, a frequency step-sweep is conducted that encompasses one or more eigenmodes of a sample. At each step the rectified amplitude is recorded. The driving level is increased, and the measurement is repeated, and so on. The next figure shows NRUS results obtained from the same material as that illustrating harmonic generation shown in Figure 13.11b and using the same experimental setup shown in Figure 13.11a [26]. Results are shown before and after damage was done by three point bending. As noted in previous chapters, the dependence of the resonance frequency shift Δf normalized to its linear value, f_0, is predicted by the P-M space theory to be linear with the strain amplitude of the driving frequency: f_1, $\Delta f / f_0 = \varepsilon_{f_1}$. The scaling for acceleration $\ddot{u}$ is shown in Figure 13.18c. Acceleration and strain are related by a constant ω^2: $\ddot{u}/\omega^2 = \varepsilon$, and $\varepsilon = du/dx$, where ω is $2\pi f$. As expected, as damage increases, the space is traversed upward to the left, indicating a larger nonlinear coefficient, α, meaning as damage increases, the nonlinear coefficient increases.

13.4.6
Nonlinear Ringdown Spectroscopy (NRS)

An elegant and quick way to conduct NRUS (and simultaneously obtain the parameter α) is to conduct a moving window analysis of a time signal during the signal ringdown. The pump signal, which also is used as the probe signal, is a continuous-

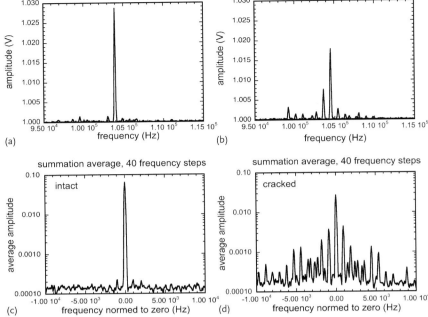

Fig. 13.16 Modulation applying an impact source with pure tone, for multiple frequencies over a band encompassing 95 to 115 kHz, in 500-Hz steps. (a). Spectral plot at *cw* single frequency at 105 kHz with impact where the pure tone is located at or near a node (only frequency band around pure tone is shown). (b) Spectral plot at a slightly higher frequency (105.5 kHz), where strong mixing takes place. (c) Summation average of spectra transformed to zero frequency for intact sample. (d) Summation average in otherwise identical, cracked sample. Note that the difference in the amplitudes of the *cw* are due to differences in coupling efficiency and linear Q in the two samples, as well as energy leakage to the sidebands in the cracked sample (from A. Sutin and P. Johnson, unpublished).

wave or long tone burst or mechanical tap used to soften the sample. One can then take advantage of the material ringdown when the probe signal is terminated in order to extract the relation between the resonance frequency shift and its strain amplitude in a single experiment. In order to accomplish this, one can conduct a moving-window analysis of the ringdown signal (the probe), for progressively later increments of the time signal, to extract the resonance amplitude and frequency. Specifically, the detected time signal for a tone burst at fixed source frequency f_s and at fixed distance r, $u \cos(\omega t + kx)$ has a ringdown amplitude of $u(t) \propto u_0 e^{-\omega r/(2Qc)}$, where U_0, c are the source amplitude and wave speed, respectively. The relation is not an equality due to nonlinear attenuation and apparent nonlinear attenuation due to the generation of harmonics. The deviation from this equation to elastic nonlinearity is used to advantage: for n time windows, and window length m points, n successive fits applying the equation $u(n, t) = u_0 \sin(\omega t + kx) e^{-\omega(u)t/(2Qc)}$, where a $u(n, t)$ and $\omega(u)$ are extracted and plotted. Note that the procedure works

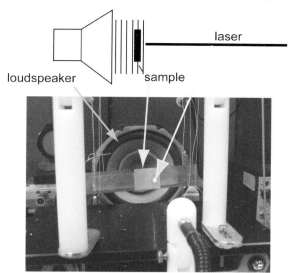

Fig. 13.17 Experimental configuration for NRS. Top shows cartoon of setup, showing speaker, sample (side view), and laser vibrometer beam. The arrows show the location of each in the photo below. Notch in sample is where the crack initiates during tension cycling (courtesy of P.-Y. LeBas).

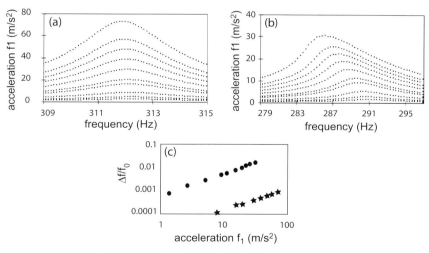

Fig. 13.18 NRUS in synthetic slate sample before (a) and after damage is induced (b). (c) The relation of f_1, $\Delta f / f_0 = \varepsilon_{f_1}$. Experimental setup shown previously in Figure 13.11a. The acceleration amplitude is shown rather than the strain. Modified from [26].

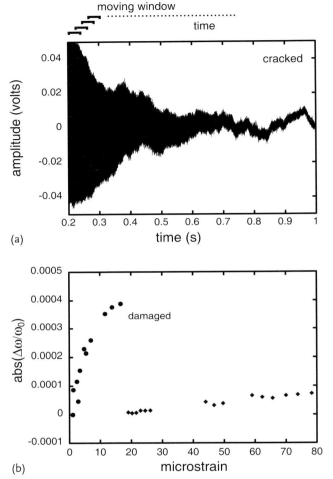

Fig. 13.19 Example of NRS. (a) Time signal measured from a cracked sample. Above the figure the concept of the moving-window analysis is illustrated. (b) NRS analysis in an intact (diamonds) and cracked (circles) sample (courtesy of P.-Y. LeBas).

despite the fact that the experiment is conducted at a single frequency because of the finite width of the frequency peak. As the frequency decreases with amplitude, energy will be transferred to the softening frequency from the exciting frequency pump. It is the softened-state resonance frequency that dominates the ringdown.

The experiment that illustrates the technique is shown in Figure 13.17. A steel sample is fatigued in tension and suspended in front of a speaker. An identical sample is left intact for comparison. A signal is sent to the loud speaker via an amplifier, comprised of a tone burst of 2000 cycles at a resonance frequency. A laser vibrometer detects the response. The resonance frequency was selected by sweeping the speaker through a low-frequency band beforehand and looking for a strong mode, in this case the first flexural mode. The frequency was 1300 Hz for the cracked sample and 1380 Hz for the intact sample. $n = 40$ nonoverlapping time windows were used in the analysis.

Figure 13.19a shows a time signal obtained from the cracked part. Just above the figure, the moving-window analysis is shown conceptually. Figure 13.17b shows results from the identical cracked and intact samples. There is some frequency shift in the intact sample, but clearly, that in the damaged sample is much larger. α can be extracted from the slope of this curve.

13.4.7
Slow Dynamics Diagnostics (SDD)

Slow dynamics (SD), the recovery process back to equilibrium after large-amplitude wave excitation described in the chapter regarding fast and slow dynamics, can be used as a damage diagnostic as well. The approach is based on probing a sample in order to see if slow dynamics exist at an early time after large-amplitude disturbance. An advantage is that the method can be applied relatively quickly. The existence of SD indicates that damage is present [28]. Specifically, the diagnostic employs a low-amplitude pure-tone signal near an eigenmode of a sample that probes the eigenmode slope change. Upon applying an impulse or high-amplitude pump tone burst, the presence of a lasting amplitude change in the probe, corresponding to a modal shift and the onset of SD, is the diagnostic. Figure 13.20 describes the concept. The eigenmode is normally determined by first sweeping the frequency in a band appropriate for the sample, based on the equipment used as well as such issues as the sample Q, geometry, and size. In Figure 13.20a a mode is selected arbitrarily or, if necessary, modeling studies are conducted so that the mode can be identified. Figure 13.20b shows that a low-amplitude (linear) probe is input at the frequency denoted by $A - A'$, with an amplitude proportional to A'. When the sample is disturbed by a large-amplitude signal (point source for instance), the eigenmode abruptly shifts due to nonlinear induced softening, if mechanical damage is present. This has the effect of changing the probe amplitude from A to A', as seen in Figure 13.20c. Depending on the frequency of the probe in relation to the frequency of the mode, the amplitude may shift upward or downward. For instance, a probe with a frequency that corresponds to the line at $B - B'$ will increase in amplitude from B' to B upon large-amplitude excitation. The probe wave recovers slowly back to its equilibrium amplitude (this can take tens of minutes to many hours, depending on the material), as shown in Figure 13.20b; however, it is only necessary to monitor the onset of the SD in the time signal for use as a diagnostic. SDD can be extremely sensitive due to the fact that it measures the change in slope.

An example of SDD is shown in Figure 13.21, from an bearing cap identical to that shown in Figure 13.15. The crack is located at the arch top as shown in Figure 13.21c. Here, a relatively high-amplitude impulse (strain ~ 5×10^{-5}) is used to produce nonlinear material softening and simultaneously inducing the material

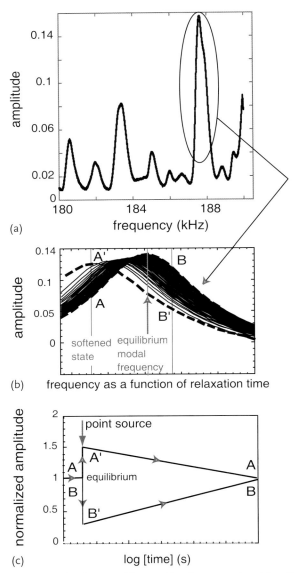

(a)

(b) frequency as a function of relaxation time

(c)

Fig. 13.20 The slope amplifier used for slow dynamics diagnostics (SDD). (a) Arbitrary portion of the spectrum from a sample, showing eigenmodes. The spectrum was obtained from a resonance sweep over the frequency band (rectified amplitude vs. frequency). The circled mode is shown in (b) in expanded view. (c) Cartoon of the rectified amplitude of a pure-tone probe during the process (from A. Sutin and P. Johnson, unpublished).

(localized) softening/slow dynamics. The impulse is delivered by a mechanical excitation in this case (equivalent in energy to a tap with a pencil). A low-amplitude (strain~ 10^{-7}), pure-tone probe is input into the sample to monitor material change before, during, and after the impulse. Figure 13.21 a shows that at the time of the tap (time 0 s in the figure), the probe amplitude changes. In an undamaged, but otherwise identical, sample, no amplitude change is observed after the time of the strike (again time 0).

What is observed in the experiment is dependent on the frequency of the probe with respect to the eigenmode peak. By judicious choice of the fixed frequency (e.g., near the mode inflections), one gains significantly, or not, in one's ability to observe the onset of SD. The drawback is that the modes must first be characterized by conducting a modal analysis from the spectrum of an impact or by applying a resonance sweep in order to select the probe frequency.

In a simplistic manner, the amplification is proportional to the specific dissipation Q : $\Delta A = \text{abs}(C Q \Delta f)$, where C is a constant relating Q and Δf, meaning there is an amplification by $C Q$ in sensitivity over the frequency shift. With increasing Q the slope of the resonance curve steepens, making the slope amplifier more effective. At low Q, the method works less well because ΔA becomes relatively smaller.

13.5
Progressive Mechanical Damage Probed by NEWS Techniques

A number of experiments have shown empirically the relation between linear wave speed, Q, and nonlinear response as a function of mechanical damage induced by oscillating or bending (e.g., [18, 29–31]). To our knowledge, the first work on this topic was conducted by Buck *et al.* [18, 29], where flexural fatigue damage was studied in aluminum [29] as a function of surface stress, employing surface acoustic waves. In their experiment, a tapered flexural-fatigue sample geometry was used in order to generate a uniform surface stress and therefore a homogeneous density of fatigue cracks (see Figure 13.22a for sample geometry). In the experiment, a strain gauge was used to measure the bending moment of the sample, and surface stress was calculated from this. The apparatus consists of a tuned quartz transducer (30 MHz) as probe source, and a capacity microphone and heterodyne receiver were used to detect harmonics. Measurements were made on an aluminum single crystal of [100] orientation. Deformation was obtained in compression in an MTS system between two parallel flat plates. Harmonic generation was determined at several intervals of compression cycling, outside of the MTS [17].

Figure 13.22 shows the results, where the ratio of the harmonic displacements, U_2/U_1, versus fatigue cycles are seen (recall that in one dimension, at fixed distance, U_2/U_1 is proportional to β, by the classical relation $U_2/U_1 \simeq \omega_{U1}^2 \beta$). The sensitivity of U_2/U_1 to increased dislocation density is clear, by a factor of two or so.

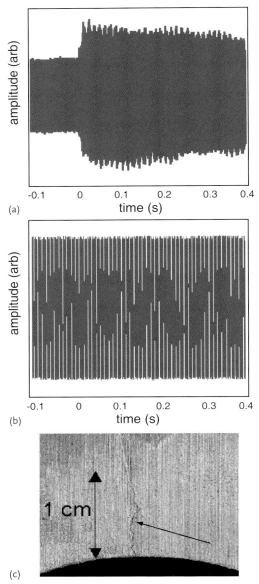

(a)

(b)

(c)

Fig. 13.21 Application of the slope amplifier in a bearing cap identical to that shown in Figure 13.15. Results for a damaged (a) and undamaged (b) sample are shown. The impact time was at 0 s in both cases. A high pass filter was used in the experiment in order to damp the free vibrations of the sample. This is why no signal is apparent in the undamaged sample shown in (b). (c) A photograph of an expanded view of the sample showing the surface-breaking crack located at the top of the sample arch.

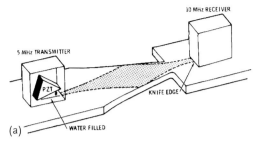

(a)

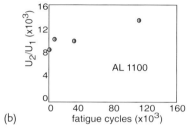

(b)

Fig. 13.22 Progressive damage in a sample of aluminum.
(a) Sample setup showing flexural fatigue specimen with transducer locations. The shaded region is the high-amplitude region of second harmonic. (b) Normalized second harmonic
displacement as a function of fatigue in aluminum (modified
from [17]).

A follow-on measurement by the same group for the same type of sample geometry and experiment is shown in Figure 13.23. The figure shows the peak of the
second harmonic amplitude as a function of the percentage of the "expended life in
the initiation phase of fatigue." This quantity is defined as that part of the total life
necessary to produce the first surface crack having a length of order 0.5 mm. The
results for three applied surface stresses are shown. Harmonic signals are detected
as early as 10 to 20% of the fatigue life, and change by up to a factor of 4 to 4.5.

Cantrell and Yost have conducted numerous studies of fatigue damage. In what
follows, results are described in aluminum alloy 2024 [32]. In the tests, ASTM
standard dogbone specimens were fatigued at a rate of 10 Hz under uniaxial,
stress-controlled load at 276 MPa. Bulk waves of frequency 5 MHz were applied
by a lithium-niobate transducer. After propagating through the solid, both the
fundamental and harmonic signals were detected by an air gap capacitance transducer at the opposite end, as indicated. Actual measurements of β are shown in
Figure 13.24 and were extracted from absolute amplitude measurements of the
fundamental and second harmonic signals. The β increases monotonically with
increasing fatigue cycles and is related to an increase in the volume fraction of
persistent slip bands throughout the fatigue life.

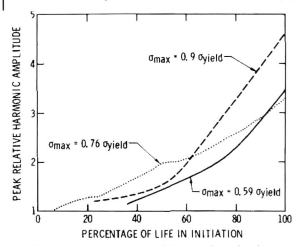

Fig. 13.23 Progressive mechanical damage due to bending in aluminum 7075-T6. The sample geometry is that shown in Figure 13.22a. Relationship between harmonic amplitude and fundamental amplitude U_2/U_1 for three values of surface stress as noted. The stress values refer to the maximum cyclic surface stress of the material's yield strength. Higher surface stress produces larger harmonics (from [29]).

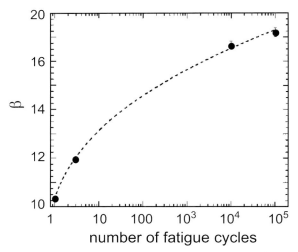

Fig. 13.24 Progressive mechanical damage showing β as function of number of fatigue cycles for aluminum alloy 2024-T4 (from [32]).

A very nice study by Nagy [31], shown in Figure 13.25, relates the linear elastic quantities of wave speed and attenuation to the nonlinear parameter in a sample of polymer. In the study, a sample fixed at the bottom end and free at the other was

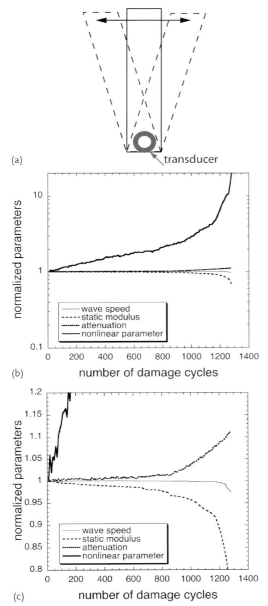

Fig. 13.25 Progressive mechanical damage due to bending on a polymer sample. (a) Experimental sample showing deformation. (b) Comparison of linear and nonlinear parameters, semilog plot. (c) Zoom of (b) showing the linear parameters (linear axes). The nonlinear response is affected early on in the damage process and becomes enormous very quickly (data courtesy of P. Nagy; experimental procedure described in [31]).

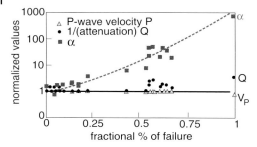

Fig. 13.26 Progressive mechanical damage in concrete. α, wave speed, and $1/Q$ were monitored at each damage step. Data are normalized to their intact values (modified from [26]).

shaken in constant shear displacement (equivalent to a strain of 7×10^{-3}), at a period of about 10 s until failure (Figure 13.25a). A pulse-echo probe measured the linear ultrasonic longitudinal velocity, and attenuation was measured during each loading cycle, while the sample was in its central (undeformed) position, across the base of the sample. In addition, a number of velocity measurements were made through each cycle of excitation. During an oscillation, half of the sample was in compression while the other half was in extension when not in the vertical position. During each bending cycle, the observed velocity modulation was twice the loading frequency. Thus the first-order nonlinear effect cancelled out and only the second-order effect was observed, as a second-harmonic modulation of the instantaneous velocity. The nonlinearity parameter plotted in Figure 13.25b is this quadratic coefficient. The quasistatic Young's modulus was calculated in each bending cycle from the bending force required to produce a given end deflection. One can see that the linear responses [$1/Q$ (dashed line), wave speed (solid line), and modulus (dashed line in Figure 13.25c) Figure 13.25b,c] are relatively insensitive to induced damage until just before the sample fails. The nonlinear parameter shows a change early on in the cycling and, just before failure, is about six times that of its initial value.

A similar procedure was applied to damage detection in thin, synthetic slate strips, described previously, using the same experimental setup shown in Figure 13.11a [26], where samples were subjected to cyclic fatigue loading under three-point bending. Before cycling, and after each loading session, the linear parameters (wave speed and wave dissipation) and nonlinear parameter (α obtained from the modal frequency shift in resonance) were measured. Figure 13.26 summarizes the evolution of the linear and nonlinear parameters relative to their initial values as a function of fractional percent to failure.

13.6
Mechanical Damage Location and Imaging

Isolating and/or imaging the source of elastic nonlinear response, and thereby mechanical damage, has been addressed in a few studies but is still in an early stage except in limited cases.

13.6.1
Harmonic Imaging

Second harmonic imaging is a well-known and frequently applied method that works well in a variety of circumstances provided one can separate system-related harmonic generation from that of the test specimen, as is the case with all nonlinear methods. At present second harmonic imaging is widespread. For instance, nearly every medical ultrasound imaging system offers second harmonic imaging. The primary application of these systems is to boost frequency rather than image nonlinearity, but they are used to create images from the generation of a second harmonic either based on tissue nonlinearity or on injected bubble-dynamics-induced nonlinearity.

There are many examples of second harmonic diagnostics and imaging for NDE in the literature. In Figure 13.27 we describe one example of second and third har-

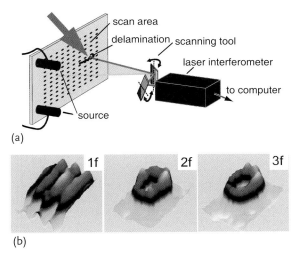

Fig. 13.27 Harmonic images. (a) Experimental configuration showing location of delamination (arrow), as well as location of sources and scanning laser. Large arrow points to scan region and defect. (b) Fundamental and harmonic images, $20 \times 40\ mm^2$ in dimension, obtained in the region of an oval-shaped delamination in a fiber-reinforced composite plate. The $1f$ image shows a standing wave pattern, but in the $2f$ and $3f$ images, the oval-shaped pattern is clear (used with permission from I. Solodov). (Please find a color version of this figure on the color plates.)

monic imaging of a delamination. In the study, the source emits a very high strain amplitude at a frequency of 50 kHz, and after detection, the signal is used to extract the second and third harmonics at 100 and 150 kHz, respectively. This is done at each point in a two-dimensional rastor-type scan in order to create an image. Note that the $1f$ image shows no hint of the delamination, and the largest nonlinear response in the $2f$ and $3f$ images is on the edges of the delamination. This observation underscores the point that the nonlinear response here is showing the edges rather than the center of the delamination. Later, we will describe second harmonic imaging in the context of time reversal used with nonlinear waves that also shows that the nonlinear response does not necessarily originate in all of a crack/delamination (*TR NEWS*). This important point means that extracting the damage volume from a nonlinear response may be ultimately possible only in some instances.

An example of second harmonic imaging in a spot weld follows in Figure 13.28 from [33]. In the experiment, a C-scan acoustic microscope imaging system, was modified to generate high-power tone bursts and to extract the second harmonics of the reflected waves. The frequency of the C-scan was 30 MHz, with a 25-mm focal point size. In the experiment, spot welds were created with artificial, small defects of order 0.1 mm inside. The C-scan is an image of acoustic impedance contrast (density times velocity) shown as dark regions. A cross section of the region is shown as well. The second harmonic image is an image of where the nonlinear response is highest (e.g., bright regions at border between black and gray regions). Dark regions correspond to regions of high acoustic impedance.

13.6.2
Modulation Imaging

A vibration/modulation approach is described next [34–36]. The method is based on application of ultrasonic tone bursts in the presence of low-frequency, continuous-wave vibration. The presence of wave modulation in a tone burst sequence provides the means to locate a crack and to distinguish cracks from other wave scatterers. Following Figure 13.29, the technique can be thought of as follows. A low-frequency, continuous-wave excitation is applied to a thin, rectangular-shaped specimen containing a crack and a hole, simultaneous with a group of high-frequency tone bursts. A first pair of tone bursts is reflected at a point where the low-frequency excitation stress reaches a maximum. At this time the crack is relatively compressed and the reflected pulse has a reduced amplitude, due to a transient, reduced acoustic impedance. The second pair of tone bursts is shown when the low-frequency wave stress is zero, meaning there is slightly larger amplitude in the reflected signal. The third pair is shown at the moment of wave extension, when the crack is maximally opened and the reflection coefficient is maximized. Through all phases of vibration the amplitude of the tone burst reflected from the hole remains constant, essentially independent of the changing stress field. In this manner, the amplitude modulation of the ultrasonic tone burst due to the crack flexing as a result

(a) linear C-Scan (b) second harmonic

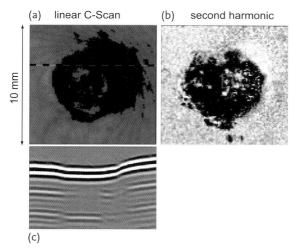

(c)

Fig. 13.28 Two-dimensional image of a spot weld. Dark regions correspond to high impedance contrast. (a) C-scan image. (b) Second harmonic image of same region. Bright regions correspond to highest nonlinear response. The small rectangle shows one of the small defects and the high associated harmonic response. (c) Cross-section of slice noted by dashed line in (a). Vertical dimension is time (increasing downward vertically), horizontal dimension is space, and the relative impedance is shown as gray tones, where light shade is higher impedance [33].

of the low-frequency excitation distinguishes cracklike defects from other inhomogeneities.

The procedure is as follows. The source located at x on the sample face launches a tone burst of 3 MHz and duration 0.66 μs at a sequence of times, $T_1 \ldots T_N(x)$, $N = 512$, where

$$T_n(x) = \frac{1}{366}(n-1)s, \quad n = 1 \ldots N.$$ (13.4)

The detected signals are composed of the tone bursts and low-frequency excitation in the time intervals $T_n \leq t \leq T_n + \Delta T$ shown in step 1 in the figure. The tone burst signals are rectified, the low-frequency excitation is eliminated by a high pass filter, and this signal is recorded and digitized at 0.6 μs, that is, $\Delta T = 256 \times 0.6$ μs.

As a consequence, the output is an "amplitude" matrix

$$V_{nm}(x) = V(T_n, \tau_m),$$ (13.5)

where $\tau_m = m\tau_0$, $256\tau_0 = \Delta T$, $m = 1 \ldots 256$, and $n = 1 \ldots N$, as illustrated in step 2 in the figure. The basic procedure is to study the Fourier structure (step 3) of $V(T_n, \tau_m)$ at fixed τ_m,

$$A(\tau_m, x) = \sum_{n=1}^{N} V(T_n, \tau_m) \sin \Omega T_n,$$ (13.6)

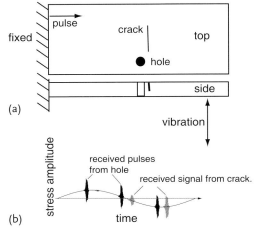

Fig. 13.29 Figure showing experiment for modulation imaging. (a) Sample top and side view showing location of crack and hole, as well as fixed end and location of vibration source. Pulses applied from fixed end. (b) During compression the crack is closed and therefore almost transparent, so there is very little reflection; during dilation, there is an impedance contrast, and thus the signal is reflected. The reflection from the hole is always the same due to the fact that stresses have no influence on it (the hole shunts the stress) (modified from [36]).

where the low-frequency modulation signal, the 10-Hz signal, is $\sin \Omega t$. This procedure will identify those τ_m at which the detected signal has evidence of the nonlinear scatterer. The association of time T_n with space Z_n can be carried out independently of the experimental procedure (if the wave speed is known), and therefore the above steps do not depend on this identification. Thus if $T \propto Z$ in some known way, the output can be used to identify the location of the nonlinear scatterer. This procedure is carried out for 26 source locations across the sample face, $x_1 \ldots x_{26}$, and in this manner the full wave profile is obtained by iterating the above steps at each source position (x).

The processed data set is the amplitude of the 10-Hz signal at each x_i, $i = 1 \ldots NX$, $NX = 26$ and at each $\tau_m = m\tau_0$, $m = 1 \ldots M$, $M = 256$, that is, the $NX \times M$ matrix:

$$A_{10}(i, m) .\tag{13.7}$$

The location of the damage is found by solving the inverse problem associated with this matrix. Here the nonlinear scatterers are point sources to within the resolution $\Delta z = c\tau_0$. The amplitude $A_{10}(i, m)$ comes from a source on the line (x_i, z) at

$$z = mc\tau_0 \equiv z_m .\tag{13.8}$$

The output data from the experimental and processing procedure are an $A - T$ grid of amplitude values V, much like using the absolute value of amplitude in a seismic reflection profile. The difference with a seismic reflection profile is that we have no off-axis detection in this experiment. Detection is only done at each source location – it is truly a pulse-echo-type experiment in that sense.

One of the plate ends was clamped to a support composed of a large mass. The ultrasonic tone burst wave was radiated at the plate face, as shown in the figure. The diameter of the piston radiator was 9 mm, and the resonance frequency of the transducer was 3 MHz. The tone burst signal applied to the radiator had a duration of about 0.66 ms. The radiated signal was about three times longer due to the transducer resonance properties. The repetition frequency of the tone burst signals was 366 Hz. The tone bursts were rectified and then digitized with a sampling rate of 0.6 ms. Triggering was done using the electrical tone burst applied to the transducer.

The far end of the plate was connected to a mechanical shaker. The shaker excited flexural vibration of the plate at a frequency of 10 Hz. Vibration amplitudes were detected and feedback-controlled for amplitude stability by use of an accelerometer located on the plate.

The two-dimensional reflection profiles of the "linear" response and of the nonlinear response (modulation level) as a function of time are shown in Figure 13.30. In Figure 13.30b, it is clear that the amplitude of the linear signal reflected from the hole is higher than the amplitude of the signal reflected from the crack (Figure 13.30a). Both features are imaged, but the nature of the features is unknown without a priori knowledge or some other means of testing. In the case of the nonlinear reflection profile, the level of the modulation in the signal reflected from a crack is large. The hole is a linear scatter and therefore should not produce modulation of the reflected signal. This experiment clearly demonstrates that a crack can be discriminated within the background of linear scattering from inhomogeneities other than cracks, and including sidewall and backwall reflections.

13.6.3
Imaging Applying Time Reversal Nonlinear Elastic Wave Spectroscopy (TR NEWS)

Much of the seminal research in time reverse acoustics (TRA) has been carried out by a group in Paris at the University of Paris VII (Laboratoire Ondes et Acoustique, ESPCI) beginning in the late 1980s [37–41]. Time reversal offers tremendous spatial and temporal focusing of sound regardless of the heterogeneity of the medium in which the wave propagates. TRA systems have a range of applications in development at this writing, including destruction of tumors and kidney stones and long-distance communication in the ocean. The NDE applications of TRA in development to date include detection of small, low-contrast defects within titanium alloys [39, 40] and detection of cracks in a thin air-filled hollow cylinder [41]. A review of TRA applications to NDE is given in [39].

The time reversal (TR)-induced focusing of wave energy at a point in space and time is ideal from the perspective of inducing an elastic-wave, nonlinear response because of the large amplitudes one can obtain. The first studies to do this applying broad-frequency band waves and a laser vibrometer detector in samples that were not submerged were conducted in sandstone [42]. Follow-on studies showed that one could induce a localized nonlinear response at the focal volume in rock. Based on this work, time reversal methods have been developed for investigating the nature of a nonlinear scatterer, such as a crack on the surface of a solid [43]. In addition, methods are in development to isolate internal nonlinear scatterers in an otherwise elastically linear solid. The general category of methods that combines time reversal and elastic nonlinearity is termed time reversal nonlinear elastic wave spectroscopy (TR NEWS), because nonlinear induced spectral components are used in identification and localization [44].

In what follows, we describe a study aimed at imaging a surficial crack in a solid taken from [45]. The experiments, shown in Figure 13.31a, were conducted in a glass block, shown in Figure 13.31b. Two piezoelectric ceramics were used as sources bonded to the face of the glass block opposite the crack location. Waves were detected on the cracked face using a laser vibrometer. The two sources were driven with separate frequencies ($f_1 = 170\,\text{kHz}$ and $f_2 = 255\,\text{kHz}$) in order to observe nonlinear wave mixing (sidebands) at the crack, and to isolate it from the surrounding, intact material.

The procedure for obtaining a one-dimensional line scan across the crack is as follows [45]. A signal $s_n(t)$ is transmitted from a source n where f_n is the frequency transmitted from the nth transducer with a maximum amplitude U_n and duration τ_n. The signal is received $r_{m,n}(t)$ at a selected point (x_m, i.e., focal point of the laser

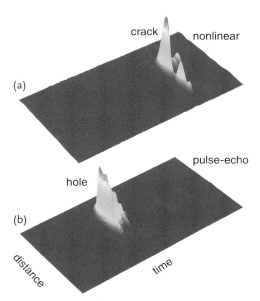

Fig. 13.30 Results from modulation imaging method. (a) Results of method showing isolation of crack only. (b) Results from standard pulse echo showing that the hole dominates the image. (please find a color version of this figure on the color plates)

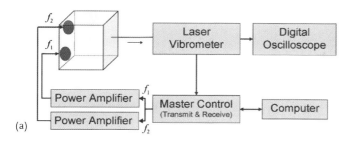

(a)

(b)

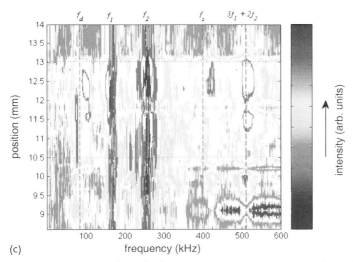

(c)

Fig. 13.31   Line scan of a surficial crack applying TR NEWS. (a) Experimental configuration for TR experiment. (b) $101 \times 89 \times 89 \, \text{mm}^3$ glass block on which the experiments were conducted. The source transducers, bonded with epoxy. The crack was located on the opposite surface to the sources.

(c) Line scan across the crack showing spectrogram. Primary frequencies f_1, f_2 as well as sum f_s and difference f_d are noted, as well as $3f_1 + 2f_2$. The region of the crack is denoted by dotted lines (modified from [45]). (please find a color version of this figure on the color plates)

vibrometer) on the sample for each f_n independently. The signal is time reversed $r_{m,n}(t) \longrightarrow r_{m,n}(-t)$. Simultaneously the time-reversed, amplified signals $S_{m,n}$ are transmitted from their original source locations. The signal is recorded at the TR focused signal R_m at the mth focal point of the laser vibrometer (x_m). N is the total number of source transducers, x_n is the location of the nth transducer, and $G(x_m t | x_n t')$ is the Green's function taking the signal transmitted from location x_n and time t' to point x_m at a later time t. The laser is stepped to the next desired location (x_{m+1}) and the process is repeated for M total focal points.

The above steps are the generalized procedure followed for the scan. In this experiment a scan (in one dimension) was conducted across the cracked area. At each of the M = 22 points (x_m) the same two f_1 and f_2 sources were used. Additionally, the drive amplitudes and durations remained constant, as did the source locations x_1 and x_2. In the step interval in and near the cracked region, the step size was 200 μm, while away from the crack the step size varied from 200 to 500 μm. The focal-point resolution was ~ 10 μm. This experiment differs from standard time reversal experiments in that the recorded (and time reversed) signals were then input back at the original source locations. This method, known as reciprocal TR, is based on the fact that reciprocity dictates that the resulting signal will focus at the point of the original detector [42, 46].

The primary frequencies f_1, f_2 show amplification at the crack as manifested by the high amplitudes in Figure 13.31c. This behavior is nearly always observed in TR measurements with cracked solids. It is due to energy that is trapped in the cracked region (it is a low-velocity zone), and therefore, due to energy conservation, wave speed decreases, and amplitudes increase. The effect is enhanced by the softening induced by the high-intensity focused wave. Tangentially, the same effect is observed in sandstone, where the high-intensity wave in the focal volume induces nonlinear softening, and wave speed decreases (A. Sutin, pers. comm.). Note the very different amplitude response of the sum and difference frequencies, related to their different spatial wavelengths, and therefore their different spatial resolution as well as complex interaction with the crack. Note that the figure also shows $3f_1 + 2f_2$.

In a follow-on experiment a scan in two dimensions n, m was conducted across the cracked area [45]. The results are shown in Figure 13.32.

The difference in the drive amplitudes $U1$ and $U2$ shown in Figure 13.32a,b arises due to the use of two different amplifiers with differing gains for the two channels. As in the line scan, the sum and difference frequencies are not identical, that is, the difference frequency is relatively constant in the damaged region while the sum frequency is only seen in one portion of the crack. The differences between the different nonlinear frequencies may ultimatley provide the ability to characterize the crack.

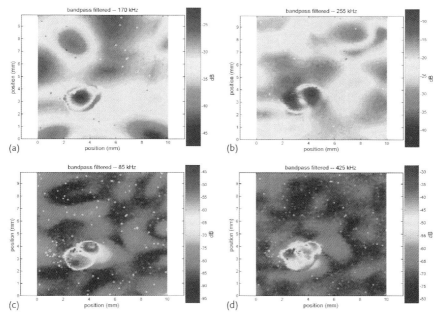

Fig. 13.32 Time reversal nonlinear elastic wave spectroscopy images in a glass block with a surficial crack (Figure 13.31). (a) The linear response showing only f_1 and (b) f_2 in the scan area. The high amplitudes in the lower left-hand quadrant indicate the crack location. (c) The nonlinear response (difference frequency $f_2 - f_1$) and (d) sum frequency ($f_1 + f_2$) in the scan area. The images are constructed by first bandpass filtering the focused signals about the desired frequency and then extracting the maximum amplitude of the filtered signals at each scan point (from [45]). (please find a color version of this figure on the color plates)

13.7
Other Methods for Extracting the Elastic Nonlinearity

13.7.1
Time Reversal + Phase Inversion

This method was initially proposed for ultrasonic imaging of microbubbles (contrast agents) in medical ultrasound [50] and was first employed in TR NEWS initially by Sutin [47, 48]. The method is not restricted to time reversal applications; however, we will describe it in this context.

If an isolated nonlinear scatterer located in a solid (or fluid) is excited simultaneously at two frequencies ω_1 and ω_2, then the backscattered signal will contain spectral content at the sum and difference frequencies, as well as higher-order terms as described previously. The idea behind the phase-inversion method is that the linear portion of a signal is canceled, leaving nonlinear terms for analysis. In one implementation described in [49], two time-reversed pulses that are sent out in sequence, the first, $r_k(t)$, with a sign opposite that of the second, $-r_k(t - T)$, where T is the period of the signal. The response from each of these signals is recorded, producing signals $p_{tot}(t)$ and $-p_{tot}(t)$, respectively. The two linear responses at the measurement point will have opposite signs, and shifting the second signal in phase by T and summing will cancel the response. If a nonlinear component is present, then the summed signal will contain energy. In the latter case, the sum of the two received signals is

$$p_{sum} = \left[p_{tot} + \phi(e_k(t)) \right] + \left[-p_{tot} + \phi(-e_k(t)) \right] , \tag{13.9}$$

where $\phi(e_k(t))$ is the nonlinear portion of the detected signal [49]. As usual, the signal amplitude is proportional to the elastic nonlinearity of the medium, as well as the source amplitudes and frequencies. The fact that the signal is phase inverted means that even components of the elastic nonlinearity are canceled, however (or the odd components, depending on the sign of the second wave). Thus the signal amplitude $\varepsilon_{p_{sum}} \propto \phi(e_k(t))$, where $\phi(e_k(t) \propto (\omega^3 \delta \varepsilon^3 + \omega^2 \alpha \varepsilon^2)$, and δ is the classic, cubic parameter and α is the hysteretic parameter. An advantage of the method is the inclusion of more of the spectrum than a single-frequency component, as is frequently the case in modulation or harmonic analysis, for instance.

In [49] a robust, hybrid version of the method was developed for antipersonnel mine location, but has since been generally applied. The hybrid version of the method is essentially a form of matched filtering where an approximation of a system impulse response is broadcast through the real system. Following the development in [49], first, a wide-frequency bandwidth signal obtained by sweeping the source in frequency as a function of time is sent to each detector sequentially. The measured signal can be represented as

$$v_k(t) = e(t) \star h_k(t) , \tag{13.10}$$

where $v_k(t)$ is the detected signal, $h_k(t)$ is the system impulse response, $e(t)$ is the source term, and $\star$ denotes the convolution operator. The electronic drive signal is cross-correlated with the radiated signal resulting in a short pulse, $s_k(t)$,

$$s_k(t) = v_k(t) \star e(-t) , \tag{13.11}$$

which approximates the system impulse response $h_k(t)$ (it is a limited-frequency band impulse response). This can be seen as follows:

$$S_k(\omega) = E(\omega) H_k(\omega) E^*(\omega) \tag{13.12}$$

where $E(\omega)$, $H_k(\omega)$, and $S_k(\omega)$ are the Fourier spectra of the initial signal $e(t)$, the transfer function $h_k(t)$, and the cross correlation of the received and radiated signals $s_k(t)$, respectively. For an initial signal with constant spectral density in the frequency band of interest $E(\omega) = E_0$ for $\omega_1 \le \omega \le \omega_2$, the cross correlation of the received signal is proportional to the system impulse response,

$$S_k(\omega) = E_0^2 H_k(\omega) . \tag{13.13}$$

Thus the system impulse response in a limited frequency band can be approximated in the time domain as

$$\hat{h}_k(t) = S_k(\omega)/E_0^2 \,. \tag{13.14}$$

The partial-band system impulse response is time reversed and normalized for all channels. The normalization maintains the same peak amplitude for all radiated signals. The signals have the form

$$r_k(t) = s_k(t_{\text{delay}} - t)/\max(s_k(t)) \,, \tag{13.15}$$

where t_{delay} is an arbitrary delay chosen to ensure causality. A second measurement records the time-reversed responses, $p_k(t)$, broadcast over all sources simultaneously. The measured signal associated with sources k can be expressed as

$$p_k(t) = r_k(t) \star h_k(t) \,. \tag{13.16}$$

Because the time-reversed signals are applied simultaneously to all sources, the response is the superposition of the individual responses:

$$p_{\text{tot}}(t) = \sum r_k(t) \star h_k(t) \tag{13.17}$$

over all sources.

In what follows we describe an application of the impulse response/phase inversion technique applied to locating landmines. Mines are known to be more elastically nonlinear in comparison with the nonlinear surrounding geomaterial, for example, [49]. This is apparently due to the highly nonlinear interface between the mine and the soil, rather than the landmine itself (A. Sutin, pers. comm.). In the experiment shown in Figure 13.33, a multichannel source array composed of loudspeakers was used to generate the signals and a laser vibrometer was used for detection. The experiments were performed in sand and soil containing clay, sand, small rocks, and fine organic matter. An antipersonnel mine was buried at a depth of 2 cm and surface vibration measurements were taken in a line scan at positions directly above as well as adjacent to the mine applying a laser vibrometer. A sum and difference frequency were generated by the array of loudspeakers and the signal was coupled into the ground via the air. The signal was detected by the laser vibrometer. In the simplest case, the signal was then time reversed and reemitted by the loudspeakers and detected again by the laser vibrometer, for both the signal and its phase inverse applying the concept of reciprical time reversal. This signal was again detected by the vibrometer and the two signals subtracted, leaving only the nonlinear response of the focal region.

In the hybrid version of this technique, the cross correlation of a swept sine wave input signal with the detected signal is created, giving the approximate impulse response of the material. The impulse response is time reversed and phase inverted, and both signals are reemitted.

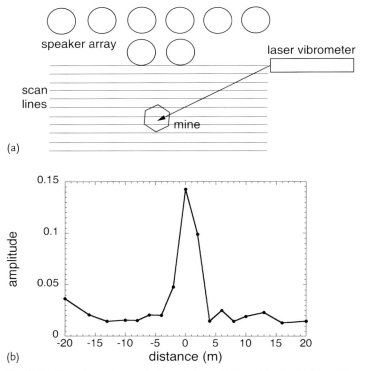

Fig. 13.33 Phase inversion method applied to antipersonnel mines. (a) Experimental configuration. An array of loudspeakers is suspended over the surface. The signals they emit are recorded with the laser vibrometer. The vibrometer can scan a large surface area, as indicated by the horizontal lines. (b) Step-scan over landmine, showing the strong nonlinear response (at 0 m) obtained from the phase inversion method. The closed circles represent the actual measurements (from A. Sutin, Pers. comm.).

13.7.2
Scaling Subtraction/Variable Amplitude Method

A method, known as the variable amplitude method (A. Sutin, pers. comm.), and alternatively the scaling subtraction method [51], is a highly effective approach in isolating the elastic nonlinear behavior of a sample. The approach relies on recording a low-amplitude, assumed linear elastic waveform. The signal could be a time-reversed, focused signal; alternatively, it could be applied with any of the methods described above. Following this, a high-amplitude, nonlinear signal is collected. The low-amplitude signal is artificially scaled to match in amplitude the nonlinear signal, and the two are subtracted, leaving only the nonlinear portion for analysis.

13.8
Summary

In this section we began by describing the history of developing nonlinear diagnostics to applications. We followed this with a list of diagnostic methods showing examples. This was followed by several examples of imaging. We have by no means described all of the experimental methods, but we have touched on most of them. In addition, we have attempted to sketch many of the signal-processing methods in use, but again, not all of them, as new methods are appearing all the time. There is no "best" diagnostic/imaging method or signal-processing method because it is application dependent. It is also dependent on the order of type of elastic nonlinearity one wishes to probe. NRUS only gives the hysteretic elastic nonlinearity parameter. The second harmonic gives only the second-order term proportional to β. In many instances, it does not matter. For instance, a yes/no result for quick diagnostics may be obtained from SDD, NWMS, or ringdown spectroscopy.

References

1 Read, T.A. (1941) Internal friction of single crystals of copper and zinc. *Transact. Am. Inst. Min. Metall. Eng.*, **143**, 30–44.

2 Zener, C. (1946) *Anelasticty of metals*, American Institute of Mining and Metallurgical Engineers Technical Publications, Chicago, University of Chicago, no. 1992, p. 35.

3 Read, T.A. and Tyndall, E.P.T. (1946) Internal friction and plastic extension of zinc single crystals. *J. Appl. Phys.*, **17**, 713–720.

4 Nowick, A.S. (1950) Carnegie Institute of Technology Symposium on the Plastic Deformation of Crystalline Solids, ONR. Nowick, A.S. (1950) Carnegie Institute of Technology Symposium on the Plastic Deformation of Crystalline Solids, ONR.

5 Koehler, J.S. (1952) *Imperfections in nearly perfect crystals*, John Wiley and Sons, Inc., New York, p. 197.

6 Weertman, J. and Salkovitz, E. (1955) Internal friction of dilute alloys of lead. *Acta Metallurg.*, **3**, 1–9.

7 Granato, A. and Lücke, K. (1956) Application of dislocation theory to internal friction phenomena at high frequencies. *J. Appl. Phys.*, **27**, 789–805.

8 Chambers, R.H. and Smoluchowski, R. (1960) Time-dependent internal friction in aluminum and magnesium single crystals. *Phys. Rev.*, **117**, 725–731.

9 Strumane, R.R., De Batist, R., and Amelinckx, S. (1963) Charged dislocations in sodium chloride single crystals. *Phys. Status Solidi (b)*, **3**, 1379–1386.

10 Gedroits, A.A. and Krasilnikov, V.A. (1963) Finite-amplitude elastic waves in solids and deviations from Hooke's law. *Sov. Phys. JETP-USSR*, **16**, 1122–1124.

11 Hikata, A., Chick, B.B., and Elbaum, C. (1963) Effect of dislocations on finite amplitude ultrasonic waves in aluminum. *Appl. Phys. Lett.*, **11**, 195–196.

12 Hikata, A. and Elbaum, C. (1966) Generation of ultrasonic second and third harmonics due to dislocations. I, *Phys. Rev.*, **144**, 469–477.

13 Hikada, A., Sewell, F.A., and Elbaum, C. (1966) Generation of ultrasonic second and third harmonics due to dislocations. II, *Phys. Rev.*, **144**, 442–449.

14 Scorey, C.R. (1970) Recovery of an ultrasonic third harmonic amplitude changes in sodium chloride following small plastic deformation. *Philos. Mag.*, **21**, 723–734.

15 Yermilin, K.K., Krasilnikov, V.A., Mezintsev, Y.D., Prokhorov, V.M., Khilkov, K.V., and Zarembo, L.K. (1973) Variation in the second harmonic of a shear ultrasonic

wave in metals under cyclic alternating load. *Phys. Met. Metallogr.*, **36**, 174–176.

16 De Batist, R. (1972) *Internal friction of structural defects in crystalline solids*, North Holland, London.

17 Buck, O. (1976) Harmonic generation for measurement of internal stresses as produced by dislocations. *IEEE Trans. Sonics Ultrasonics*, **SU-23**, 346–350.

18 Buck, O., Morris, W.L., and Richardson, J. (1978) Acoustic harmonic generation at unbonded interfaces and fatigue cracks. *Appl. Phys. Lett.*, **33**, 371–373.

19 Granato, A., and Lücke, K. (1956) Theory of mechanical damping due to dislocations. *J. Appl. Phys.*, **27**, 583–593.

20 Zarembo, L.K., Krasilnikov, V.A., Sluch, V.N., and Serdobolskaya, O.Y. (1966) Certain effects in the forced nonlinear vibration of acoustic resonators. *Akust. Zhournal*, **12**, 486–487.

21 Johnson, P.A. (1999) The new wave in acoustic testing, Materials World. *J. Inst. Mater.*, **7**, 544–546.

22 Solodov, I.Y., Pfleiderer, K., and Busse, G. (2007) Nonlinear acoustic NDE: Inherit potential of complete nonclassical spectra, in: *The Universality of Nonclassical Nonlinearity, with Applications to NDE and Ultrasonics*, (eds P.P. Delsanto and S. Hirshekorn) Springer, New York, pp. 467–485.

23 Richardson, J.M. (1979) Harmonic generation at an unbonded interface-I. Planar interface between semi-infinite elastic media, *Int. J. Engng. Sci.* Nonlinear Acoustic NDE: Inherit potential of complete nonclassical spectra, **17**, pp. 73–85.

24 Solodov, I.Y. (1998) Ultrasonics of nonlinear contacts: Propagation, reflection and NDE applications. *Ultrasonics*, **36**, 383–390.

25 Van Den Abeele, K.E.-A., Johnson, P., and Sutin, A., Non-linear elastic wave spectroscopy (NEWS) techniques to discern material damage. Part I: Non-linear wave modulation spectroscopy. *Res. Nondestruct. Eval.*, **12**, 17–30 (2000).

26 Van Den Abeele, K.E.-A., Carmeliet, J., Sutin, A., and Johnson, P.A. (2001) Micro-damage diagnostics using non-linear elastic wave spectroscopy (NEWS). *NDT&E International*, **34**, 239–248.

27 Korotkov A.S. and Sutin A.M. (1994) Modulation of ultrasound by vibrations in metal constructions with cracks. *Acoust. Lett.*, **18**, 59–62.

28 Johnson, P.A. and Sutin, A. (2005) QNDE 2004, Nonlinear elastic wave NDE I. Nonlinear resonant ultrasound spectroscopy (NRUS) and slow dynamics diagnostics (SDD), Review of Progress in: *Quantitative Nondestructive Evaluation*, vol. 24B (eds D. Thompson and D. Chimenti), pp. 377–384.

29 Morris, W.L., Buck, O., and Inman, R.V. (1979) Acoustic harmonic generation due to fatigue damage in high strength aluminum. *J. Appl. Phys.*, **50**, 6737–6741.

30 Korotkov, A.S., Slavinskii, M.M., and Sutin, A.M. (1994) Variations in nonlinear parameters with the concentration of defects in steel. *Acoust. Phys.*, **40**, 71–74.

31 Nagy, P. (1998) Fatigue damage assessment by nonlinear ultrasonic materials characterization. *Ultrasonics*, **36**, 375–382.

32 Cantrell, J.H. and Yost, W.T. (2001) Nonlinear ultrasonic characterization of fatigue microstructures. *Int. J. Fatigue*, **23**, S487–S490.

33 Kawashima, K., Murase, M., Yamada, R., Matsushima, M., Uematsu, M., and Fujita, F. (2006) Nonlinear ulstrasonic imaging of imperfectly bonded interfaces. *Ultrasonics*, **44**, e1329–e1333.

34 Zaitsev, Y.V., Sutin, A., Belyaeva, Y.I., and Nazarov, V.E. (1995) Nonlinear interaction of acoustical waves due to cracks and its possible usage for cracks detection. *J. Vib. Control*, **1**, 3355–344.

35 Didenkulov, I.N., Sutin, A., Kazakov, V.V., Ekimov, A.E., and Yoon, S.W. (2000) Nonlinear acoustic technique of crack location, in: *Nonlinear Acoustics at the Turn of the Millennium, the 15th International Symposium on Nonlinear Acoustics*, Gottingen, Germany, (eds W. Lauterborn and T. Kurz), AIP Conference Proceedings, vol. 524, pp. 329–332.

36 Kazakov, V.V., Sutin, A., and Johnson, P. (2002) Sensitive imaging of an elastic nonlinear wave source in a solid. *Appl. Phys. Lett.*, **81**, 646–648.

37 Fink, M., Cassereau, D., Derode, A., Prada, C., Roux, P., Tanter, M., Thomas, J.L., and Wu, F. (2000) Time-reversed acoustics. *Rep. Prog. Phys.*, **63**, 1933–1995.

38 Fink, M. (1999) Time reversed acoustics. *Sci. Am.*, **281**, 91–97.

39 Prada, C., Kerbat, E., Cassereau, D., and Fink, M. (2002) Time reversal techniques in ultrasonic nondestructive testing of scattering media. *Inverse Probl.*, **18**, 1761–1773.

40 Chakroun, N., Fink, M., and Wu, F. (1995) Time reversal processing in nondestructive testing. *EEE Trans. Ultrasonic Ferroelectr. Freq. Contr.*, **42**, 1087–1089.

41 Kerbat, E., Clorennec, D., Prada, C., Royer, D., Cassereau, D., and Fink, M. (2002) Detection of cracks in a thin air-filled hollow cylinder by application of the d.o.r.t. method to elastic components of the echo. *Ultrasonics Int.*, **40**, 715–720.

42 Sutin, A., TenCate, J., and Johnson, P.A. (2004) Single-channel time reversal in elastic solids. *J. Acoust. Soc. Am.*, 116, 2779–2784.

43 Sutin, A., Johnson, P., and TenCate, J. (2003) Development of nonlinear time reverse acoustics (NLTRA) method for crack detection in solids, Proceedings of the World Congress on Ultrasonics (Paris), pp. 121–124.

44 Ulrich, T.J., Johnson, P., and Guyer, R.A. (2007) Investigating interaction dynamics of elastic waves with a complex nonlinear scatterer applying the time reversal mirror. *Phys. Rev. Lett.*, doi:10.1103/PhysRevLett.98.10430.

45 Ulrich, T.J., Sutin, A., and Johnson, P.A. (2007) Imaging and characterizing damage using time reversed acoustics, Quantitative Nondestructive Evaluation, 894

(eds D. Thompson and D. Chimenti), American Institute of Physics, pp. 650–656.

46 Draeger, C., and Fink, M. (1997) One-channel time reversal of elastic waves in a chaotic 2d-silicon cavity. *Phys. Rev. Lett.*, **79**, 407–410.

47 Sutin, A., Sarvazyan, A., Johnson, P., and TenCate, J. (2004) Land mine detection by time reversal acousto-seismic method. *J. Acoust. Soc. Am.*, **115**, 2384, (abstract).

48 Sutin, A., Johnson, P., TenCate, J., and Sarvazyan, A. (2005) Time reversal acousto-seismic method for land mine detection. *Detection and Remediation Technologies for Mines and Minelike Targets X*, (eds R.S. Harmon, J.T. Broach, J.H. Holloway), Proc. of SPIE 5794, pp. 706–716.

49 Sutin, A., Libby, B., Kurtenoks, V., Fenneman, D., and Sarvazyan, A. (2006) Nonlinear detection of land mines using wide bandwidth time-reversal techniques, Proceedings of the SPIE – The International Society for Optical Engineering 6217, 62171B-1-12. Sutin, A., Libby, B., Kurtenoks, V., Fenneman, D., and Sarvazyan, A. (2006) Nonlinear detection of land mines using wide bandwidth time-reversal techniques, Proceedings of the SPIE – The International Society for Optical Engineering 6217, 62171B-1-12.

50 Krishnan, S. and O'Donnell, M. (1996) Transmit aperature processing for nonlinear contrast agent imaging. *Ultrasonic Imaging*, **18**, 77–105.

51 Scaleranti, M., Gliozzi, A., Bruno, C., and Van Den Abecle, K. (2008) Nonlinear acoustic time reversal imaging using the scaling substraction method. *J. Phys. D: Appl. Phys*, **41**, 215404–215414.

Color Plates (courtesy of S. Levy)

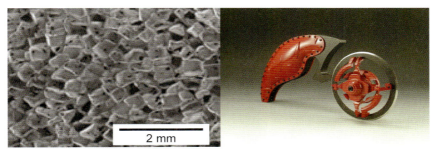

Fig. 1.1 Porous aluminum powder. (This figure also appears on page 1)

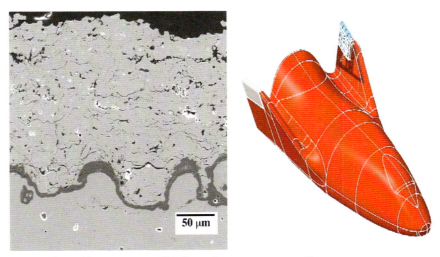

Fig. 1.2 Thermal barrier coating. (This figure also appears on page 2)

Nonlinear Mesoscopic Elasticity: The Complex Behaviour of Granular Media including Rocks and Soil. Robert A. Guyer and Paul A. Johnson
Copyright © 2009 WILEY-VCH Verlag GmbH & Co. KGaA, Weinheim
ISBN: 978-3-527-40703-3

Fig. 1.3 Sandstone (typical grain size 100 μm). (This figure also appears on page 2)

Fig. 1.4 Cement. (This figure also appears on page 3)

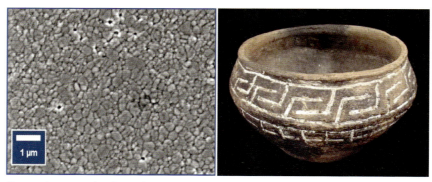

Fig. 1.5 Ceramic. (This figure also appears on page 3)

Fig. 1.6 Soil (sieved, typical grain size 1 mm). (This figure also appears on page 3)

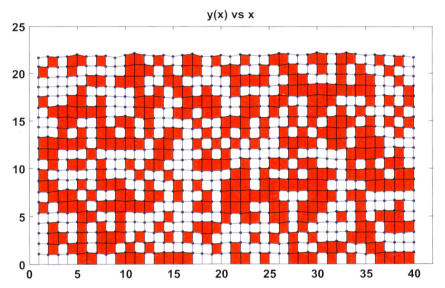

Fig. 5.3 Thermal configuration. An anisotropic thermal expansion, $\lambda_\alpha = 1 + 0.5\mathrm{sign}(r_\alpha)$, is assigned to each elastic element ($\lambda > 0$ red and $\lambda < 0$ white). This produces an average expansion of the system, as well as compres-sional and shear forces on individual elastic elements. The spectrum of these internal forces is shown in Figures 5.4 and 5.5. (This figure also appears on page 100)

S$_w$ = 0.80 (BT)

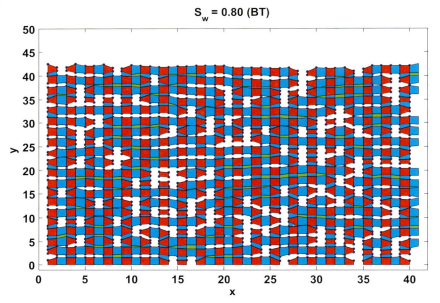

Fig. 5.9 System configuration, S_w = 0.70, filling. Displacement of the elastic elements in a 20 cell by 20 cell realization of the elastic system with S_w = 0.70. The bricks are red. The space allocated to the mortar and contacts is treated as the pore space and filled blue (mortar) and green (contact) if there is fluid in the associated pore and white otherwise. All elastic elements are shown as if they are of the same size, see the caption to Figure 5.6.

This configuration, formed on chemical potential increase, has homogeneous filling of the pore space. Elastic elements adjacent to filled vertical pores feel forces of tension tending to elongate the elastic element. Elastic elements adjacent to filled horizontal pores feel forces tending to pull the elastic element into the pore space. This appears as the pulling of bricks toward the pore. See Figure 5.12. (This figure also appears on page 106)

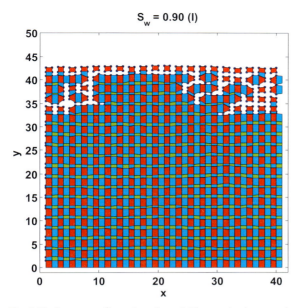

Fig. 5.10 System configuration, $S_w = 0.90$, emptying. Displacement of the elastic elements in a 20 cell by 20 cell realization of the elastic system with $S_w = 0.90$ as the chemical potential is being lowered. This configuration occurs near the onset of invasion percolation. As the vapor invades the pore space from the surface, well below the interface between full and partially empty pores, the pore space is uniformly filled. Shear forces occur only near the outer edge of the system. See Figure 5.13. (This figure also appears on page 107)

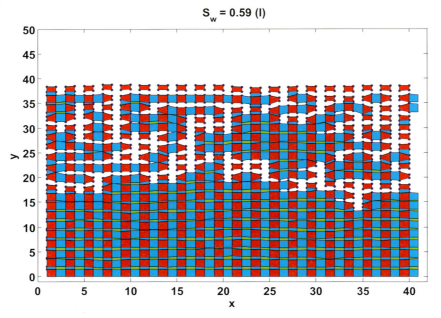

Fig. 5.11 System configuration, $S_w = 0.59$, emptying. Displacement of the elastic elements in a 20 cell by 20 cell realization of the elastic system with $S_w = 0.59$ as the chemical potential is being lowered. See Figure 5.14. (This figure also appears on page 108)

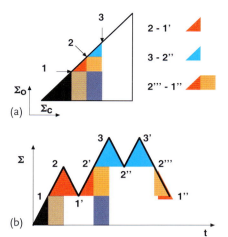

(a)

(b)

Fig. 6.9 Inversion in Preisach space; elaborate stress protocols. The regions of Preisach space, shown with color coding in (a), are swept over by the stress protocol in (b) with the same color coding. The elastic elements in the red region (of Preisach space) are responsible for the strain as the stress evolves $2 \rightarrow 1'$ ($1' \rightarrow 2'$). The elastic elements in the blue region are responsible for the strain as the stress evolves $3 \rightarrow 2''$ ($2'' \rightarrow 3'$).

The elastic elements in the blue, yellow, and red regions are responsible for the strain as the stress evolves $3' \rightarrow 2''' \rightarrow 1''$. By manipulating the stress protocol the strain due to elastic elements in a particular region of Preisach space can be found. In more physical terms, the strains due to elastic elements that respond to particular stresses can be determined. (This figure also appears on page 124)

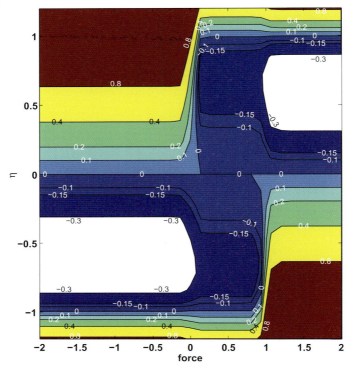

Fig. 7.7 Energy landscape. The energy landscape for $(f_c, f_o) =$ $(1.0, 0.0)$, see Eq. 7.20 and the discussion below, Eq. 7.22. (This figure also appears on page 155)

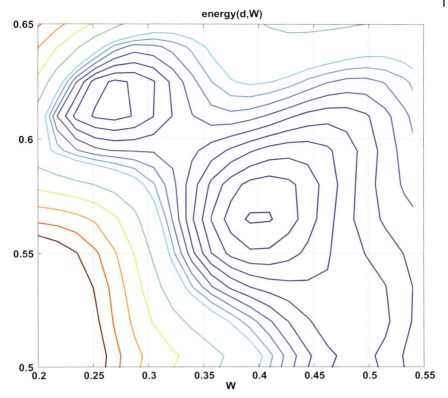

Fig. 9.5 Energy Landscape. The energy in Eq. 9.44, due to the difference between the data vector e_X and the model vector e_M, as a function of d and W, the parameters of the model form factor $f_M(x)$. The energy minimum is at $d = 0.565$ and $W = 0.40$. See Figure 9.3. (This figure also appears on page 209)

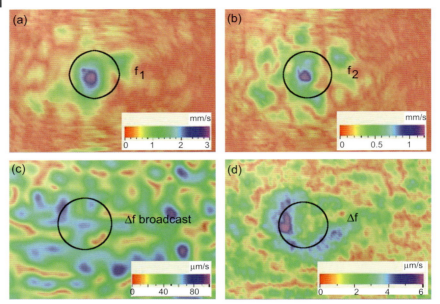

Fig. 11.11 Parametric array results. (a) Amplitude of direct arrival on the scan area of signal $f_1 = 65$ kHz, broadcast from 10 transducers (open circles) in the array, Figure 11.10. (b) Amplitude of direct arrival on the scan area of signal $f_1 = 60$ kHz, broadcast from 9 transducers (shaded circles) in the array, Figure 11.10. (c) Amplitude of direct arrival on the scan area of signal $\Delta f = f_2 - f_1 = 5$ kHz, broadcast from **all** 19 transducers in the array. (d) Amplitude of arrival on the scan area of signal at $\Delta f = f_2 - f_1 = 5$ kHz when $f_1 = 65$ kHz is broadcast from 10 transducers (open circles) in the array and $f_2 = 60$ kHz is broadcast from 9 transducers (open circles) in the array. The black circle on each panel is the projection of the source array onto the scan area. (This figure also appears on page 273)

Northridge Earthquake

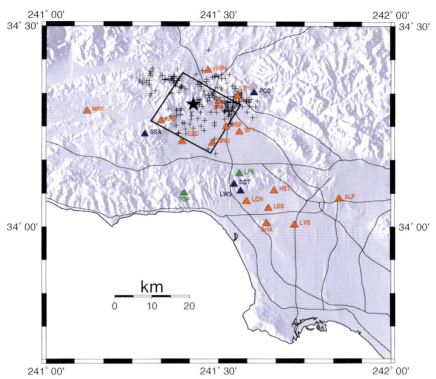

Fig. 12.5 Relief map of the study region in southern California, showing the location of 1994 Northridge earthquake (star). The alluvium recording sites are shown as red triangles and the hard rock sites as blue triangles. Aftershock epicenters are shown with black crosses and the mainshock rupture distribution is outlined by the box. The fault plane dips to the southwest, with the top edge at a depth of 5 km and the bottom edge at a depth of 20.4 km. The location of maximum earthquake slip is marked with the black star (from [8]). (This figure also appears on page 319)

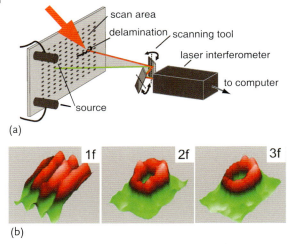

(a)

(b)

Fig. 13.27 Harmonic images. (a) Experimental configuration showing location of delamination (arrow), as well as location of sources and scanning laser. Large arrow points to scan region and defect. (b) Fundamental and harmonic images, $20 \times 40 \, \text{mm}^2$ in dimension, obtained in the region of an oval-shaped delamination in a fiber-reinforced composite plate. The $1f$ image shows a standing wave pattern, but in the $2f$ and $3f$ images, the oval-shaped pattern is clear (used with permission from I. Solodov). (This figure also appears on page 353)

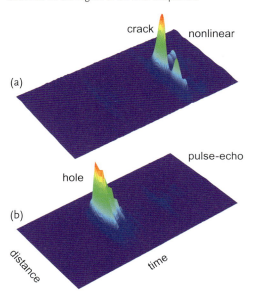

Fig. 13.30 Results from modulation imaging method. (a) Results of method showing isolation of crack only. (b) Results from standard pulse echo showing that the hole dominates the image. (This figure also appears on page 358)

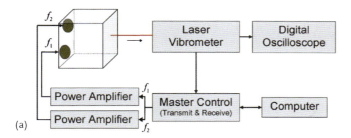

(a)

(b)

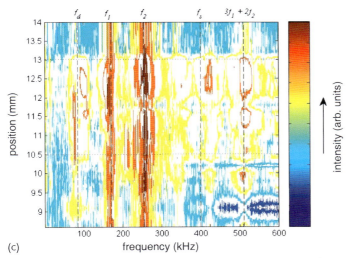

(c)

Fig. 13.31 Line scan of a surficial crack applying TR NEWS. (a) Experimental configuration for TR experiment. (b) $101 \times 89 \times 89\,\text{mm}^3$ glass block on which the experiments were conducted. The source transducers, bonded with epoxy. The crack was located on the opposite surface to the sources.

(c) Line scan across the crack showing spectrogram. Primary frequencies f_1, f_2 as well as sum f_s and difference f_d are noted, as well as $3f_1 + 2f_2$. The region of the crack is denoted by dotted lines (modified from [45]). (This figure also appears on page 359)

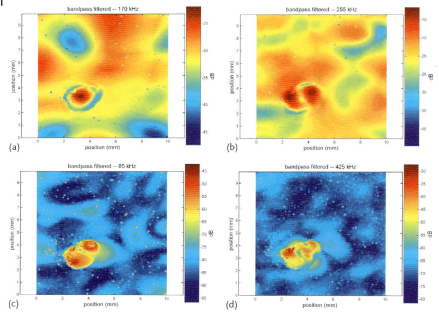

Fig. 13.32 Time reversal nonlinear elastic wave spectroscopy images in a glass block with a surficial crack (Figure 13.31). (a) The linear response showing only f_1 and (b) f_2 in the scan area. The high amplitudes in the lower left-hand quadrant indicate the crack location. (c) The nonlinear response (difference frequency $f_2 - f_1$) and (d) sum frequency $(f_1 + f_2)$ in the scan area. The images are constructed by first bandpass filtering the focused signals about the desired frequency and then extracting the maximum amplitude of the filtered signals at each scan point (from [45]). (This figure also appears on page 361)

Plate 1 Carrara marble, Tuscany, Italy. Plane-polarized light. Long dimension is 0.94 mm. Note the distinct grain boundaries leading to the highly elastically nonlinear response of this material.

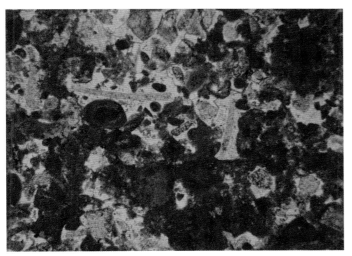

Plate 2 Lavoux oölitic limestone, France. Plane-polarized light. Long dimension is 2.35 mm. Some grain boundaries are apparent, but the dominant features are the fossils, the oöliths. Lavoux limestone is highly elastically nonlinear due to both grain boundaries and perhaps differential compressibility between fossilized and nonfossilized regions. The mechanical behavior of this rock is featured throughout this book.

Plate 3 Berea sandstone, Ohio, USA. Forty-five-degree polarizers with quartz plate. Long dimension is 2.35 mm. Note grain contacts. Berea sandstone is also highly elastically nonlinear due primarily to the grain boundaries. The mechanical behavior of this rock is featured throughout this book.

Plate 4 Meule sandstone, Alsace, France. Forty-five-degree polarizers with quartz plate. Long dimension is 2.35 mm. Meule and Vosges sandstones are very similar, and their composition is also similar to Berea sandstone. The presence of the grain contacts leads to highly elastic nonlinear behavior. The mechanical behavior of this rock is featured throughout this book.

Plate 5 Granodiorite, Jemez Mountains, New Mexico, USA. Parallel polarizers with quartz plate. Long dimension is 2.35 mm. Granodiorite is a crystalline rock whose elastic nonlinear behavior is due to micro- and macrocracks. A large number of microcracks are evident.

Plate 6 Quartz-cemented Fontainebleu sandstone, France. Forty-five-degree polarizers with quartz plate. Long dimension is 2.35 mm. Fontainbleau sandstone is one of the most elastically nonlinear rocks we are aware of. Its behavior is due to very soft contacts between grains. The mechanical behavior of this rock is featured throughout this book.

Plate 7 Vosges sandstone, Alsace, France, showing microcracks in sand grains. Forty-five-degree polarizers with quartz plate. Long dimension is 2.35 mm. Vosges* sandstone is similar in composition to Meule sandstone and, as a result, has similar mechanical behaviors. Its elastic nonlinear behavior is due primarily to soft grain contacts.

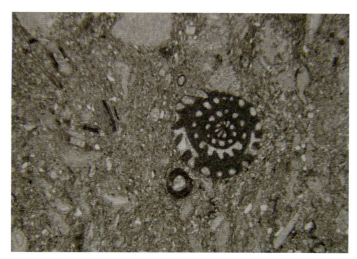

Plate 8 Madera limestone with marine fossils, Valle Grande, New Mexico, USA. Plane-polarized light. Long dimension is 2.35 mm. Madera limestone is another fossiliferous limestone (see the sample of Lavoux limestone), e.g., fossil nautilus is clearly apparent. The elastic nonlinear behavior may be due to both grain boundaries and perhaps differential compressibility between fossilized and nonfossilized regions.

Plate 9 Porphyritic basalt, Los Alamos, New Mexico, USA. Plane-polarized light. Long dimension is 2.35 mm. Basalt is a volcanic rock with, in this case, a large mix of grain sizes from very small (dark regions) to very large. Its elastic nonlinearity is likely due to micro- and macrocracks.

Plate 10 Fractured and altered peridotite, Mashaba Igneous Complex, Zimbabwe. Plane-polarized light. Long dimension is 2.35 mm. The elastic nonlinearity of this rock is due to the intense network of fractures, clearly evident. The material would be expected to have a highly anisotropic elastic nonlinear response, strong perpendicular to the cracks.

Plate 11 Immature sandstone (highly angular grains, wide range of grain sizes, abundant nonquartz grains), Pigeon Point, California, USA. Plane-polarized light. Long dimension is 2.35 mm. The elastic nonlinearity is due to the grain contacts.

Plate 12 Chlorite schist, Franciscan Formation, Oakland, California, USA. Plane-polarized light. Long dimension is 2.35 mm. Schist, a metamorphic rock, is strongly anisotropic and would be expected to have strong nonlinear response across the grain boundaries (dark regions).

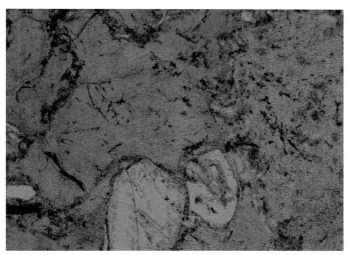

Plate 13 Rhyolite pebble in Puye Conglomerate, Los Alamos, New Mexico, USA. Plane-polarized light. Long dimension is 2.35 mm. Rhyolite is a volcanic rock containing much glass. We have not tested its elastic nonlinear response but would expect it to be small unless microfractured.

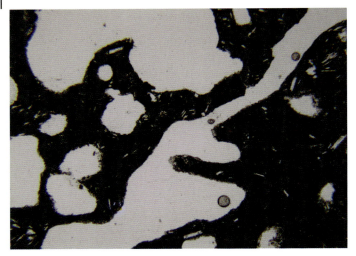

Plate 14 Highly vesicular basalt, Lathrop Wells, Nevada, USA. Plane-polarized light. Long dimension is 2.35 mm. Basalt is a volcanic rock (compare to the other basalt sample). Its elastic nonlinearity would be dominated by microcracks.

Plate 15 Novaculite, Oauchita Mountains, Arkansas, USA. Crossed polarizers. Long dimension is 0.94 mm. Novaculite is formed from low-grade metamorphism of chert. It is one of the few rocks we are aware of that exhibits no mesoscopic nonlinear response.

Index

Nonlinear Mesoscopic Elasticity: The Complex Behaviour of Granular Media including Rocks and Soil. Robert A. Guyer and Paul A. Johnson.
Copyright © 2009 WILEY-VCH Verlag GmbH & Co. KGaA, Weinheim
ISBN: 978-3-527-40703-3